D0648501

ADVANCES IN CLINICAL CHEMISTRY

VOLUME 2

Advances in CLINICAL CHEMISTRY

Edited by

HARRY SOBOTKA
Department of Chemistry, Mount Sinai Hospital, New York, New York

C. P. STEWART
Department of Clinical Chemistry, University of Edinburgh, Royal Infirmary, Edinburgh, Scotland

VOLUME 2

ACADEMIC PRESS, INC.
111 Fifth Avenue, New York, New York 10003

United Kingdom Edition published by
ACADEMIC PRESS, INC. (LONDON) LTD.
Berkeley Square House, London W1X 6BA

LIBRARY OF CONGRESS CATALOG CARD NUMBER: 58-12341

Second Printing, 1970

PRINTED IN THE UNITED STATES OF AMERICA

CONTRIBUTORS TO VOLUME 2

SAMUEL P. BESSMAN, *Departments of Pediatrics and Biochemistry, School of Medicine, University of Maryland, Baltimore, Maryland*

E. J. BIGWOOD, *Department of Biochemistry, Faculty of Medicine, Brussels University, Brussels, Belgium*

BARBARA H. BILLING, *Department of Surgery, Postgraduate Medical School of London, London, England**

R. CROKAERT, *Department of Biochemistry, Faculty of Medicine, Brussels University, Brussels, Belgium*

JOHN O. FORFAR, *Department of Pediatrics, Edinburgh Northern Group of Hospitals, and Department of Child Life and Health, University of Edinburgh, Edinburgh, Scotland*

WALTON H. MARSH, *Department of Pathology, State University of New York, Downstate Medical Center, Brooklyn, New York*

H. PEETERS, *St. Jans Hospitaal, Brugge, Belgium*

E. SCHRAM, *Department of Biochemistry, Faculty of Medicine, Brussels University, Brussels, Belgium*

P. SOUPART, *Department of Biochemistry, Faculty of Medicine, Brussels University, Brussels, Belgium*

S. L. TOMPSETT, *Department of Biochemistry, Edinburgh Northern Group of Hospitals and Department of Clinical Chemistry, University of Edinburgh, Edinburgh, Scotland*

H. VIS, *Department of Biochemistry, Faculty of Medicine, Brussels University, Brussels, Belgium*

* Present address: Department of Medicine, Royal Free Hospital, London, England.

FOREWORD TO THE SERIES

A historian of science in years to come may well be astonished at the explosive burst of scientific activity round about the middle of the twentieth century of our era. He will be puzzled by the interrelationship between the growth of population and the rise of the standard of living; he will be interested in the increased percentage of scientists among the population, their greater specialization and the resulting fragmentation of science; he will analyze the economic and the psychological motivation of scientists; he will compare the progress of knowledge with the broadness of the current of scientific publication.

Living as we do in the midst of these events, we are hardly aware of their relatively rapid rate. What we notice is a doubling of the scientific output every ten years, regardless of contemporary political events. It is this climate which has engendered the appearance of series of reviews in dozens of disciplines. It may be with yearning or with a feeling of superiority, that we look back at such annual compendia as "Maly's Jahresberichte der Thierchemie" of one hundred years ago, which encompassed the annual progress in the zoological half of biochemistry within 300 to 400 pages.

Nowadays, that number of pages would not suffice to record the complete annual increment of knowledge in a single specialized division of the subject such as Clinical Chemistry. Media already existing furnish a comprehensive list of publications and an encyclopedic summarization of their contents; the present series of "Advances in Clinical Chemistry"—like other "Advances" series—attempts something different. Its aim is to provide a readable account of selected important developments, of their roots in the allied fundamental disciplines, and of their impact upon the progress of medical science. The articles will be written by experts who are actually working in the field which they describe; they will be objectively critical discussions and not mere annoted bibliographies; and the presentation of the subjects will be unbiased as the utterances of scientists are expected to be—*sine ira et studio.*

The bibliography appended to each chapter will not only serve to document the author's statements, it will lead the reader to those original publications in which techniques are described in full detail or in which viewpoints and opinions are discussed at greater length than is possible in the text.

The selection of the subjects in the present and in future volumes will include discussion of methods and of their rationale, critical and comparative evaluation of techniques, automation in Clinical Chemistry, and microanalytical procedures; the contents will comprise those borderline subjects, such as blood coagulation or complement chemistry, which are becoming more chemical with increasing knowledge of the underlying reactions; in some instances the discussion of a subject will center around a metabolic mechanism or even around a disease entity.

While recognizing that the elaboration and testing of methods is of the greatest importance in a subject, part of whose function is to provide reliable, accurate diagnostic and prognostic procedures, the new series will take cognizance of the fact that Clinical Chemistry plays an essential part in the progress of medical science in general by assisting in elucidating the fundamental biochemical abnormalities which underlie disease. The Editors hope that this program will stimulate the thinking of Clinical Chemists and of workers in related fields.

May 1958

HARRY SOBOTKA
C. P. STEWART

PREFACE TO VOLUME 2

The material for this volume has again been selected with the object of presenting a cross section through the growing points of clinical chemistry. It ranges from a review of the hypercalcemia of infancy, a clinical entity which can only be understood on the basis of the chemical situation involved, to a consideration of means by which the principles of automation can be applied in the hospital biochemical laboratory. Between these extremes lie the other subjects discussed: the principles and technique of zone electrophoresis as a complement to Owen's paper (in the first volume) on the clinical value and interpretation of paper electropherograms; the development of our knowledge concerning the elaboration and excretion of the bile pigments; the determination and significance of ammonia in blood; and amino aciduria, a vast subject related to a variety of congenital and acquired metabolic abnormalities.

Automation opens a new chapter in our discipline. Apart from the growing use of labor-saving devices in the organization of the laboratory and in such preparative work as the cleaning of glassware, certain trends have become apparent in the development of automatic or semiautomatic methods. As an important principle, one may discern the replacement of absolute volumes, obtained by pipetting, with the mixing of sample, diluent, and reagents in definite proportions kept constant by an ingenious mechanism. The chemistry of the methods, thus automatized, remains essentially unchanged although minor modifications may be introduced. In these processes it is the method as a whole which is calibrated (by standard solutions) rather than the individual measuring instruments. We may expect that in the future new principles will be applied, which lend themselves preferentially to serial analysis.

Automation will accelerate the change in the relationship between the clinician and the chemist. In the past, analytical procedures were worked out by the chemist but were simple enough to remain within the scope of the practical knowledge and skill that could be acquired by the medical student in the course of his curriculum. The increased range and complexity of the analyses which can yield clinically useful information, as well as the refinements and variety of the instruments involved, have meant that the new techniques cannot be adequately covered in a general preclinical course primarily intended to train the future clinician in chemical thinking and designed to acquaint him with the fundamental laws of animal metabolism.

This inevitable process of specialization will be considered by many as dangerous for the future of medicine. Even now the physician is removing himself progressively from the inspection of the patient's excreta and relying on a technician's written report; he cannot draw more information from this report than has been put in it within the compass of a dozen adjectives. These conditions among others induce the clinician to order a great many unnecessary analyses, for he naturally wishes to obtain the maximum of information, even though much of it may be of limited significance. This development is predicated upon economic and scientific factors beyond our control, and an important corollary is that the medical schools must combine in their teaching the proper utilization of laboratory data with the avoidance of mechanization in the practice of medicine. These considerations also hold for the application of digital computing methods to the study of disease, a subject of growing applicability which may be discussed in a future volume of this series.

We are grateful to many colleagues who have sent us constructive criticisms or have proposed topics for treatment in future volumes, and although it has not proved possible to include reviews on all these topics we hope that some, of continuing interest, will appear in Volume III. Meanwhile we invite further suggestions of subjects which are of immediate or developing importance.

October, 1959

HARRY SOBOTKA
C. P. STEWART

CONTENTS

LIST OF TABLES

PAPER ELECTROPHORESIS: PRINCIPLES AND TECHNIQUES

H. Peeters

St. Jans Hospitaal, Brugge, Belgium

Page

Introduction

Theoretical considerations of paper electrophoretic methods are few in comparison to the numerous papers dealing with the practical use and the clinical applications.

Probably no physicochemical method has ever been used so extensively with so little knowledge of its fundamentals. The reason seems to be that the apparatus and its manipulation are simple, and results are consequently so readily obtained that no theory is considered necessary. It is obvious, however, that many clinical results suffer from experimental inaccuracy owing to the lack of fundamental knowledge and even of practice in manipulating paper electrophoretic equipment.

A survey of techniques helps one to understand the principles on which electrophoretic separation on paper is based, enables one to appreciate the possibilities, and makes one aware of the limits and errors of the method.

New possibilities and safer results as well in the different fields of medicine as of other biological sciences are open with this method if the outline of its principles becomes better understood.

In this survey both the one-dimensional and the two-dimensional methods will be reviewed.

The unidimensional type of paper electrophoresis is an extension of free boundary electrophoresis, the method developed by Tiselius (T1). There are several differences between the two systems. One is the presence of a substrate (supporting medium) as anticonvectional medium during the electrophoretic separation. Another important difference is the starting point. In paper electrophoresis the entire load of material due to be separated is collected on the starting line, whereas in free boundary electrophoresis the material is present in equal concentration over one leg of the electrophoretic cell. Fortunately these differences simplify the qualitative and quantitative appraisal of separation after the run on paper, and for practical work both prove to be true inherent qualities and go far to account for the success of the method (K1, V1, W1).

In two-dimensional electrophoresis two forces are acting simultaneously at right angles to one another (G1, G2, H1, S1, S2). The first force is a vertical field due to gravity. Its origin is a vertical flow of buffer solution running along the paper from top to bottom. The second force is a horizontal field due to an electrical potential between two long electrodes applied vertically along the right and left edges of the paper.

The originality of this method has not been fully exploited in practice partly because of lack of appreciation of the theoretical principles and partly because of the practical difficulties in running the experiment. This accounts for the as yet limited success of this brilliant and new technique.

In this critical review the two methods will be considered successively. For each of them a description of the available apparatus with special regard to its structure in promoting and controlling the key function of migration will be followed by a study of the anticonvectional medium (also called the stabilizing medium or substrate), the buffer, and the electrical field. Next, the run itself will be considered with the factors involved. Under the third heading, the quantitation of results will be surveyed.

Finally, a consideration of the new possibilities will describe the trend of electrophoretic methods and indicate their probable development in the future.

If the electrophoretic tool is called upon to bring a definite advance to medical science, only accurate techniques and reproducible results under standardized conditions can warrant an important future for these hopeful but still imperfect beginnings.

References

G1. Grassmann, W., and Hannig, K., Ein einfaches Verfahren zur Analyse der Serumproteine und anderer Proteingemische. *Naturwiss.* **37**, 496 (1950).

G2. Grassmann, W., and Hannig, K., Ein einfaches Verfahren zur kontinuierlichen Trennung von Stoffgemischen auf Filtrierpapier durch Elektrophorese. *Naturwiss.* **17**, 397 (1950).

H1. Haugaard, G., and Kroner, T. D., U.S. Patent 2,555,487 (1948).

K1. König, P., *Actas e trabalhos do Terceiro Congresso Sud-Americano de Quimica Rio de Janeiro e São Paulo.* **2**, 334 (1937).

S1. Strain, H. H., and Sullivan, J. C., Analysis by Electromigration plus chromatography. *Anal. Chem.* **23**, 816 (1951).

S2. Svensson, H., and Brattsten, I., *Arkiv Kemi* **1**, 401 (1949).

T1. Tiselius, A., The moving boundary method of studying the electrophoresis of proteins. *Nova Acta Regiae Soc. Sci. Upsaliensis* [4] **7** (4), 1-107 (1930).

V1. von Klobusitzky, D., and König, P., Biochemische Studien über die Gifte der Schlangengattung Bothrops. *Arch. exptl. Pathol. Pharmakol.* **192**, 271 (1939).

W1. Wieland, T., and Fischer, E., Ueber Elektrophorese auf Filtrierpapier. Trennung von Aminosäuren und ihren Kupferkomplexen. *Naturwiss.* **35**, 29 (1948).

Part I. Critical Review of One-Dimensional Methods

1. General Conditions of the Electrophoretic Separation, i.e., "The Run"

The conditions for a satisfactory run include a series of physical factors in connection with the *electrophoretic chamber, the paper, the buffer, and the electrical field.* While each of these has its separate effect, the interaction among them must be considered, for it is the electrical current passing through a buffered paper strip that causes electrophoretic migration.

1.1. THE FACTORS CONSIDERED IN ISOLATION

1.1.1. *Apparatus*

The general description of any satisfactory apparatus can be given as follows. Filter paper is soaked in buffer and stretched in such a way that each extremity is in contact with a buffer vessel, which contains an electrode receiving current from a DC source. In this way an electrical field lies along the paper. Probably *evaporation* of water is one of the most important physical forces active during the electrophoretic run, apart from the electrical field itself, and this may be compensated by a flow of buffer solution originating in both buffer vessels. *Gravity* may also cause shifts of water between different parts of the apparatus so that hydrodynamic forces are an important cause of difficulties during the electrophoretic separation.

The power supply, the electrodes, and the buffer vessels, as far as electrical conditions are concerned, will be described under the heading of electrical field (Section 1.1.4.).

1.1.1.1. *Classification According to Evaporation.* Any moist paper goes on losing water through evaporation until the surrounding atmosphere is vapor saturated. Some few types of apparatus suppress any evaporation by covering the paper with an impermeable surface. Most designs, however, do allow some sort of evaporation, but some of them favor, while others try to avoid, the phenomenon, the former being designated "electrorheophoresis equipment." However, all apparatus in which the surface of the paper is not sealed presents some rheophoretic component, which is often far larger than is commonly supposed.

(*a*) *Significance of evaporation.* Evaporation is easily demonstrated when paper, soaked in buffer solution, but without any current passing, is exposed to air during a certain time. At the end of the experiment the paper is cut up into fragments which are eluted in triple-distilled water. Sodium, one of the buffer components, is determined by means of flame photometry (D5). The quantity of sodium grows higher toward the middle of the paper and reaches a maximum halfway between the buffer vessels. When the experimental period is prolonged, the concentration rises with time (Fig. 1)

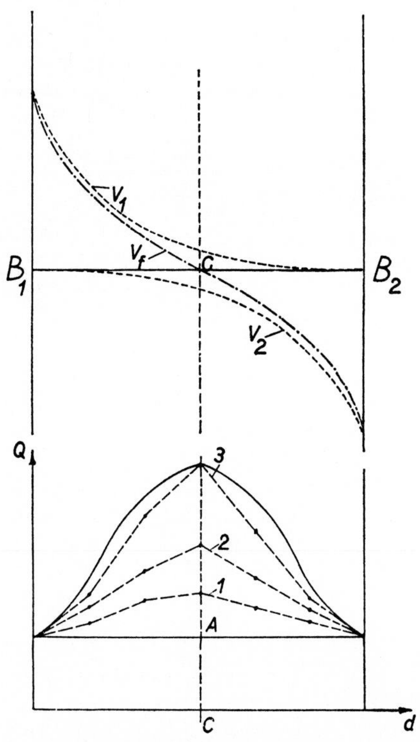

FIG. 1. Influence of evaporation on hydrodynamics (upper half) and salt distribution in the strip (no current applied) (lower half). A flow of buffer originates in each buffer compartment (B_1 and B_2) with a velocity V_1 and V_2. The resultant V_f has a velocity $V_f = 0$ in the center of the strip (C). There is a corresponding increase in the quantity of salt. Quantity of Na (Q) plotted against distance d increases above the original quantity (A) toward the center of the strip (C). The different curves 1, 2, and 3 give the results after 1, 2, and 3 hours evaporation (A8).

Relatively higher values are found along the edges of the strip due to evaporation from the central path toward the edges (Fig. 2). The additional effect of the electrical field on the distribution of salt will be considered later (see Section 1.1.3).

Inside a sealed box there will be evaporation as long as the atmosphere is not saturated with vapor. After that no further loss by evaporation is possible. But any condensation of water on the lid or on the walls of the chamber will call for further evaporation. It is clear that evaporation, and consequently the salt concentration in the paper will increase when the electrical energy provided by the current is converted to heat while passing through the strip.

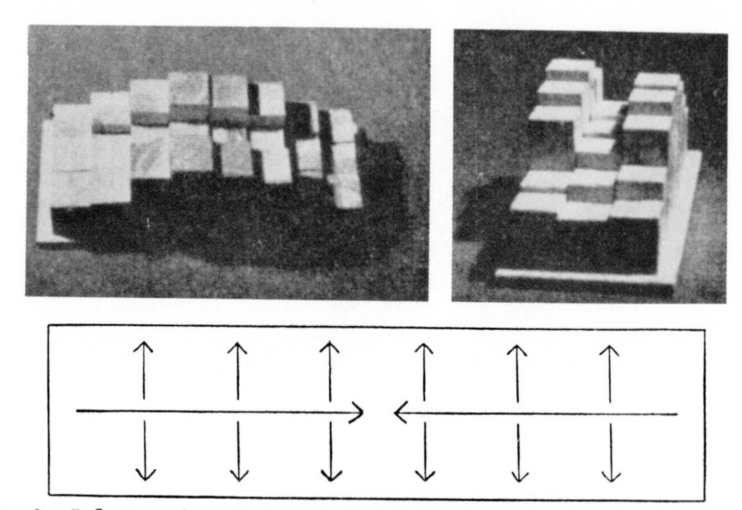

FIG. 2. Influence of evaporation on salt distribution according to the length and the width of the strip (no current applied). Apart from increasing toward the center of the strip, the salt concentration also increases toward the edges. Schematic representation of the two simultaneous evaporation flows explains the model.

A theoretical account of evaporation has been given by several authors (A8, D6, D18) since the first description and application of the phenomenon by Macheboeuf in 1953 (M2). A good mathematical approach is given by Audubert and de Mende, using the following expression for the evaporation flow v_f:

$$v_f = v_0^{-kx}$$

where v_0 is the speed of flow at the point leaving the buffer vessel and x is the distance from the buffer vessel (A8). Water evaporating at the surface is compensated by water flowing through the paper capillaries from the buffer vessel. As water is lost in each segment, there is an exponential decrease in the quantity of water arriving in the following segment.

The buffer salts follow the same rule, and there is an exponential increase in the quantity of salt present in the strip. This is proved by the shape of the curves in Fig. 1. If evaporation were a linear phenomenon, salt distribution would show a linear increase.

The evaporation flow is important in connection with migration, as the experimental velocity V_e of a particle moving on the strip in a given direction is the absolute velocity of that particle V modified by the evaporation flow v_f.

$$V_e = V \pm v_f$$

The resulting experimental speed is shown in Fig. 3. Special attention to the electroendosmotic flow will be given later.

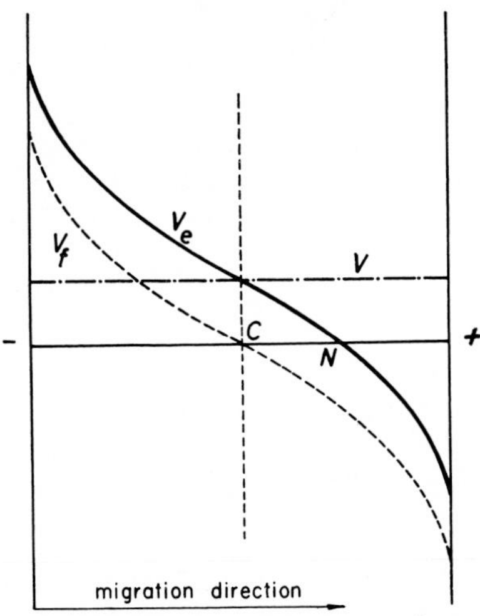

FIG. 3. The experimental velocity V_e is the resultant of the electrophoretic velocity V and the velocity of the hydrodynamic flow V_f in each point of the strip. There is a point N where $V_e = 0$. The migrating particle is unable to leave this point (A8).

If evaporation velocity is equal to the electrophoretic velocity but opposite in direction, the result of the interaction is the immobilization of the migrating ion on a definite point of the strip. If we consider the diagram, it is easy to understand why a particle must find its restpoint N independent of the actual spot of the paper where it was originally applied. This phenomenon explains the experiment of Fig. 4.

Among the applications of this principle is the possibility of applying a diluted sample over a large surface of paper and ending with a concentrated fraction suffering but little from diffusion.

(*b*) *Types of apparatus.*

(1) *Apparatus without evaporation* (sealed strip). To seal the paper, either a nonconducting solid (K22) or a nonconducting liquid immiscible with water can be used.

The use of glass plates (B8, D14, K22), greased or siliconed, or plastic-coated metal plates (M25) has been proposed. Complete suppression of evaporation and an excellent possibility for cooling, by immersion of the whole system into an organic liquid (C6) are two procedures which allow rapid work under high potential. In practice the trapping

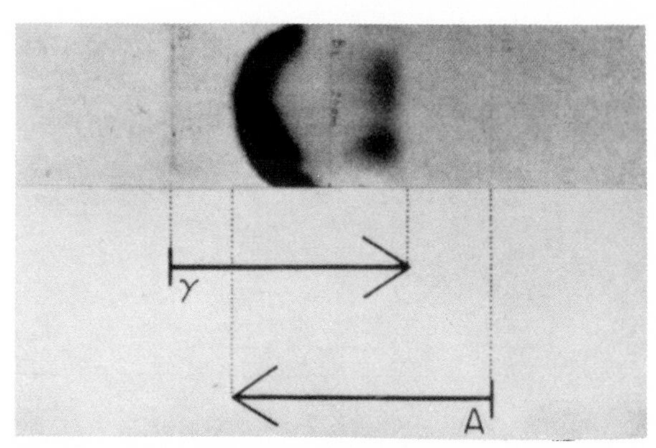

FIG. 4. Influence of evaporation on migration distance (rheophoresis). Albumin (A) and γ-globulins (γ) can be applied so as to migrate across themselves in opposite directions toward their points of immobilization.

of air bubbles and the existence of a buffer layer between paper and support have led to the use of pressure, which is, however, difficult to maintain uniformly over a long surface. Only thick glass affords good uniform pressure (K21), but then cooling poses a new problem.

The use of organic liquids, such as hexane (H5), heptane (D15), toluene (M24), chlorobenzene (C3), carbon tetrachloride (C4, S4) in direct contact with the paper was less successful because interaction between the organic solvent and the buffer material under investigation cannot be avoided.

In my own laboratory we have tried a new solution to this problem. The paper is sealed by means of a layer of plastic (*Nobecutane*) which is sprayed over the moist substrate, and this gave satisfactory results in both one- and two-dimensional techniques (unpublished).

(2) *Apparatus with evaporation* (open strip). The use of air as adjacent phase to the paper remains the easiest solution, but the surrounding air enclosed in the chamber becomes humid at the cost of the paper.

Helium and hydrogen (M8) have been used because these gases are better conductors than air, but in practice they provide little improvement. Some investigators have tried to humidify the air by introducing a flow of hot water (H15), and we ourselves tried hot vapor, but without satisfactory results. Under these conditions cooling became a real problem, for condensation occurred on any colder wall of the chamber and could not easily be controlled.

Usually the strip hangs free in a closed chamber where vapor saturation eventually occurs spontaneously. Not until Macheboeuf (M2) proposed to accelerate and regulate evaporation was much attention given to the problem, although spontaneous migration of icteric fractions (i.e., without application of electrical potential) was observed by many investigators. This spontaneous migration goes toward the center of the strip and is of course very fast during the first minutes after the lid of the chamber has been closed, especially if current is applied immediately afterward.

Evaporation forces the whole sample rapidly towards a straight and narrow zone at the center. This can easily be observed if the serum sample is stained with a trace of bromophenol blue (V7). Only after some time do the β- and γ-globulins migrate backward toward the starting line, while the albumin and α-globulin migrate normally forward for some time (Fig. 5). This migration of the fractions is slower than the first rheophoretic movement.

To avoid the loss of vapor-saturated air when applying the sample, a slot in the top of the box should allow the application of the sample without disturbing the atmosphere inside (V1). Only a few designs are satisfactory with respect to this very important point.

For further description it is convenient to divide this type of apparatus into three classes according to moderate, forced, or reduced evaporation.

Moderate evaporation. In describing the chamber of most apparatus, whether commercially produced or not, no mention of volume is made. The multiplicity of designs is further evidence that no importance is generally ascribed to the actual size or shape (D13, G17, K10, O1, S4, S21, W5, W11).

The sealing of the chamber is usually by no means complete. Often a lid merely lies on the box, but as it does not always fit exactly and is commonly of plastic material, there is often warping (perhaps after some months), and a gap occurs between lid and box. Any chamber can be sealed correctly if a water-filled moat runs around the top edge of the box and the lid fits into it. It is practical to add some dye to the fluid in the moat in order to watch easily any drying out. An adaptation of this basic design allows for easy and perfect sealing of any existing equipment (P4).

The length and the number of strips is also varied in an arbitrary manner, although in determining the rate of evaporation the paper surface area is an important factor.

Forced evaporation. In this type of equipment, introduced by Macheboeuf (M2), evaporation is favored and also controlled by means of some porous adsorbent inside the lid, or by having a porous material form the lid or the wall of the apparatus (D18). Some designers make a hole in the roof through which the vapor-saturated air escapes regu-

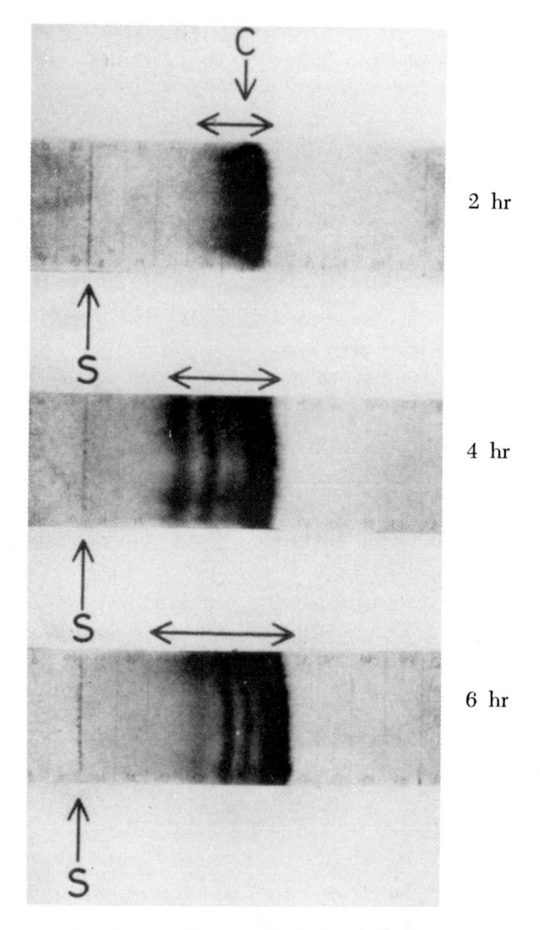

FIG. 5. In the beginning of the run the evaporation flow pushes the unseparated fractions from the application zone (S) toward the center of the strip (2 hours). After 4 hours there is some opposite migration of globulins, and albumin progresses only slowly as they get nearer their immobilization points (6 hours).

larly (M21). This type of equipment, however, is very sensitive to drafts, which cause a sudden increase of evaporation.

It must be stressed that there is no fundamental difference between this so-called "rheophoresis" equipment and the usual types, but these authors emphasize evaporation as having a real bearing on migration.

Reduced evaporation. Evaporation can be kept down by various means. One is to use a small and easily saturated chamber and this can be obtained merely by good design (K18, L6). To achieve rapid saturation a damp cardboard may be fixed just beneath and above the strip (F7).

Another means of keeping evaporation low is to avoid condensation on lid and walls as much as possible by having a very thin-walled chamber. The usual glass or Plexiglass or other plastic is not satisfactory, but a thin plastic sheet, self-supporting through the angle folds or stretched on a very light plastic or metal frame is an easy solution (P4, S21). It must be noted that the authors describing thin plastic boxes do not mention low distillation as an argument.

Low evaporation can also be obtained if a cooling coil is arranged inside the chamber, eliminating the major part of the heat normally dissipated through evaporation (O4).

Placing the entire equipment inside the refrigerator gives only a relative advantage, because the current itself creates a rise in temperature. At low temperatures, however, conductivity is smaller, less current passes, and as an indirect result, less evaporation occurs. In practice, working at room temperature is the simplest, and the runs are longer because adsorption of protein to paper is lower at higher temperatures. Even in the oven at 36° C perfect runs can be made (Fig. 6).

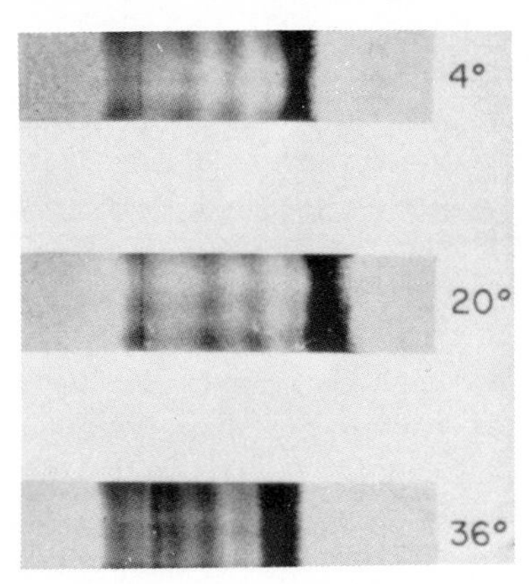

FIG. 6. Influence of temperature on migration. At any temperature a fair pattern can be obtained (P6).

When using very high potentials, as in high voltage electrophoresis (K6, M24, M25, V8, W4), positive cooling is of foremost importance and is best achieved by means of a refrigerator system with a cooling coil on which the paper rests in the bottom of the chamber.

1.1.1.2. *Classification According to Hydrodynamics.* Water moves through the capillary system of the paper as a result of several physical forces: capillarity, evaporation, and gravity.

The third of these can be eliminated by leveling the two electrode compartments, thus preventing the buffer solution from siphoning through the paper strip. This can be done by tipping the apparatus slightly or by connecting the two buffer compartments by a narrow tube (which, because of the low conductivity of the buffer, may be left open during the run).

The strip itself is stretched between the two buffer compartments. In most cases the strip is fixed on a separate frame which can be lowered into the buffer vessel. If it is stretched *horizontally* as recommended by Wieland and Fischer (W5) and many others, at a level only slightly higher than that of the compartments, it will contain too much water in proportion to the water capacity of the paper and will tend to sag unduly. Excessive water content causes reduced anticonvectional power together with greater electrical conductivity, i.e., more loss of energy as heat. With some experience water saturation of the strip can be judged by the sheen of the paper. It is evident that a horizontal strip stretched at a rather high level above the buffer vessels will have a low water content which is easily kept constant, at least if evaporation is small (V1), because the opposing effect of gravity on capillarity is minimized.

When the strip is hung over a rod in the form of an inverted V in the so-called "tent" or "church" type (D13, F3, W21), excess fluid is avoided in the highest part of the paper, but the lower ends are very wet and there is a steep water gradient along the paper. The water gradient is not linear but follows an S-curve (L10; Fig. 7). Moreover, about half the length of the strip is of no use.

A few authors (G7) support the paper at both ends, but with one end higher than the other. The shape of the strip is intended to counteract electroendosmosis.

The best way of keeping the strip taut is by means of appropriate weights; other designs stretch the strip by means of the lid of the chamber (S23) or by means of holders with adjustable tension (O4), but the commonest method is by adhesion of its descending parts to the walls of the buffer vessel. The use of a support such as nylon thread (D18) should be avoided because it spoils the paper structure and so interferes with the reading. A fakir type supporting board proves satisfactory (D5, V1).

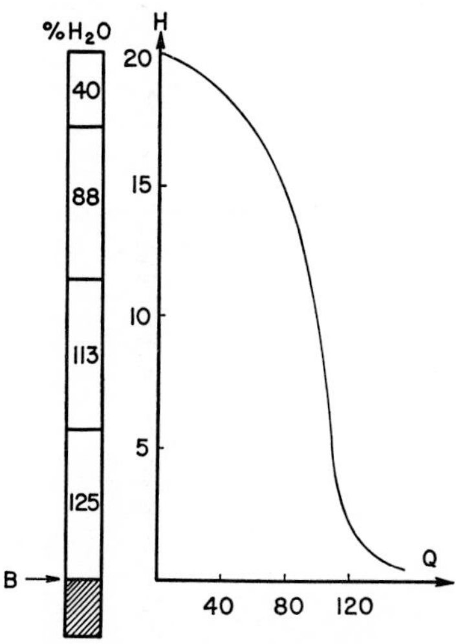

FIG. 7. Water content of paper in ascending chromatography. Paper, Schleicher and Schüll 598 G (data according to Grüne, G23); *B*, water level; *H*, height above water level; *Q*, water content of the strip expressed in percentage of the paper (w/w).

1.1.1.3. *Summary.* All types are summarized in Fig. 8a and b where the general layout of each system is given schematically.

The design of a good apparatus should have the following features: a small chamber with thin walls, hermetic seal, no break of the seal when applying the sample after a preliminary saturation period, easy leveling of buffer compartments, easy stretching of the strip, a horizontal strip at a moderate distance above the buffer level. The design of such an apparatus is given schematically in Fig. 9.

1.1.2. *Paper*

König (K18) published the first work on paper electrophoresis, but it seems that the success of paper chromatography many years later gave the real start to the general use of paper electrophoresis. Other anticonvectional media, such as agar gels, starch gel, and cellulose acetate, present advantages for some types of work, but the ease of handling paper strips is difficult to surpass.

Paper is nearly pure cellulose which is a polymer of β-glucopyranose, the glucose molecules (numbering up to 3000 in the short cotton fibers, "linters" used for filter paper) being linked by β-glucoside bonds with cross linkages involving secondary valences or hydrogen bonds. Some breakage of the glucose rings provides free aldehyde groups (and thence carboxyl groups by oxidation), but this occurs much less in filter paper from "linters" than in paper from wood pulp. Filter paper is hygroscopic (though much of the lyophilic property of glucose is lost in the condensation to cellulose). Consequences of this structure are: (*a*) positively charged particles tend to be adsorbed because of the cation-exchanging power of the carboxyl groups; (*b*) in the alkaline pH range the buffer tends to flow toward the cathode, the immobile paper phase

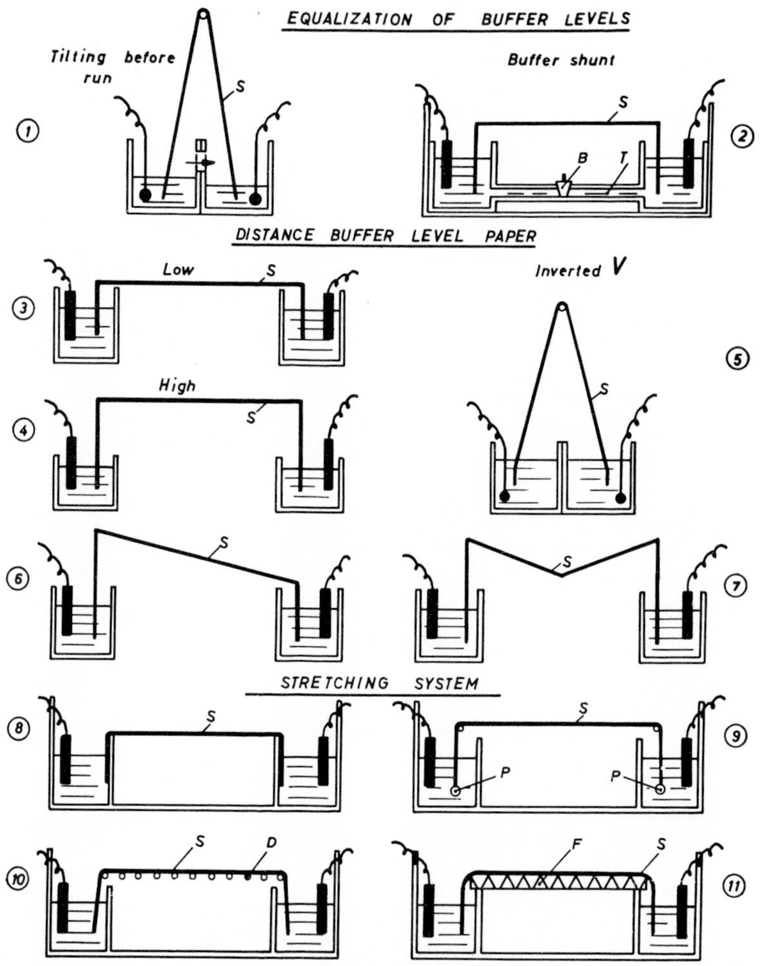

FIG. 8a. Devices according to hydrodynamics. For key to letters see Fig. 8b.

being anionic (endosmotic buffer flow); (*c*) in the acid pH range, the carboxyl groups being relatively undissociated, there is less interference with the migrating protein molecules (G23). The lacunar spaces inside the cellulose network are 10–600 Å wide (H10, S2), and the structure of cellulose consequently provides an enormous surface (500 or 600 m^2/cm^3 of cotton fiber).

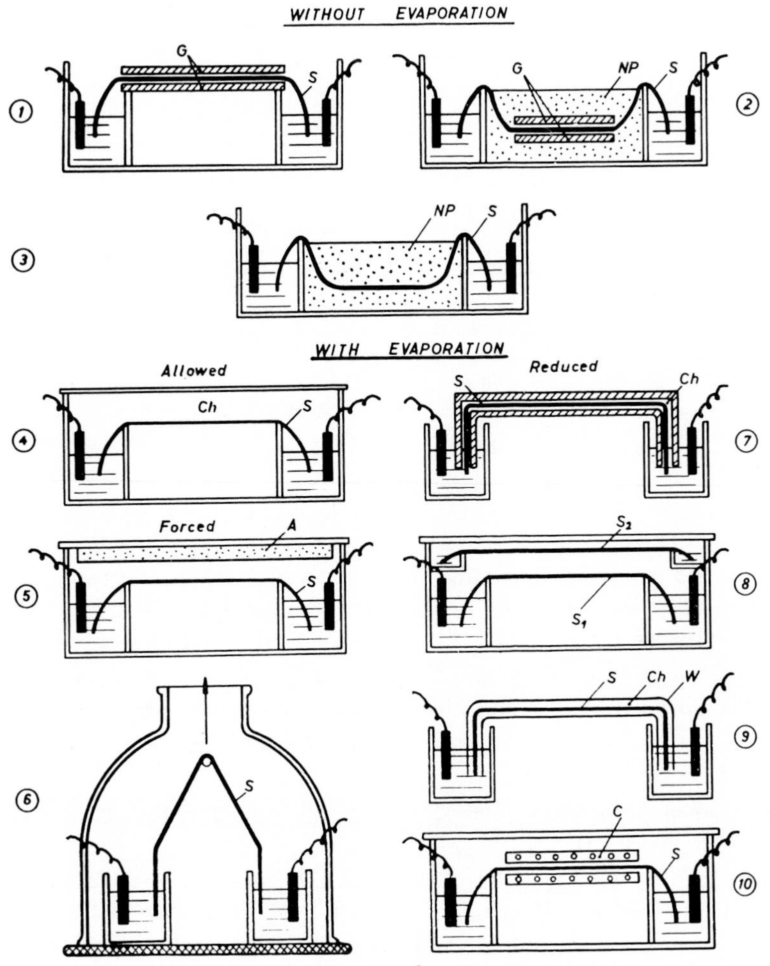

FIG. 8b. Devices according to evaporation.

KEY: *A*, vapor-adsorbing medium; *B*, stopcock; *C*, cooling system; *Ch*, chamber air; *D*, nylon threads; *F*, fakir-type supporting board with nylon tabs; *G*, glass plates; *NP*, nonpolar liquid; *P*, plastic-coated weights; S, substrate; S_1, electrophoretic strip; S_2, saturating strip; *T*, tubing; *W*, thin plastic wall.

Structural qualities. The sheets from which strips are cut usually measure 58 by 58 cm or 58 by 60 cm. *Thickness* has a bearing on evaporation, as the surface is larger at the edges for thicker than for thinner sheets.

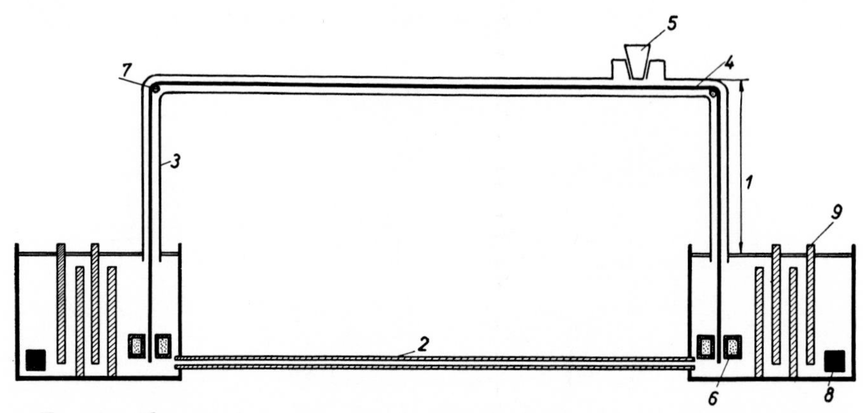

FIG. 9. Theoretical arrangement for an ideal zone electrophoresis chamber. Hydrodynamics: strip high above buffer level (1) and equalization of buffer level (2). Low evaporation: thin wall (3), small chamber (4), no loss of vapor-saturated air (5). Taut paper: plastic-coated magnets (6) and strip support (7). Electrical field: electrodes over the full width of the vessel (8). Electrolysis products: efficient baffle system (9).

The *weight* expressed in g/m^2 of the absolutely dry sheet is correlated with the mesh and solidity, loose papers being lighter and often less resistant to stretch. Normally paper contains some moisture in accordance with the humidity of the atmosphere, about 7 % of dry weight in normal room air and up to 20 % in vapor-saturated atmosphere.

The *strength* of paper is much less when the paper is wet because the water tends to pull the hydrophilic paper fibers apart. The *width* of the pores is 1–12 μ. The macroscopic structure cannot be expressed in mathematically accurate figures.

A paper of perfectly *regular structure* and without clouds is important for direct photometric readings.

Machine direction (in manufacture) is the direction of the faster flow and is to be found in all papers, although it is far smaller in the special quality made for electrophoresis or chromatography, where this machine effect is deliberately kept as small as possible. It is often indicated by an arrow on the wrappings or printed as a watermark in the sheet itself.

It may be determined by some simple method. A drop of buffer becomes oval when falling on dry paper; the longer axis of the oval is the machine direction. A better method is the following. When folding a small square of paper (5×5 cm) between the fingers, the paper folds along and cracks against the machine direction. This is easy to demonstrate with corrugated paper, which has a macroscopic machine direction (Fig. 10). It is important to cut the strips always in the same direc-

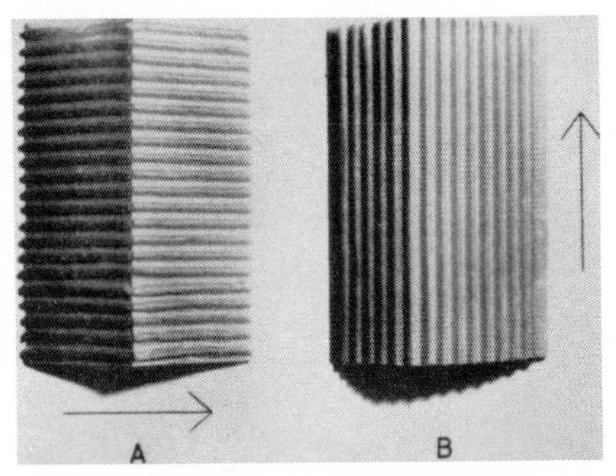

FIG. 10. Determination of machine direction. Paper folded against machine direction cracks (A); paper folded along machine direction bends (B). From reference P7.

tion to obtain identical stretch after wetting the paper and uniformity in the length of the migration path as well as constancy of the anticonvectional effect due to the paper structure. Diffusion is greater along the machine direction, as the flow has a tendency to be faster. Strips cut against the machine direction give the best results in zone electrophoresis.

Functional qualities. There is first of all the relationship between paper and buffer, and this has several aspects.

Capillarity of the paper with respect to thickness is called the *filtering capacity* and expressed, according to Herzberg, as the time in seconds needed for 100 ml of water at 20–21° C to filter through a surface of 10 cm^2 under a constant pressure of 5 cm of water. The corresponding value may be given for a filter with a diameter of 15 cm.

Capillarity in relation to the length of the paper or "*water-imbibing capacity*" may be expressed, according to Klemm, as the height in millimeters to which water rises in a vertical strip 15 mm wide by 250 mm long, dipping into water at 20°C, after 10 minutes of imbibition. In Table 1 a few figures are given for commercially available papers.

The water (or buffer solution) content of the paper is an important factor in determining the current and thus the rate of evaporation and the protein migration. Of the total, which may be 200 % of the dry

TABLE 1
COMPARATIVE TABLE OF ANALOGOUS COMMERCIAL PAPERS

Schleicher and Schuell	Binzer "Ederol"	Whatman	Machery, Nagel *et al.*	Weight g/m²	Sucking velocity mm/10 min	Sucking velocity mm/30 min
2040 a	201	1	60/261	85–90	55–60	115–125
2040 b	202	1	60/261	120–125	55–60	115–125
2040 b/Gl	202/S			120–125	55	115–125
2043 a	207	4	60	85–90	45–50	75–85
2043 a/Gl	207/S			90–95	45	90–100
2043 b	208	4	60	120–125	45–50	90–100
2043 b/Gl	208/S			120–125	45	90–100
2045 a	214		63/263	90–95	30–35	60–70
2045 b	215			120–125	30–35	60–70
2045 b/Gl	215/S			120–125	35	60–70
598 G				140-145		145–155
2071				630–680		65–70
2230				700–750		160–170
	225			180	150	
	226		214	140	50–60	
		3MM		160		

weight of the paper, about 7 % of the dry weight is fixed to the hydroxyl groups, about 25 % is held firmly by capillarity, and the remainder is held loosely in the pores and on the surface of the paper. The water content will be most nearly constant throughout if the paper is held horizontally well above the buffer level.

The interaction between the *paper and the buffer when the field is applied* causes the electroendosmotic flow.

It is easy to demonstrate this phenomenon by adding a test substance (e.g. glucose) to the buffer vessels or at any point on the paper before the run and testing the distribution on the paper afterward (B8, C7, D6, D18, J2, P8, T2, W15, W16; Fig. 11).

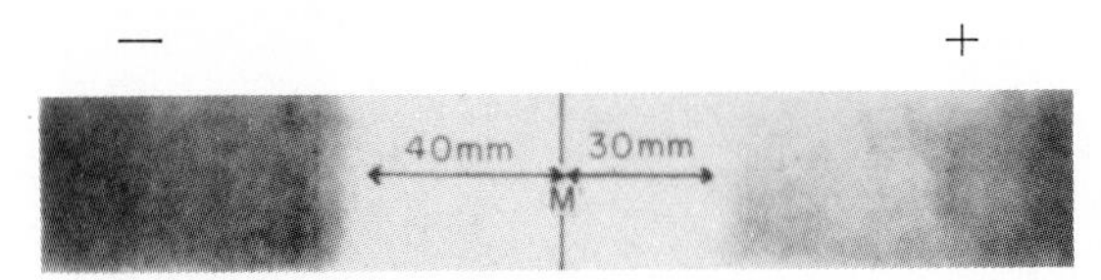

FIG. 11. Zone electrophoresis: 110 volts, barbiturate buffer, pH 8.6, $\mu = 0.1$, 4 ma per strip during 4 hours. Glucose was added to the buffer in each electrode compartment. Rheophoresis causes movement of glucose toward the center (M). The effect of electroendosmosis can be seen as a flow toward the cathode. The strip was developed by means of heat. From reference P8.

It is clear that evaporation is superimposed on the electroendosmotic flow and that partial measurements on the open strip account for both phenomena at once (Fig. 12).

Moreover, methods of this type give only an approximate indication, the results being strictly applicable only to the test substance used, since molecular shape and volume influence the electroendosmotic effect. Theoretically the movement of water molecules and not that of dissolved migrant should be measured (M12).

None of the methods proposed for neutralizing the electro-endosmotic flow is effective and in practice the phenomenon is tolerated. Cellophane barriers (M25), still used in two-dimensional techniques, are inefficient because cellophane itself has a negative potential. Other suggestions include a hydrodynamic counterflow, obtained by arranging the cathode at a higher level than the anode and saturating some of the active groups of the paper by an adsorbable ion (S28) or modifying the polarity of the paper by chemical action (J5).

There is finally the *interaction between the paper and the protein or other ions migrating on the buffered strip.* Adsorption of the migrating ion on the cellulose resists the driving force of the electrical field. As adsorption is different for each component of a mixture, it may alter the electrophoretic velocity in such a way that two particles with a different mobility happen to migrate together as an apparently homogeneous fraction. On the other hand two particles with an identical electrophoretic mobility may actually move at different rates because of a different adsorption and may show up as two fractions.

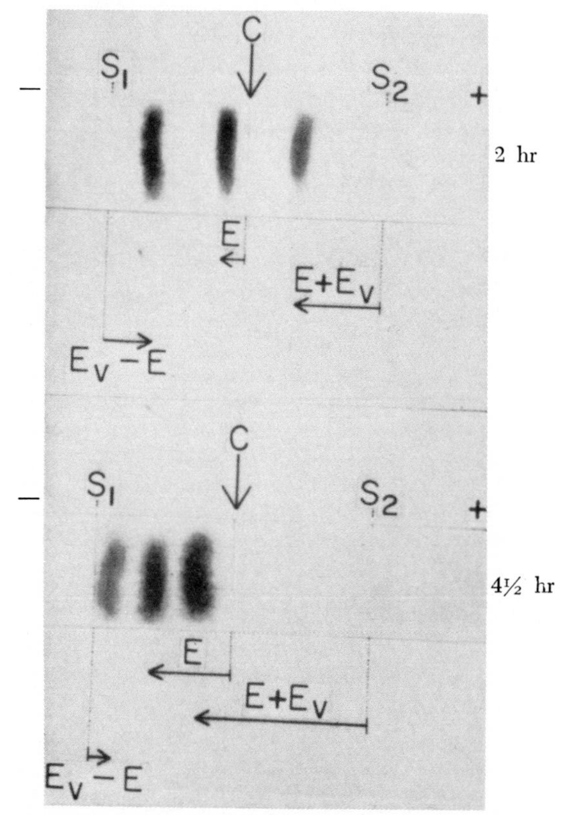

FIG. 12. Influence of evaporation and electroendosmosis. Glucose applied at S_1, C, and S_2; E_v, evaporation flow; E, electroendosmotic flow. Upper run stopped after 2 hours; lower run stopped after 4½ hours. Strips developed by heat.

For albumin there is an irreversible adsorption which is the cause of "trailing" and thus of a quantitative error in the estimation of protein. About 0.7 μg of protein is said to be lost for each square centimeter of paper traversed (J4, M18), but this estimate seems too high (A1). However, the globulins running on a protein-saturated path do not suffer any loss through adsorption during migration. Buffer composition and pH have a great influence on this effect, and fibrinogen, which does not leave the starting zone in barbiturate at pH 8.6, migrates normally in phosphate at pH 9.2 (B7). The preliminary impregnation of paper by a protein was used by Kallee (K1) to prevent the adsorption of iodinated insulin. For work on serum the impregnation of the paper with albumin gives a practical means for correcting the trailing of albumin (H4, K20; Fig. 13; Tables 2 and 3).

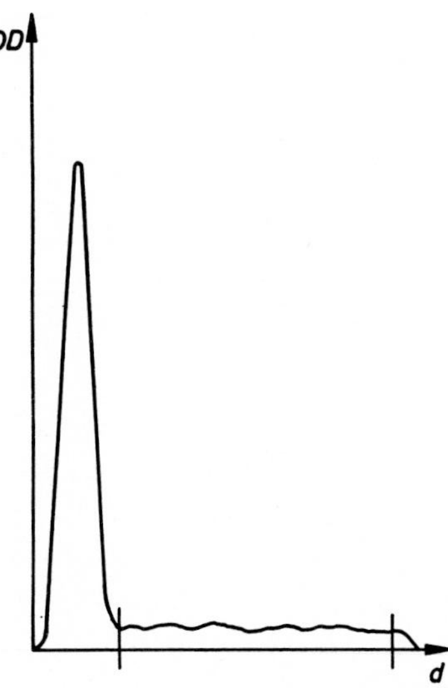

FIG. 13. Scanning (OD/*d*) of a run of pure albumin (K20). OD, optical density; *d*, distance of migration.

TABLE 2
SERUM PROTEIN LOSSES THROUGH ADSORPTION ON THE STRIP (K20)

Fraction	Number of strips	Temperature, ° C	Migration distance (cm)	Surface of adsorbed fraction (planimetric units)	Adsorbed protein (planimetric units/cm migration distance)
Albumin	10	20	6.6	0.24	0.036
		6	6.7	0.53	0.079[a]
α_1-Glycoprotein	10	20	5.6	0.19	0.034
α_2-Glycoprotein	10	20	4.6	0.18	0.039
β_1-Globulin	10	20	3.2	0.12	0.037
γ-Globulin	10	20			
					0.037 average

[a] Not included in average.

TABLE 3
ELECTROPHORESIS OF NORMAL SERUM ON PAPER TREATED WITH 0.01 % ALBUMIN (K20)[a]

Treatment	Albumin	Globulin			
		α_1	α_2	β	γ
Without impregnation	47.8	5.5	10.6	11.8	24.3
With impregnation	56.7	4.6	8.5	10.8	19.6

[a] Results are given as percentage of total protein.

Impregnation of the paper with alumina is reported for lipoproteins (B15, D2). Some authors have accentuated the disadvantages of the paper substrate in order to obtain some specific effect (D15), but in practice the results in work on serum proteins were small. Rather than complicating the problem by specific adsorption, a new and neutral substrate showing neither physical nor ion exchange adsorption seems to be the ideal goal.

The shorter migration distance on paper than in free solution has been explained on the basis of a longer migration path along the tortuous capillaries of the paper (A8, D15, K22), with a field actually smaller than that calculated from the potential measured at the ends of the strip. It seems impossible to make absolute mobility measurements on paper, although relative mobilities may be obtained for clinical use (M22).

Although it is possible, for a given apparatus, paper, and set of ions, to calculate on the basis of capillary length a relationship between migration distance on paper and in free electrophoresis, any such formula must be applicable only to constant working conditions, since a variation in paper wetness, for instance, must affect the buffer concentration and hence the conductivity and migration rate. Other factors include ionophoretic movement of the buffer ions (M14, M15), the dimensions of the migrating ions (P23), and hydrodynamic factors. No theretical formula can at present take into account all these variables, but a practical conclusion from the simple consideration of capillary length is of interest. It appears that measurement of the resistance or conductivity of the strip is more important than determination of potential. A good general idea of the conditions is given by following voltage and amperage simultaneously, keeping one constant and paying particular attention to the other.

1.1.3. *Buffer*

Both the solvent and the solutes of the buffer solution must be discussed in relation to their functional properties. The pH is important in connection with mobility of the protein, and the ionic strength affects the conductivity of the buffer. A new and probably important approach is the study of the ionophoretic pattern of the buffer ions inside the supporting medium; this result of the electrical field is, together with evaporation, a major cause of local disturbances in buffer ion distribution.

Solvent. Water is the usual solvent, but because it evaporates very easily some workers (D15) have used glycerol or analogous substances to reduce the vapor tension. These, however, have higher viscosities than water and will affect the mobility u of the migrating particle according to the equation

$$u = \frac{Q}{6\pi r \eta}$$

where Q is the net charge in *esu* (electrostatic units), r the radius of the particle in cm, and η the viscosity coefficient of the solution. Electro-endosmosis is also inversely proportional to viscosity, and this is an interesting feature.

For clinical application in paper electrophoresis, however, the use of additives which decrease vapor pressure has not become established usage, but may receive renewed interest with new substrates (supporting media) where it is difficult to maintain the water content at a sufficiently high level.

Buffer salts. An ideal buffer should have the following qualities. The pH should be near the p*K* of the acid to give good buffering capacity; the two ions should have the same migration velocity to avoid irregular distribution of the buffer ions due to ionophoresis; only monovalent ions should be present to avoid heavy double layers and thus provide good mobility; conductivity of the buffer should be low to avoid high current and undesirable effects of heat production (Table 4). Organic substances and ampholytes are interesting from this point of view, and barium ions seem to have the same effect (J7). If direct evaluation in the ultraviolet is to be done, buffer salts which do not absorb at the same wavelength as proteins are desirable. Glycine (W17), borate, and phosphate (B5) are satisfactory in this respect, but Veronal is not. Adsorption of buffer salts by the proteins is different according to the salt and to the protein fraction (L13). The splitting up of β-globulins into β_1- and β_2-globulins (L6) is partially due to the presence of calcium ions (L5). In general the buffers used in electrophoretic work have a low buffering capacity, due to the need for low ionic strength and conductivity.

pH. The pH of the buffer of course determines the net charge of the migrating particle. For electrophoretic work on proteins it is usual to employ an alkaline pH of 8.6 at which all proteins are anions and at

TABLE 4
BUFFERS GROUPED ACCORDING TO THE INTENDED SEPARATION

Kind of buffer	pH	μ	Components in 1 liter	Voltage	Time, hours	Reference
For proteins Elphor (Na)	8.6	0.1	9.809 g Na barbiturate 3.9 g Na acetate 60 ml 0.1 *N* HCl	5 volts/cm	16	(M23)
Barbital	8.6	0.05	10.3 g Na barbiturate 1.84 g Barbital			(M26)
Barbital	8.6	0.075	15.45 g Na barbiturate 2.76 g Barbital	270 volts	4.5	(D15)
Elphor (K)	8.6		120.7 ml *N* KOH 4.77 ml Acetic acid 100% 8.75 g Barbital 60 ml 0.1 *N* HCl			
Rheophor	8.6	0.05	10.05 g Na barbiturate 90 ml 0.1 *N* HCl			(D18)
Complexon	8.6		5.88 g Na barbiturate 3.88 g Na acetate 43.6 ml 0.1 *N* HCl 40 mg EDTA (Complexon III)			
Phosphate	7.4		0.6 g $NaH_2PO_4 \cdot H_2O$ 2.2 g Na_2HPO_4			(D15)
Borate	9.0		7.63 g Na borate 0.62 g Boric acid			(D15)
Borate	8.6		8.80 g Na borate 4.65 g Boric acid			(D15)

TABLE 4 (*continued*)

Kind of buffer	pH	μ	Components in 1 liter	Voltage	Time, hours	Reference
For mucoproteins						
Acetate	4.5	0.1	3.51 g Sodium chloride 3.28 g Sodium acetate adjust to pH 4.5 with *N* HCl	1.6 volts/cm	16	(S22)
Acetate	4.5	0.28	7.2 g Magnesium sulfate 3.28 g Sodium acetate adjust to pH 4.5 with *N* HCl	1.6 volts/cm	16	(M4)
Acetate	4.5	0.19	7.14 g Sodium sulfate 3.28 g Sodium acetate adjust to pH 4.5 with *N* HCl	1.6 volts/cm	16	(M4)
Phosphate	4.4	0.2	9.44 g Disodium phosphate 10.3 g Citric acid	1.6 volts/cm	16	(M4)
Citrate	4.5	0.133	28.46 g Sodium citrate 20.6 g Citric acid	1.6 volts/cm	16	
Barbiturate	8.6	0.1	20.6 g Sodium barbiturate 3.68 g Barbital			(S22)
For hemoglobin						
Barbital	8.5	0.05	5.0 g Sodium barbiturate 3.33 g Hydrated sodium acetate 1.0 g Hydrated copper sulfate 34.2 ml 0.1 *N* Sulfuric acid	250–300 volts	16–20	(M26)
Barbital	8.6	0.05	10.30 g Sodium barbital 1.84 g Barbital	270 volts	4.5	(G6)
Phosphate	7.8	0.12	0.294 g $NaH_2PO_4 \cdot H_2O$ 3.25 g Na_2HPO_4	6.3 volts/cm	17	(G6)

TABLE 4 (*continued*)

Kind of buffer	pH	μ	Components in 1 liter	Voltage	Time, hours	Reference
For hemoglobin (*continued*)						
Phosphate	6.5	0.1	3.11 g KH_2PO_4 1.49 g Na_2HPO_4	6.3 volts/cm	17	(G6)
For amino acids						
Michaelis	8.6	0.1	9.80 g Sodium barbiturate 3.9 g Sodium acetate 60 ml 0.1 *N* HCl	300 volts	1.5	(V5)
Phosphate	7.2	1/15*M*	2.196 g NaH_2PO_4 8.622 g $Na_2HPO_4 \cdot 2H_2O$			(B8)
Phosphate	4.6	0.15	20.4 g KH_2PO_4	300 volts	2	(M19)
Phthalate	5.9		5.1 g Potassium acid phthalate 0.86 g Sodium hydroxide			(W20)
Borate	9.2	0.05	3.3 g Sodium borate	90 volts/cm	1	(G22)
Acid	2.0		500 ml 1.5 *N* Formic acid 500 ml 2 *N* Acetic acid.	90 volts/cm	1.25	(G22)
Acetic	2.2		120 ml Acetic acid	500 volts	3	(V5)
Acetate	2.9	0.012	54.5 ml Acetic acid 1 g Sodium acetate	110–200 volts/cm	1	(G22)
Formic acid	1.8		1 liter 1.5 *N* Formic acid	65 volts/cm	1.25	(G22)
Ammonium formate	3.08		375 ml 2 *M* Formic acid 200 ml 1 *M* NH_4OH			(H13)
Ammonium formate	3.40		263 ml 2 *M* Formic acid 200 ml 1 *M* NH_4OH			(H13)

TABLE 4 (*continued*)

Kind of buffer	pH	μ	Components in 1 liter	Voltage	Time, hours	Reference
For amino acids (*continued*)						
Ammonium formate	4.10		138 ml 2 *M* Formic acid 200 ml 1 *M* NH_4OH			(H13)
Ammonium acetate	4.48		220 ml 2 *M* Acetic acid 200 ml 1 *M* NH_4OH			(H13)
Ammonium acetate	5.00		147 ml 2 *M* Acetic acid 200 ml 1 *M* NH_4OH			(H13)
Ammonium acetate	5.46		114 ml 2 *M* Acetic acid 200 ml 1 *M* NH_4OH			(H13)
Ammonium acetate	5.60		228 ml 2 *M* Acetic acid 400 ml 1 *M* NH_4OH			(H13)
Ammonium acetate	6.80		260 ml 2 *M* Acetic acid 500 ml 1 *M* NH_4OH			(H13)
Collidine acetate	7.0		9.1 ml Collidine 46 ml *N* acetic acid	500 volts	3	(N1)
Pyridinium	4.0		100 ml Acetic acid and Pyridin to pH 4.0	50 volts/cm		(M24)
For insulin						
Barbital	8.6	0.05	10.3 g Sodium barbiturate 1.84 g Barbituric acid	220 volts	8	(S26)
Acetic acid	1.7		333 ml Acetic acid	150 volts	16	(S12)
For gonadotropic hormone						
Phosphate	7.0	0.1	12.50 g $Na_2HPO_4 \cdot 2H_2O$ 4.66 g $NaHPO_4$			(D19)

which their differences in mobility allow for good fractionation (E3, S30). Electroendosmosis and evaporation may however carry some anionic proteins toward the cathode.

Slightly different pH values have also been proposed (L6) and even quite different ones for special purposes. Thus Sonnet recently stressed the use of pH 4.5 (S22) for separation of mucoproteins in the native serum. The size of the charge on human proteins in relation to pH has proved to be important in a few instances and should probably be studied more carefully in many pathological conditions. For example, the many myeloma globulins seem to be identical apart from the charge, which may even be different in blood and urine of the same patient (O6). Furthermore, a double peak for human albumin was reported by Knedel (K12); in this congenital disturbance there is a difference in charge only between two immunologically identical albumins.

The use of surface active agents often modifies the charge of the protein irreversibly through formation of complexes, and because of this a careful study of lipoproteins may prove useful, since anionic detergents influence alpha-lipoproteins, and cationic detergents, beta-lipoproteins (A7, M17).

Ionic strength. The ionic strength of the buffer solution is very important because it influences mobility, and it can be calculated according to the formula:

$$\mu = \frac{1}{2} \sum_{1}^{n} C_i \, Z^2{}_i$$

where C is the concentration in moles per liter, and Z_i the valence of each type of ion i in an electrolyte solution containing n ions.

With increasing ionic strength, some charges of the protein particle are neutralized by buffer ions of the opposite sign, according to the theory of Debye and Hückel, and mobility consequently decreases. Mathematical expression of this phenomenon may be given by the equation of Audubert (A8):

$$u = u_0 \left(\frac{1}{1 + aA\sqrt{\mu}} \right)$$

where u is the observed mobility at a given ionic strength μ, u_0 is the theoretical mobility for $\mu = 0$, a is the diameter of the particle, and A is a constant.

In Fig. 14 the depressing effect of increasing ionic strength on mobility is demonstrated. The practical effect of increasing ionic strength is not only a decrease in mobility, even followed by salting out of proteins for very high values, but also a decrease in electroendosmosis. A less favorable effect will be the larger quantity of heat developed in the system because an increase in ionic strength necessarily means an increase in conductivity.

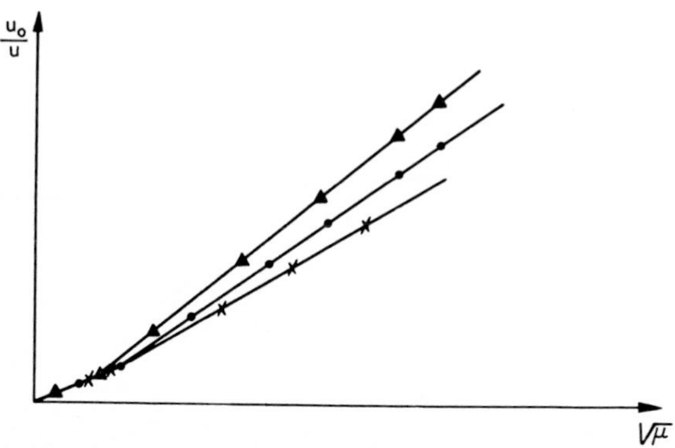

FIG. 14. Influence of ionic strength on mobility: u_0, mobility at $\mu = 0$; u, mobility at a given μ; x, monovalent ions; ●, divalent ions; ▲, trivalent ions. See reference A8.

If the ionic strength is lowered, mobility increases (Fig. 15), but there is a limit to this favorable effect, as fractionation becomes less sharp and some proteins may even lose their solubility. The practical range of μ for electrophoretic work on proteins is about 0.05–0.075. It is easy to calculate μ for a given buffer, but there is no means for determining the actual ionic strength of the buffer inside the strip and consequently no means of measuring absolute mobilities. One reason is that

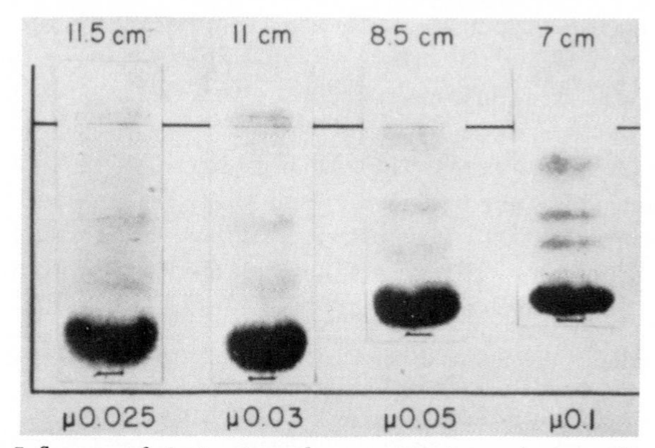

FIG. 15. Influence of ionic strength on migration velocity. The velocity increases with decreasing ionic strength, but fractionation is impaired (barbiturate buffer).

the exact water content is difficult to determine, since some water may be bound by the cellulose and this, as with evaporation, is accompanied by increasing concentration of the buffer salts (Section 1.1.1.1). Variation of ionic strength is also produced by ionophoresis during the application of the field (*vide infra*). The protein sample itself seriously influences the ionic strength at the point of application (A8, P6), as previous dialysis of the sample is infrequent in paper electrophoresis.

These factors together produce quite considerable variations in ionic strength throughout the strip and must produce local variations in field strength and consequently in mobilities. The occasional separation of β-globulins into two groups (β_1 and β_2) and the separation of α_1-globulin some time after the beginning of a run are to be ascribed largely to these causes. It is thus impossible to measure absolute mobilities on paper.

Ionophoresis of the buffer salts. When considering evaporation (Section 1.1.1.1) mention was made of sodium determination in eluates of the strip after the run. Measurements of potential on the strip have confirmed the presence of salt zones (unpublished and P21). When current is applied, the sodium peak due to evaporation moves partially toward the cathode and diminishes at the same time. A second new and unexpected sodium peak appears near the anode (Fig. 16). Determinations of both sodium and barbiturate proved the presence of both ions together on the same spot (P8). Experiments with sodium barbiturate as anodic electrolyte and potassium barbiturate as cathodic electrolyte proved that it is an accumulation of the cation leaving the anodic buffer vessel which causes this new and unexpected peak (Fig. 17). The reason appears to lie in the unequal mobilities of the two ions which tend to occupy a zone from which the faster ion (sodium) needs the same time to reach the cathode as the slower ion (barbiturate) needs to reach the anode (G26, P7). It is, of course, necessary that one anion be discharged at the same moment as one cation.

Once the existence of this abnormal distribution of buffer salts is realized, it becomes easy to see why ionic strength is unmeasurable and variable during the run. It is also easy to see that changes in ionic strength with unaltered pH foreshadow the appearance of pH changes which, when they occur, ruin the experiment.

FIG. 16. Sodium, Veronal, and conductivity (*G*) show a parallel increase in a para-anodic peak (P_1) as a result of ionophoresis. There is also an evaporation peak (P_2) which is moved away from the center *C* toward the cathode by electroendosmosis, and also partially broken down through ionophoresis. (a) *Q* Na is the quantity of sodium in the eluates in mg; (b) *Q* Veronal is the quantity of Veronal in the eluates in mg; (c) *G* is the conductivity of the eluates in mho $\times$ 10^{-5}; and *d* is the distance on the strip in cm.

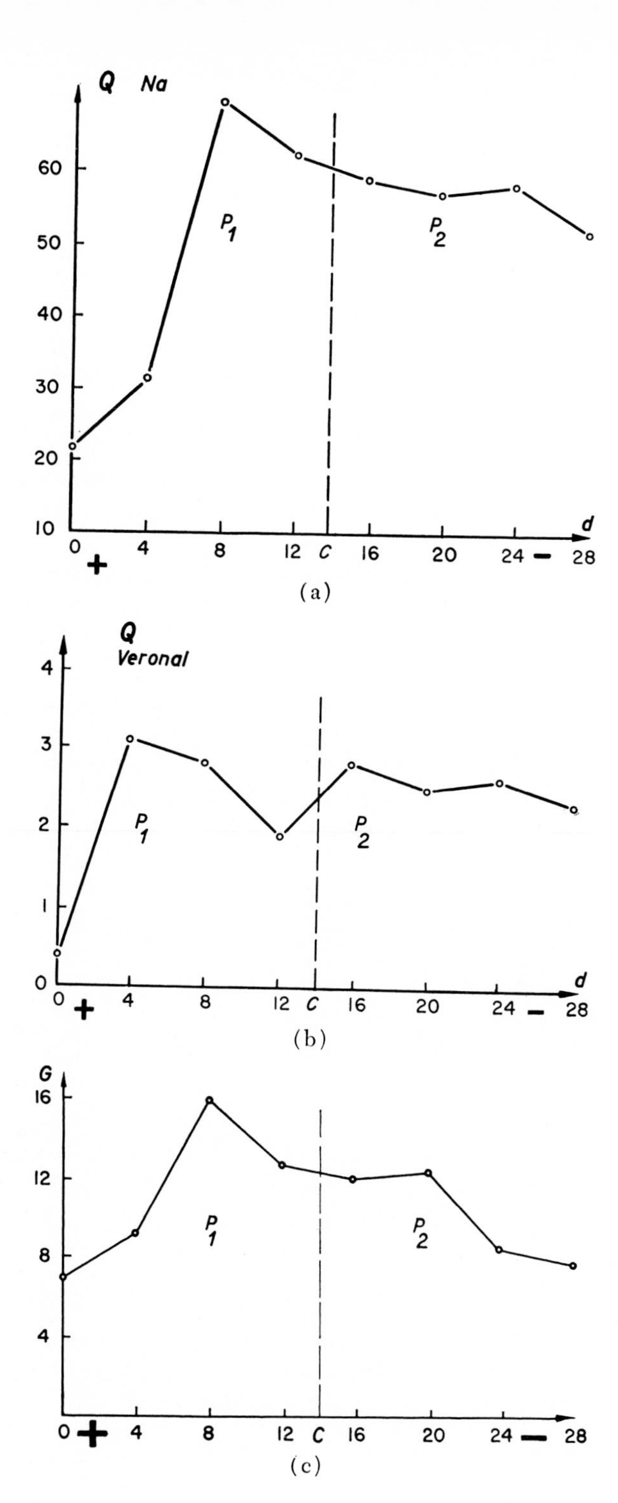

Q
Na
60
50
40
30
20
10
P1
P2
0
4
8
12
C
16
20
24
28
d
+
−
Q
Veronal
4
3
2
1
0
P1
P2
0
4
8
12
C
16
20
24
28
d
+
−
G
16
12
8
4
P1
P2
0
4
8
12
C
16
20
24
28
+
−

(a)
(b)
(c)

	BUFFER CATION			STRIP	IONIC PATTERN
	CATHODE COMPARTMENT	PAPER STRIP	ANODE COMPARTMENT	CATHODE − ANODE +	
A.	K	Na	Na	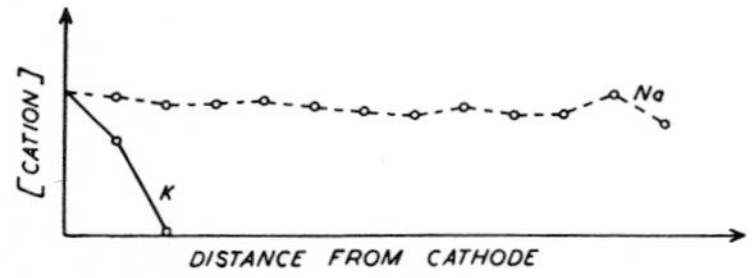 [CATION] Na K DISTANCE FROM CATHODE	K does not leave the cathode. Small para-anodic sodium peak.
B.	Na	K	K	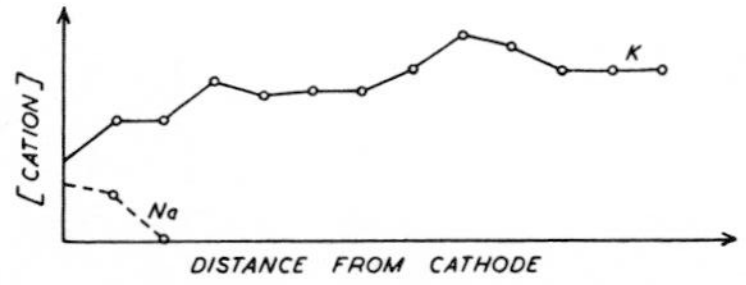 [CATION] K Na DISTANCE FROM CATHODE	Na does not leave the cathode. Para-anodic K-peak.
C.	Na	Na	K	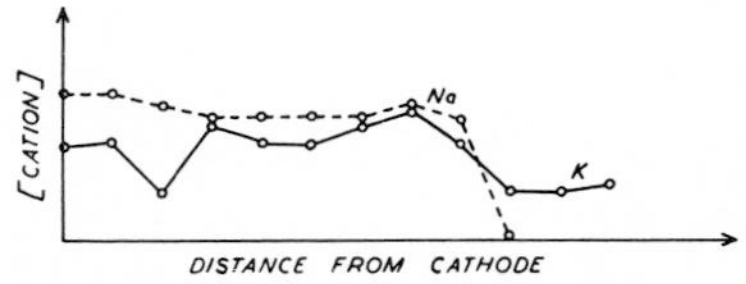 [CATION] Na K DISTANCE FROM CATHODE	Na and K migrate toward the cathode. Na has already left the anodic zone and K has already reached the cathode. The para-anodic peak is clearly visible.
D.	K	K	Na	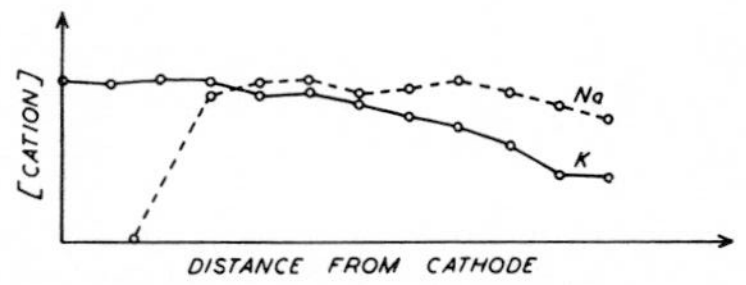 [CATION] Na K DISTANCE FROM CATHODE	Na and K migrate toward the cathode. K has not yet completely left the strip. Na has not yet reached the cathodic end of the strip.

FIG. 17. Sodium (Na) and potassium (K) measured by flame photometry of eluates of strip fragments. The strip was cut transversely into sections. Buffers are either sodium or potassium Veronalate-Veronal buffers.

For a complete picture of the situation at a given moment, accurate determinations not only of anion and cation, but also of the water content of fragments of the strip should be made. For that given moment, conductivity, ionic strength, and pH could then be calculated. We have already found a partial solution to the problem by measuring the conductivity of the eluates (Fig. 16; P8). Although these measurements are imperfect because the salts are eluted out of the strip into a given volume of distilled water, they confirm at once that ionic strength is higher at those points where high sodium barbiturate concentrations are found (P7, P9, P10).

As a general conclusion it can be stated that the strip has a nonuniform electrical field, ionic strength, and buffer concentration, but that modification in pH will only occur if these irregularities increase too quickly. In Table 4 the buffers are grouped according to the intended separation.

1.1.4. *Electrical Field*

1.1.4.1. *Power Supplies.* Most workers use direct current, usually from a rectified AC source, sometimes with a ripple in the DC to favor evaporation for rheophoretic work (D18). The use of batteries is excellent but is not very practical. Great importance has been attached to constant voltage, as providing a uniform field strength throughout the experiment, but it has already been shown (Section 1.1.3.1) that conductivity differences cause local changes in the field even if the potential between the two electrodes remains identical. Indeed the main disadvantage of constant voltage is the increase of intensity parallel to increase of conductivity, which is very likely to occur as a result of evaporation accentuated by conversion of electrical energy to heat.

A new approach to the problem of power supply for electrophoresis is the use of constant current. Durrum (D15) gives a hint in that direction. An example proves the practical value of this principle for routine work (Fig. 18), as the temperature of the laboratory has hardly any influence on the migration length. This is easy to interpret. Mobility increases with higher temperature but is automatically compensated by a lower field strength. With rising temperature the resistance in the strip becomes less and the current has a tendency to rise. But in a constant current power supply, high conductivity is automatically compensated by voltage drop, which causes lower field strength and lower migration velocity. In practice the migration distance is kept satisfactorily constant by this opposition of decreasing field strength and increasing mobility in response to a rising temperature.

Apart from routine work, experimental runs with new buffers and substrates are easily carried out without burning the paper, but an unsuccessful experiment may result if a rapid and serious drop in voltage should result from too great an original current causing too high evaporation. The automatic adaptation of a CC (constant current) power

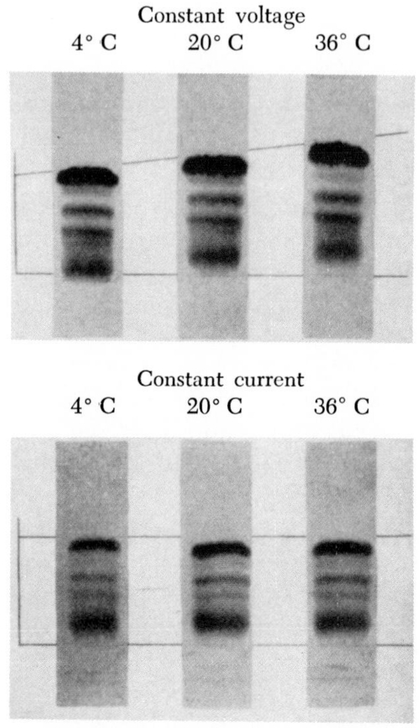

FIG. 18. Influence of temperature on migration velocity. With constant voltage, migration velocity increases with temperature; with constant current it remains constant.

supply to more "physiological" conditions, thanks to the feedback of the electrical information of the strip into the regulating device, helps to avoid destructive changes in the strip.

Circuit diagrams of CV (constant voltage) or CC sources have been published (B6, H11, H12, P11) but seldom in medical literature (A2), and therefore diagrams for CC or CV sources as used in our own laboratory are given in Figs. 19 and 20.

A thorough study of this electronic problem cannot be included here, but the subject is well reviewed by Patchett (P2).

High voltage power supplies have been used by several authors (C7, C8, D13, G20, G21, H5, K4, K5, M24, R9, S1, T5, W4) to obtain fields of

20 to 120 volts/cm; however, descriptions of the electrical parts were not given. Separation of proteins under high voltage proved unsuccessful (S27) in zone electrophoresis, but a useful two-dimensional technique will be illustrated in Part II.

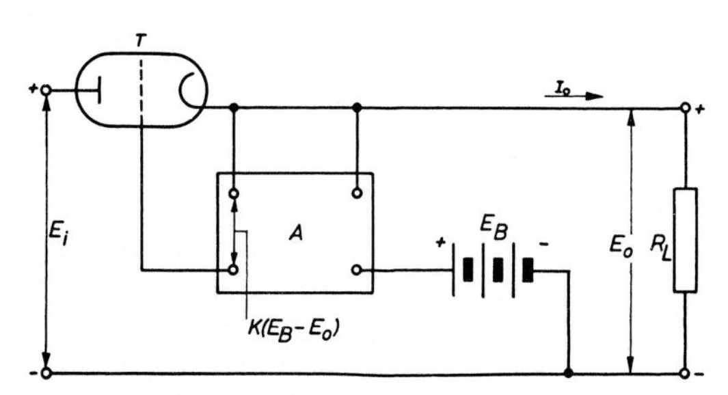

FIG. 19. Circuit of a DC power supply with CC or CV. The output potential E_0 is compared with a constant potential E_B. This battery potential is first amplified (A) and then applied to the grid of the regulating valve T which maintained E_0 constant but a little higher than the amplified potential E_B. When E_0 or also the intensity I_0 has a tendency to vary, either through a variation of the input potential E_i or of the resistance R_L, the changes are compensated by T, so that E_0 or I_0 remains constant. K is the amplification factor.

Pulsating DC at 80 kilocycles was used by Mach and Geffert (M1), and it shortened the duration of the run by about one-third, perhaps as the result of the disruption of the Debye and Hückel ion cloud. This so-called Wien effect has not yet been found by other authors (D17).

1.1.4.2. *The Electrodes.* In most commercial and homemade apparatus the electrodes are of platinum wire or platinum-coated silver. Carbon electrodes are just as satisfactory as long as the same buffer continues to be used, but washing out carbon electrodes when changing to new buffers is difficult. Silver electrodes connected with AgCl/KCl bridges give excellent results (F7), as they do in free electrophoresis.

The products of electrolysis accumulate around the electrodes and change the pH of the buffer locally. To avoid ill effects on the paper strip a baffle system is used, the paper strip dipping into one chamber, the electrode into another, the two chambers being so connected by wicks of material such as glass wool (K22) or by an agar bridge (M8) that electrical current is maintained but physical flow of buffer solution is restricted.

Wunderly (W21) proposed a baffle system consisting of three separate vessels connected by means of paper bridges. The first vessel contained the electrode, the second vessel was at a slightly higher level, and the third, where the strip started, was a little higher still. As a result, buffer tended to flow toward the electrode.

The conductivity of electrophoretic buffer is however so small (Table 5) that a great loss of potential occurs during the transport of current between the baffles. An irregular field may result if the electrode itself is confined to a given point on one side of a buffer vessel which feeds a series of parallel strips. It is better to have the openings in the baffle system long and narrow and parallel to one another and to the electrode wire (D18).

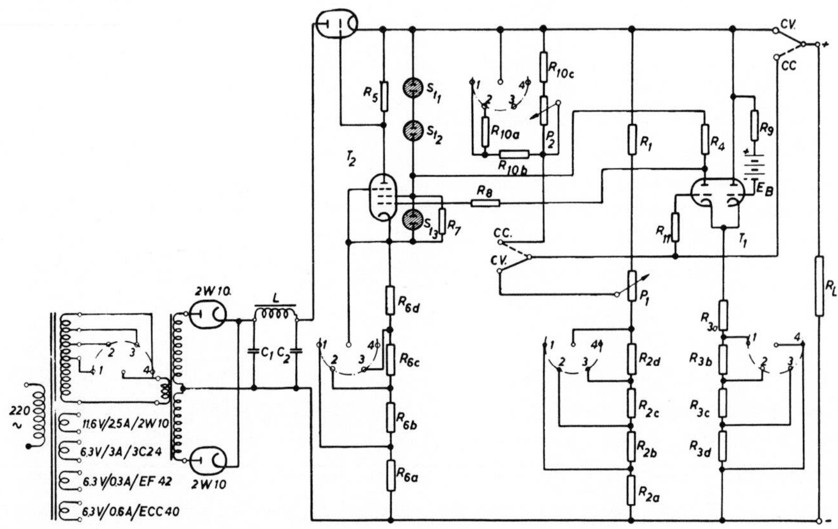

FIG. 20. Circuit of a DC power supply (N1a). Constant current: 4–30 ma, voltage 0–2000 volts; constant voltage: 800–2000 volts, intensity: 0–30 ma; ripple: 0.5 %.

R_1	= 1.4 MΩ	R_7	= 50 kΩ
R_{2a}	= 3.3 MΩ	R_8	= 100 kΩ
R_{2b}	= 1.5 MΩ	R_9	= 100 kΩ
R_{2c}	= 1.4 MΩ	R_{10a}	= 40 kΩ
R_{2d}	= 1.6 MΩ	R_{10b}	= 30 kΩ
R_{3a}	= 120 kΩ	R_{10c}	= 15 kΩ
R_{3b}	= 120 kΩ	R_{11}	= 2 MΩ
R_{3c}	= 80 kΩ	P_1	= 0.5 MΩ
R_{3d}	= 80 kΩ	P_2	= 50 kΩ
R_4	= 57 kΩ	S_{t1}	= S_{t2} = S_{t3} = 85 A1
R_5	= 40 kΩ	C_1	= 1 μF
R_{6a}	= 130 kΩ	C_2	= 8 μF
R_{6b}	= 70 kΩ	L	= 10 H
R_{6c}	= 60 kΩ	R_L	= resistance of electrophoretic substrate
R_{6d}	= 60 kΩ		

TABLE 5
CONDUCTIVITY OF SOME IMPORTANT BUFFERS

Name of buffer	pH	μ	Conductivity at 20° C, mho $\times 10^{-3}$
Elphor	8.6	0.1	6.59
Rheophor	8.6	0.05	4.35
Phosphate	7.4	0.045 *M*	3.62
Hannig	8.6	0.033	2.68
Borate	8.6	0.036	3.03
Tris[a]	8.6	0.08	4.64

[a] Tris is tris (hydroxymethyl) aminomethane.

A different procedure to avoid pH changes and also loss of potential over the baffles is to add fresh buffer continuously to the electrode vessel, which has an overflow. Thus the electrodes lie far nearer the paper without an interposed baffle system. Grabar introduced this method for agar gel electrophoresis (G9). We ourselves use a pump to circulate the buffer continuously over the two electrodes, catching the overflow in one common reservoir, where it mixes and then returns to circulation. This system proved particularly useful in continuous two-dimensional electrophoresis, as will be seen in Part II. But it will also help in the building of easily standardized electrophoretic chambers, necessary for standardization of clinical work on electrophoresis of proteins.

1.1.4.3. *The Electrical Field.* At this point it is important to emphasize the difference between the potential applied at the electrodes and the potential applied to the strip itself. It is also important to remember that because of evaporation the concentration and therefore conductivity of the buffer increase and that because of the field itself ionophoresis also modifies conductivity. As a result the potential differences per centimeter are not identical everywhere on the strip but vary according to the local changes in conductivity.

A supplementary change in conductivity is introduced by the application of the sample which often contains quantities of ions, large enough to modify seriously the original conductivity of the buffer.

It can be seen (Fig. 21) that there is a collapse of field strength where the sample is applied if it contains too many conducting ions. This is of course responsible for bad separation, as mobility decreases with increased ionic strength and velocity is impaired by the low field, until the ions are moved on by ionophoresis. If evaporation is rapid the whole sample will be pushed forward without any separation, as it is paralyzed by its own high conductivity and high ionic strength. Only after ionophoresis of the salts, will mobility and velocity increase, and only then will separation begin.

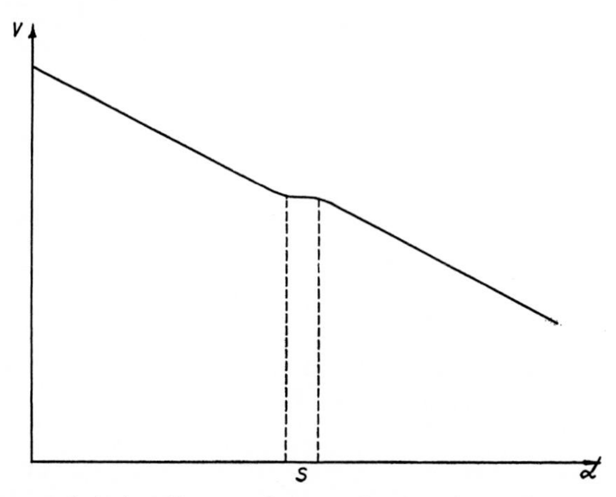

FIG. 21. Potential (*V*) falls according to the length of the strip (*d*). In the zone of application (*S*) there is greater conductivity and only a small voltage drop.

1.2. Interaction of the Several Factors

Figure 22 gives a schematic representation of the various factors involved and shows their interaction to form one comprehensive pattern.

Factors influencing migration can be summarized as follows. The electrical energy is the origin of two different physical factors. The first is the field itself which causes migration of the buffer ions, or ionophoresis, simultaneously with migration of the macromolecules. The second is heat, which increases not only the mobility of the charged particles, but also evaporation and the concentration of the buffer. The original conductivity, pH, and ionic strength of the buffer tend to be modified through ionophoresis. The buffered substrate has anticonvectional properties and also initiates processes, such as adsorption, which influence migration.

There is another important factor, namely the buffer flow inside the buffered substrate, which influences migration. This flow has three main sources: evaporation partially caused by heat production from electrical energy; the hydrodynamic result of the level of the buffer vessels with respect to one another and to the level of the paper; and the electro-endosmotic effect, the interaction between the electrical forces and the buffered substrate.

The theoretical principles given here will help toward a clear understanding of the observed phenomena, which will be described in the next section.

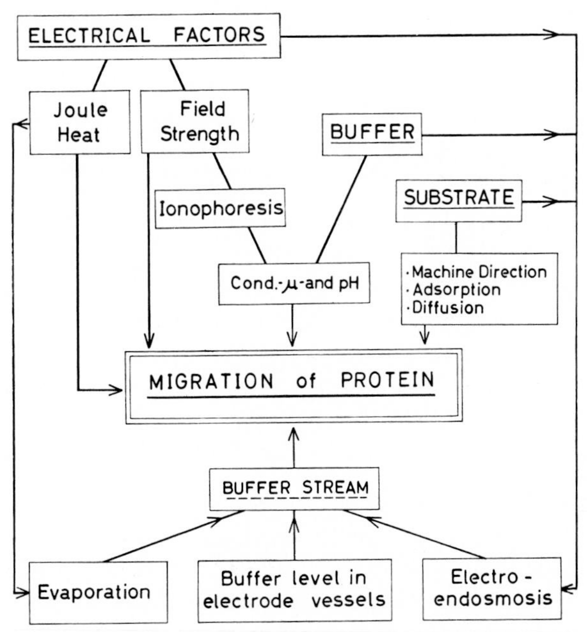

FIG. 22. Factors influencing migration.

2. The Run

This section deals with the amount, preparation, and concentration of samples; the practical conditions of the run itself and its difficulties; the fixation or elution of fractions; and, finally, the methods of quantitation.

2.1. THE SAMPLE

2.1.1. *Type of Sample*

In clinical work many biological fluids have been submitted to paper electrophoretic separation. Serum and urine have been studied extensively and the results were reviewed in Volume I of this series (p. 238). Other fluids include cerebrospinal fluid (B12, B14, E6, K18), pleural fluid (D4), gastric juice (H6), ascitic fluid (H9), synovial fluid (W3), proteins of the lens (F4, W9, W10), aqueous humor of the eye (W12, W24), edema liquid (W23), and pericardiac effusion (G2). Apart from the general separation of plasma proteins, work has been done on special protein groups, such as lipo- and glycoproteins, mucoproteins, hemoglobins (H19), coagulation factors (O5), and on other components, such as amino acids.

For clinical purposes serum is preferable to plasma because the fibrinogen band interferes with the reading of globulin fractions. Moreover, it is easier to get serum free from traces of hemoglobin liberated by hemolysis.

Many biological fluids (e.g., serum and plasma) contain sufficient protein without excessive salt concentration to be applied at once to the paper. Some other fluids, such as urine, may need preliminary dialysis because they contain too much salt, and they often need concentration also. Still other fluids, such as cerebrospinal fluid, always need concentration before electrophoresis.

Many proteins are very rapidly denatured so that ideally the fluid should be analyzed immediately. β_2-Globulin fractions are partially lost when the serum is kept for some days (O3). In work with lipoproteins, refrigeration alters the results (H1, H1a), but preservation in a deep freezer is practical for reference work.

2.1.2. *Dialysis*

Dialysis is not currently performed on samples intended to be applied on buffered supporting media. Theoretically, however, this precaution should be taken because increased conductivity, increased ionic strength, and modification of pH cause several disturbances at the application spot, the most important being local depression of the electric field intensity (G10). Some substitute for dialysis is realized if the sample is diluted with a solution of buffer and serum salts, with volume and concentration calculated to give a final salt concentration corresponding to the expected result of combined dialysis and dilution (S31). This method was successful for free electrophoresis of serum and should also be useful in zone electrophoresis on stabilizing media.

Some investigators apply the sample some distance from the desired starting line and use evaporation to move the sample toward it, thus obtaining in effect a partial dialysis. True dialysis can be achieved with the usual methods, and for micro samples Antweiler (A6) gives a successful technique. A rapid and ingenious method is dilution of the sample with buffer, followed by concentration on an ultrafilter, on which electrophoresis is subsequently performed (unpublished). Better knowledge of the physical conditions of the buffered strip will increase the use of a dialyzing technique in order to obtain better results.

2.1.3. *Concentration of Sample*

Concentration of the sample together with dialysis is a good procedure for urine and cerebrospinal fluid, although some pathological urines contain sufficient protein to be applied at once. Dialysis against dextran in concentrations of about 20–40 % at 4°C is popular for this type of work (A3, S5). Concentration through rheophoresis was proposed (D18) but has been of limited value because rheophoresis of such intensity causes ionic disturbances. Concentration of cerebrospinal fluid has been achieved by Knapp with ultrafiltration and this seems to be the best method (K8); it can also be applied to other diluted samples (H8). Treatment with acetone has also been used (B14, K2). Evaporation under reduced pressure has been proposed for urine (S25). For enzymes in gastric juice ammonium sulfate precipitation is mentioned (M20), and this method may have other applications.

2.1.4. *Application of Sample*

2.1.4.1. *Quantity.* The paper itself determines through its water-absorbing capacity the maximal volume of the sample that can be applied on a given surface, but excessive quantities of protein militate against good partition.

Narrowing the surface of application is of only limited value because diffusion is proportional to the square of the concentration or inversely proportional to the square of the width over which the sample is applied. Thus proteins, such as albumins, which are present in large quantity, always tend to diffuse over a broad zone even if the application zone is narrow. This means that large, concentrated samples are not the best for obtaining sharp resolution; their use also causes a blurring of the small α-globulin zone which lies quite near to the large albumin fraction.

In practice a volume between 0.03 and 0.005 ml of serum is applied. The sample usually recommended nowadays is 0.01 ml of serum (P6) applied over a 3-cm length on Schleicher and Schuell 2043b paper, and this amounts to approximately 0.65 mg of protein. When larger quantities of protein are present, the direct photometric scanning of large fractions becomes impossible (see Section 3.2.2). The amount of protein applied may be adjusted by determining the total quantity of protein before the run. For glycoproteins fourfold quantities of serum are applied to make reading of the stained spots easily possible.

2.1.4.2. *Application.* In most cases the paper is artificially wetted in buffer, then surface-dried between dry filter paper to eliminate excess water, mounted on a holder, charged with the sample, and fixed inside the apparatus. This avoids the use of dry paper which is a cause of difficulty, especially in horizontal methods because a paper strip is longer when wet than when dry. If, however, the paper is kept stretched with weights or with an adjustable holder (O4), this is of less importance.

In many chambers the paper tends to stick to the walls and this makes correct stretching difficult.

Samples may be applied to a dry strip and driven toward the starting zone by the capillary buffer flow, the current being applied only when moistening is complete. This is, at the same time, a partial substitute for dialysis (Section 2.1.2).

Grassmann applies the sample to a wet strip and blots the application zone with dry filter paper, thus avoiding the immediate diffusion of the freshly added solution of protein. The disadvantage of this method is that the run is started at once, before the electrophoretic chamber is vapor-saturated. Using serum stained before application, one can observe the rheophoretic movement of the entire sample toward the center before the electrophoretic migration starts. This phenomenon, together with the low mobility of γ-globulins, determines the location of the starting zone, which differs from one apparatus to another. As a general rule it is of no use to apply a sample any further from the anode than the spot where the tail of the γ-globulins will be at the end of the run. The center of the strip is also a good application zone when rheophoresis is strong, but has the disadvantage of possible paper injury.

A better method is to apply the sample through a slot in the lid of the box after a saturation period (V1). In this case the rheophoretic component of the run is reduced to a minimum.

The applicator itself may be a smooth-tipped pipet, as used by Grassmann, Hannig, Knedel and others, which is easily calibrated to determine an exact volume. It is important to distribute the sample evenly without scratching the surface of the paper. Martin and Franglen (M6) and others (W1) use a narrow paper strip soaked with the sample, but there is disturbance of the field at the application spot and often the cut edges of the applicator strip give rise to serious adsorption. We have tried application by means of cotton and nylon threads, but the quantity applied was difficult to determine and perfect contact with the substrate was not easily obtained. A glass applicator may carry the sample on a ground surface and this is stamped on the paper. In this case there is less danger of damaging the paper structure than with a pipet, but it is difficult to obtain a uniform zone because the slightest tilting of the glass surface or a lack of support under the paper causes irregular suction and the sample is not evenly distributed. A metal applicator forming a loop, devised by Durrum, is very useful, especially if the sample is applied on the top of an inverted V paper where the glass rod supports the strip. A narrow Perspex slit (D8) and a comb carrying two rows of very fine teeth (S29) are also proposed.

For purposes of quantitation the application zone is usually a straight-line. For qualitative work, a spot would also be satisfactory, but for direct photometry it gives rise to theoretical objections.

A zone of up to 3 mm is usually left free at each edge of the paper strip because protein might run along the edge (G12) and also because of the so-called "edge-phenomenon" with curving backward of each fraction (see Section 2.2.2.).

2.2. The Run

2.2.1. *Normal Working Conditions*

In general (M12) a field of about 5 volts/cm according to the free length of the strip and 0.25 ma/cm according to the width of the strip seems a fair average for work on proteins. The use of lower field strengths involves a longer duration and may be useful, for instance, for an overnight run. Some workers tend to use shorter periods of time, but although a stronger field develops more heat with its attendant difficulties, there is nevertheless a trend toward high field strengths up to 100 volts/cm and this is useful for separation of small molecules.

During the run the paper is heated above room temperature, and buffer solvent must evaporate in proportion to the size of the chamber and the quantity of condensate. This evaporation is greatest at the beginning of the run, but in a smaller chamber it may become so slight that rheophoresis falls to nil and the fraction travels into the buffer vessel (Fig. 23). Most commercial forms of apparatus are not vapor saturated before the run begins and evaporation remains throughout the experiment at a sufficiently high level to cause immobilization of the fractions at the place where buffer flow and electrophoretic velocity neutralize each other.

Many different and even contradictory examples of migration distance, measured in relation to time, are given in the literature (D15). Any such relation can be explained, but considering the theoretical origin of the buffer flow and changes in buffer concentration, with their effect on field strengths, ionic strength, and pH, migration distance on paper cannot be related to particle mobiliy only.

2.2.2. *Anomalies and Their Origin*

Table 6 lists some abnormalities of fractionation and their causes. When one sidewall *A* of the chamber is cooler than the other, the distortion shown in Fig. 24a is produced, as it is also if one single sheet of paper is used. As a result of buffer solvent evaporation more salt and less water lie along the edges than along the center (Fig. 2), mobility decreases toward the edges, and curved fractions appear (Fig. 24b). In Fig. 24c and d irregular patterns are the result of uneven distribution of sample and/or water content. While the buffer is "retiring" it runs into canals rather like water drying on land. The dryer zones have lower conductivity, become less heated, and allow for better separation than the moister ones.

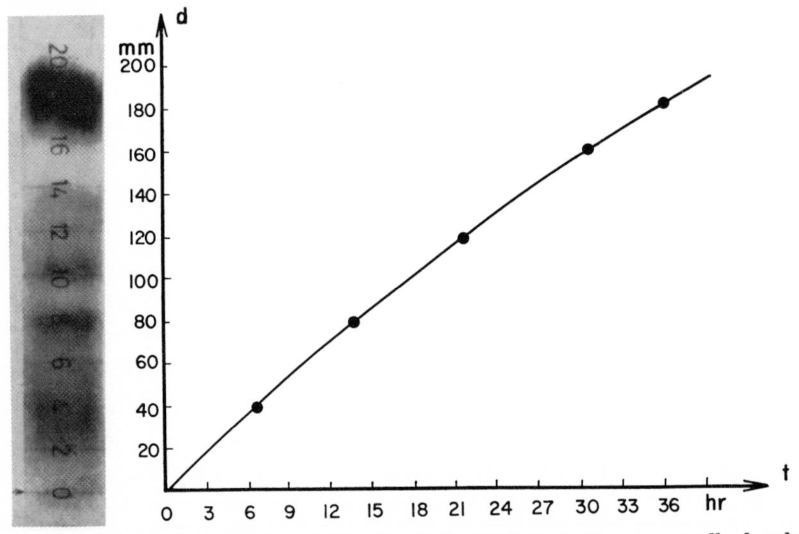

FIG. 23. Inhibition of evaporation. Zone electrophoresis in very small chamber with very thin walls. Field 2.7 volts/cm and 0.3 ma/cm strip; Veronal buffer, pH 8.6; $\mu = 0.1$; duration 36 hours. There is nearly linear relationship between migration distance and time. The fractions are widely separated. After staining of serum with bromophenol blue before the experiment, migration was followed.

2.3. Drying; Fixation

It is good practice to fix the proteins as soon as development is maximal, for blurring of the pattern by diffusion and irregular evaporation may occur if the paper strip is left exposed outside the chamber. The free-hanging ends of the paper, supersaturated with buffer, must be cut away at once.

The only objections to elution of the protein fractions from the wet paper are that it is almost impossible to locate the zones (this can be overcome by staining) and that handling and cutting wet paper are very difficult. In general, fixing, staining, and drying are necessary steps. A fast-moving infrared source can be adjusted to dry the paper almost at once without scorching and to "fix" the proteins by rendering them in-

TABLE 6
ABNORMAL FRACTIONATION IN ZONE ELECTROPHORESIS (P6)

Phenomenon	Cause	Correction
Fractions run together at paper edges	Application faulty	Apply a few mm from edge
Comet formation	Adsorption	Use a coarser paper; raise ambient temperature
Broad fraction	Diffusion	Dry starting line before application; control total concentration of proteins; after optimal duration cease electrophoretic action
Curved, oblique, jagged, irregular zones	Irregular application	Quick backward and forward movement
	Irregular evaporation	Avoid irregular temperature gradients
	Local differences of pH and of ionic strength	Change polarization after each run; mix together contents of the two electrode vessels; renew buffer at regular intervals
	Traces of drops of water	Slanting roof
	Irregular drying	Dry rapidly under infrared lamp, with rapid movement to and fro

soluble. The dye-binding capacity, however, depends on the temperature of denaturation, which in this method is difficult to control (D15, L7). Though slower, a drying oven, such as was introduced by Grassmann is more convenient and can be better controlled; the process must not be too slow, however, or irregularly "frayed" fractions are produced by convection currents (V2).

Chemical fixation, which may be combined with staining, is useful and will be considered with the latter.

3. Quantitation

The commonest methods of quantitation in general practice depend upon photometric measurements, after staining the protein fractions *in situ* on the paper. Newer methods, which will be given first are based on the measurement of properties other than the dye-binding capacity of the protein fractions. Some of these methods, which avoid the many

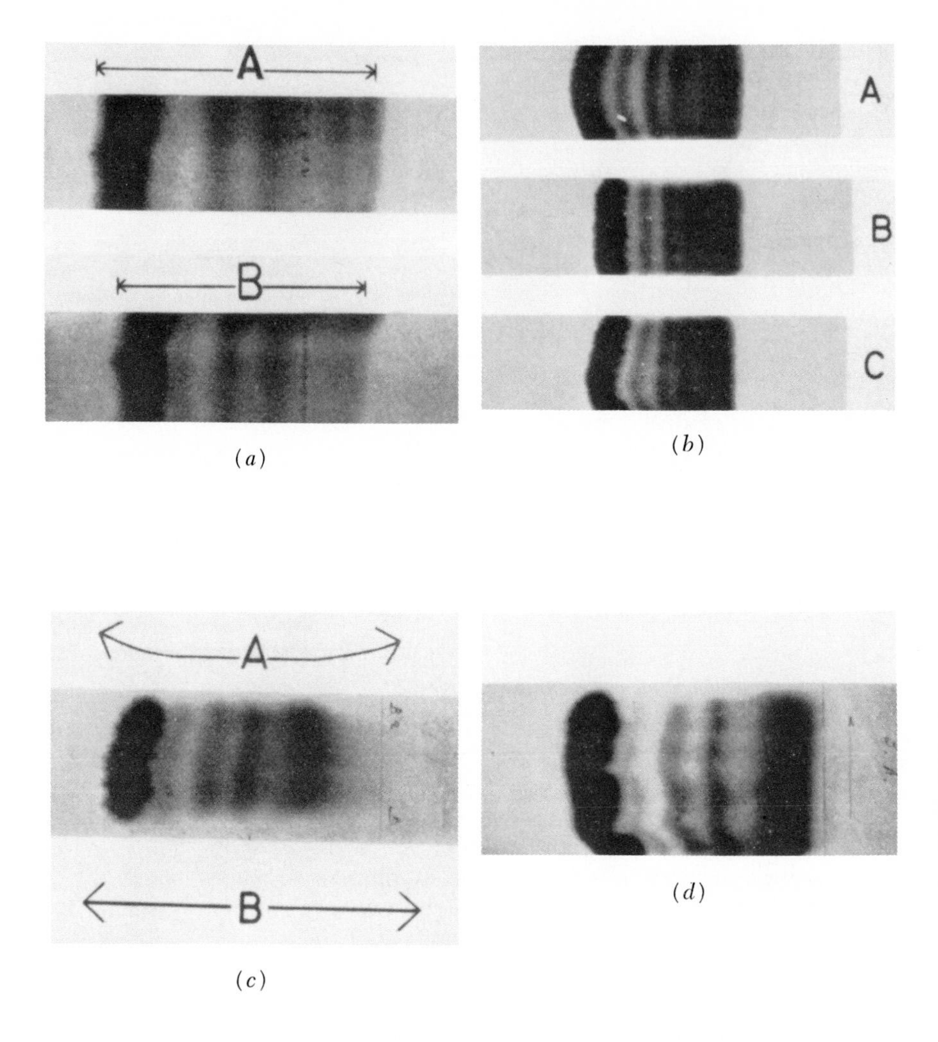

FIG. 24. (a) The two strips were running in the same chamber, but strip *A* was lying near a cooler wall and suffered less from evaporation than the warmer strip *B*. (b) Only three strips were applied in a chamber designed for seven. Strips suffered irregularly from evaporation. (c) The strip was hanging slack along edge *A* and taut along edge *B*. (d) Irregular application causes jagged migration with shorter migration of more concentrated zones.

theoretical and practical causes of error in the classic dyeing procedures and the current "reading" techniques, seem to be very promising.

3.1. Quantitation Methods without Dyeing

Among the methods which do not involve staining the fractions on the paper, two main types can be distinguished: those evaluating fractions still *in situ* on the strip and those using eluted fractions.

3.1.1. *Direct Methods*

3.1.1.1. *Isotopes.* If radioactive material is being separated, a search can be made either for the isotope itself or for the fraction of the molecule or ion in which it is incorporated or with which it is complexed. Apart from qualitative location of the radioactive fractions, the determination of radioactivity is a measure of the concentration of a larger molecule if the stoichiometric relation of the isotope to the given chemical is known. The background causes difficulty when measuring low activities, but in practice, isotopes have proved useful in determining transport of radioactive elements on and by proteins. Thus interesting work has been reported on the presence of copper as a constituent of ceruloplasmin and on the absence of a specific copper-carrying protein (W14). The antigen-antibody reaction between I^{131}-tagged serum albumin and the antibody was followed by means of autoradiography of the strip (L1). Other applications include investigations of thyroglobulins (G1, G7, M7), I^{131} in blood and thyroid extract (B3, H17, H18), radioactive gold after colloidal gold administration (S11), C^{13} in carbohydrates (J1), etc. Mobility of the protein tagged with an isotope may, however, be different from that of the native protein (G1). Reaction between separated fractions and the isotope added after the run has also been used (L14).

The use of isotopes is very promising in two-dimensional techniques.

3.1.1.2. *Ultraviolet Light.* König in 1937 (K18) published a method of detection by ultraviolet light, making the first reference to paper electrophoresis. Fluorescent fractions are easily detected (B5, H14, S10) and nonfluorescent fractions may be seen or photographed as dark zones on the slightly fluorescent background of the buffered paper.

Quantitation can be achieved by ultraviolet absorptiometry. A strong source of UV is needed (B4) and paper and buffer may not absorb light of the typical wavelength. The best wavelength is about 250 mμ (K7) though 280 mμ is also satisfactory (B5), as it lies in the tyrosine-tryptophan zone. Barbiturate buffer is not adequate for this type of work on proteins since the barbiturates themselves absorb ultraviolet light (B5), but glycocoll (W21) gives better results.

Substrates such as cellulose acetate, transparent to UV, show great promise as supporting media (see Part III).

3.1.1.3. *Retentiometric Analysis.* Wieland (W6) has proposed a different method of quantitation. After the run, the strip is stained by an acid solution of azocarmine B in methanol. The next step is to allow a solution of copper acetate in tetrahydrofuran to ascend into the paper at right angles to the migration direction. The acid groups of the protein, not being saturated by the acid azocarmine dye, react with the copper ions and give a profile all along the strip, inversely proportional in height to the quantity of protein present at each point. This negative profile of the protein is revealed by spraying with rubeanic acid (W7, W8). Unfortunately these curves are jagged, and it is almost impossible to draw a base line or measure small fractions (G12).

3.1.1.4. *Photosensitization.* Bon (B11) proposed the following method. Press the wet strip against silver bromide paper, expose for a short period to light, develop with a hard developing bath, and fix immediately with concentrated potassium metabisulfite. Protein at 1 $\mu g/mm^3$ gave a distinct spot on the photographic paper.

3.1.2. *Elution*

As a preliminary to elution of unstained fractions the zones may be located directly with ultraviolet light and the paper strip cut up accordingly.

Alternatively, one strip (or, better, half of a strip cut lengthwise) is stained and used as a test for a second strip which is cut into appropriate lengths and eluted. There is, however, no certainty as to the limits of the zones of the protein fractions, for the factors involved in application and migration of protein samples are capable of destroying the parallelism of the two strips and of thus falsifying the results. Nevertheless, this method may be used and is even indispensable for the analysis of lipo- and glycoproteins with reference to the proteins themselves.

A third method is merely to cut the strip into 0.2–0.5 cm-pieces and elute these separately (S22); but this technique cannot equal a millimeter by millimeter recording, one disadvantage being the impossibility of determining irregularities in the pherogram.

In any case the quality of the eluate is still a problem. Does it contain the complete fractions and no disturbing elements from paper or buffer? A few applications of elution are of practical interest.

(*a*) *Protein determinations.* Kjeldahl determinations on eluted fractions (A4, L9) have been superseded by the easier color reactions for proteins, especially since many buffers contain nitrogen components.

The biuret reaction (M16), the Folin and Ciocalteu reagent (K22), ultraviolet adsorption (H2) and polarographic determinations (Fig. 25) have also been proposed (B2, H16).

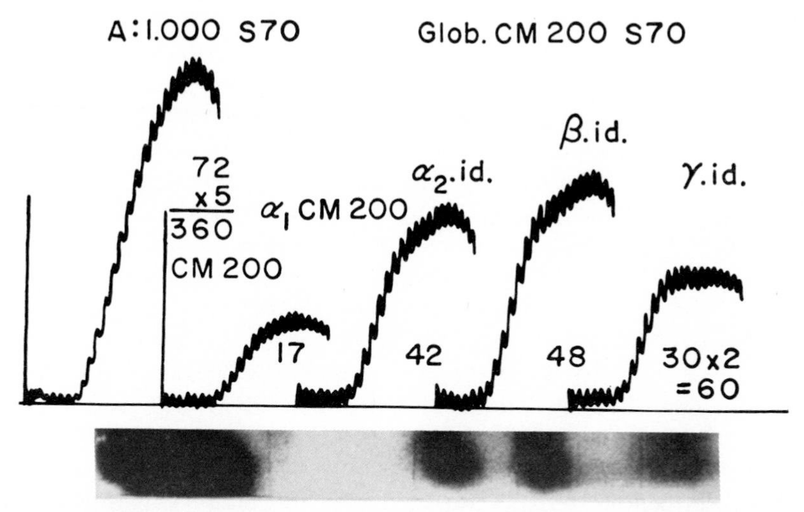

FIG. 25. Polarographic determination of the eluates of stained fractions after zone electrophoresis (B2). *A*, albumin; *CM*, current multiplication factor; *S*, sensitivity. The height of each polarogram calculated for the same current multiplication factor gives the polaroactivity of each fraction (Po_{fraction}). The estimation of the $Po_{\gamma\text{-globulin}}$ needs a coefficient of 2, e.g., $Po_{\text{albumin}}\ CM_{200} = Po_{\text{albumin}}\ CM_{1000} \times (1000/200)$. $Po\,T$ = total polaroactivity = $\Sigma\ Po_{\text{fraction}}$. Evaluation in % = $(Po_{\text{fraction}}/Po\,T) \times 100$.

(*b*) *Cholesterol.* Cholesterol can be determined on the paper itself (B1, B3) after location of the lipids in ultraviolet light; the paper dissolves in the reagent while cholesterol reacts in the normal way. Elution can be performed first (B13, L2, N2) and the reaction afterward.

(*c*) *Phospholipids.* Phospholipids (N2) and esterified fatty acids (T6) have been determined after elution in a mixture of ethanol and ether.

(*d*) *Glyco- and mucoproteins.* Staining has been carried out after elution or rather squeezing of the strip, specific carbohydrates being determined in the eluate (G5, K11, S22). But even in this field elution is used when prospecting a new field, while for routine measurements direct colorimetry is used.

3.2. Quantitation Methods after Dyeing

The primary condition for a correct dyeing procedure is the existence of a specific stoichiometric relationship between a given fraction and the quantity of dye fixed. If, as for serum, the various fractions of protein present a different dye-absorbing capacity toward the same dye, the relationship between dye and concentration of protein must be corrected by means of protein factors, which convert optical density or concentration of color into concentration of protein. Another problem is that of chemical purity of the dye, for many commercial samples are mixtures of variable composition and this accounts not only for differences in stoichiometric relationship but also in colorimetric absorption. A third problem is the fixation of the fractions on the paper before staining, and as this very often involves denaturation, the dye-binding capacity may vary according to the fixation method. Stoichiometric relationship between dye and protein therefore presupposes standard fixation methods and also standard dyeing conditions with reference to quality, concentration, solvent, pH, duration, temperature, and other factors concerning the dye.

A second condition for a correct dyeing procedure is perfect rinsing of the dye out of the paper without loss of dye fixed to the protein. If rinsing is incomplete, a "blank" of the paper must be taken into account and this may be too high for accurate evaluation of small fractions; here again standardized working conditions are essential.

A third general condition is the stability of the dye throughout staining, rinsing, and photometry.

It is difficult to achieve a dyeing procedure perfect in all these respects, and even if this were realized, the problem of colorimetry itself would remain. Either after elution or on the paper itself, the dye must be evaluated. On paper, the so-called "paper error" is a very important factor although translucency agents or reflectometry are used to reduce this error to a minimum.

3.2.1. *Staining of Fractions*

3.2.1.1. *Dyes and Dyeing Conditions.*

(*a*) *Protein-dye relationship.* Dyes which conjugate with protein have been studied extensively, while dyeing procedures for lipoproteins are less specific and less well known. Of those proposed for proteins only two, bromophenol blue and bromocresol green are chemically pure and have a definite structure; the others, such as naphthalene black, azocarmine, lissamine green, are a mixture of several dyes, and the composition differs according to the origin and even the batch. All these dyes are anionic and in fact are acid wool dyes. They react with the basic (cationic) groups of the protein and thus act best in the pH range acid to the protein isoelectric point. The influence of pH, however, is far stronger on indicator dyes such as bromophenol blue and bromocresol green (F5) than on substantive dyes such as amido black. The affinity of the different dyes for the protein is different (S7), and this explains the differences in the duration of the staining procedure.

Concentration of the dye seems to be important because if it is too high, especially with large molecule dyes, coagulation at the surface of the concentrated protein fraction hinders further diffusion of the dye into the protein spot. This is true, for instance, for naphthalene black (D3, P22, S1) and bromophenol blue (D17). The solvent is important in this connection; alcoholic solutions tend to become more watery during use, and this may have a great influence on the dye uptake (F5). Results obtained with dilute solutions and prolonged staining seem superior and more reproducible than those given by rapid staining with a concentrated dye solution (C1a).

During the staining of strips, concentration of buffer salts increases progressively in the bath, as the strips are not previously washed but are merely dried in the oven. The presence of salts depresses the dye-binding capacity of protein (F5). The use of a fixing bath before the staining may remove up to 80 % of the salts, and also assure a uniform denaturation of the proteins (F6, S20, W13).

A very interesting difficulty is the difference in dye uptake between a concentrated and a diluted protein fraction. Eberhard (E1) gave a careful account of naphthalene black evaluation of spots of chemically pure albumin and γ-globulin. He found that, in photometric measurements on the eluates of the stained fractions, those which were distributed over a large area of paper always gave lower values than fractions eluted from a smaller area of paper, although the total amount of protein was actually the same. Difficulties of photometry on paper were excluded by this technique, and Eberhard considered that there is competition between paper and protein for the same dye molecule. It is, perhaps, easier to find the key in the quantity of salt, which is greater in the diluted than in the concentrated fraction. The same phenomenon occurs with bromophenol blue (G25).

Stability of the dye itself seems to be influenced by heavy metals (D3) and may be increased by complexing agents (Fig. 26) or by the addition of very small amounts of human serum to the eluate, though there is no adequate interpretation of this latter phenomenon.

Fixation, which causes denaturation of proteins, is of course a major cause of variability in dye uptake. Heat fixation may destroy the binding capacity of part of the albumin (S20), and is also difficult to reproduce exactly (M5), but chemical fixation can be kept more constant in its application and effect; the dye is usually dissolved in the fixing solution and the two steps are performed in one.

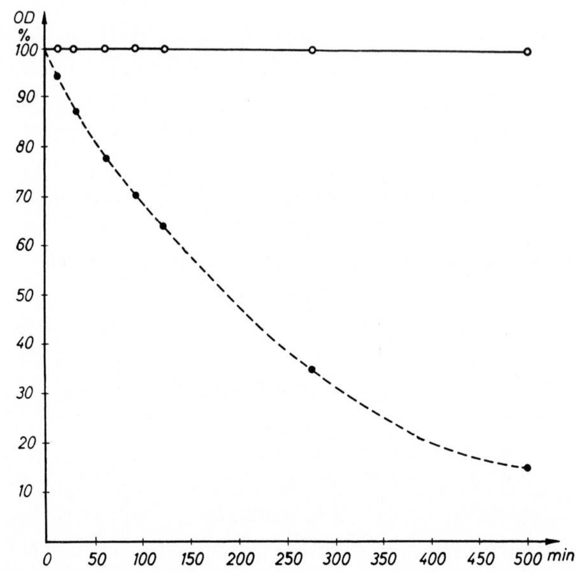

Fig. 26. Influence of complexing agents on stability of alkaline amido black solutions (D3). (- - - -) without EDTA and (———) with EDTA.

Rinsing to wash out excess dye from the paper without removing dye from the protein raises a different problem for each dye because of the differing stability of the dye-protein conjugates and must therefore be considered for each dye separately [see Section 3.2.1.1(b)].

The dye-binding capacity of albumin is generally far greater than that of the different globulins. This leads to the introduction of a "protein factor," a coefficient used to convert the optical density into concentration of protein, albumin being the reference unit. Moreover, the several globulins differ among one another in their dye-binding capacity, each fraction really requiring a separate factor, and this would again be different for pathological proteins (E4, G19) (Table 7). For most clinical purposes this refinement is ignored in practice, but it should be realized that the figures are then inaccurate.

(*b*) *Dyeing and rinsing procedures.* The different procedures will be reviewed according to the type of the pherogram, classified as pro-

TABLE 7
GLOBULIN CORRECTION FACTORS (L12)

Author	Stain	Globulin correction factors				
		α_1	α	α_2	β	γ
Esser *et al.*, 1952 (E5)	Bromophenol blue	2.36		2.09	2.20	1.62
Koïw *et al.*, 1952 (K16)	Bromophenol blue	2.8		1.7	1.6	1.4
Loeffler and Wunderly, 1953 (L12)	Bromophenol blue		2.30		1.84	1.64
	Amido black		1.71		1.79	1.50
	Azocarmine		1.18		1.30	1.04

teinogram, lipidogram, glucidogram, etc. Countless detailed variations in technique have been reported but it would be of little use to mention all of them, and we shall point out for each procedure the more usual dyes and the more important factors.

Proteinogram

1. Amido black (naphthalene black 12B.200) was introduced by Grassmann and Hannig (G13). In the original method the strip is stained with a saturated solution of amido black in methanol with 10 % acetic acid. Rinsing is done in a series of baths of methanol with 10 % acetic acid (G17). An acid reaction is required for good dye-protein combination and dye stability, but since interaction of the acetic acid and methanol with formation of methyl acetate influences the dye-protein binding, an aqueous solution is recommended (D3). The maximal light-absorbing capacity is at 620 mμ (P16). The protein factor is variously given as 1.3 (K22, S18, S19), 1.46 (B10, G15), 1.5 (G4), 1.6 (C6, P18).

The greatest problem is rinsing. Proposals have been made to modify the original Grassmann method, for instance, by the use of a Soxhlet(G3), but the only substantial technical improvement is described by Oosterhuis (O2) who uses 10 liters of an aqueous solution of 1 % acetic acid for washing 10 paper strips, with a 2-hour washing time (D3).

2. Bromophenol blue, suggested by Durrum (D13) and Jencks (J3), is used as an aqueous solution of 0.1 g per liter containing 5 % acetic acid and 5 % zinc sulfate. Here, dye fixation affects considerably the protein factor. Heat must be applied according to the standard schedule of 30 minutes at 120°C (D15).

Some authors, however, prefer an alcoholic solution of the dye, containing mercuric chloride as fixing agent (K22).

Rinsing is first done with an aqueous solution of 2 % acetic acid and finally with an acetate buffer at pH 3.6, since the influence of pH on this dye is considerable (D15).

The maximal adsorption lies between 585 and 600 mμ (P16). The protein factor is variously reported as 1.1 (G18, P18), 1.3 (J3), 1.6 (C6, K3).

3. *Azocarmine B* was originally used by Turba and Enenkel (T4) as a saturated solution and is prepared in a mixture of 50 % methanol, 10 % glacial acetic acid, and 40 % water. Rinsing is done with methanol and later on with 1 % acetic acid. The wavelength needed for colorimetry is 520 mμ. Protein factors again vary widely: 1 (T4), 1.35 (M3), 1.5 (V2), 1.6 (E4, E5), 1.8 (P18).

The most important cause of error is loss of dye during rinsing.

4. *Lissamine green* (SF 150 I.C.I. light green or lichtgrün FS) introduced by Griffiths (G18) has been used by several authors (A1, C1a, D1, D9, O4) and is read at about 600 mμ. The solution is made up with 200 mg of dye per 100 ml in 3 % (w/v) aqueous sulfosalicylic acid (D9). More dilute solutions have also been proposed, and it has been claimed that the protein factor was equal to 1 for all proteins (C1). This dye is, however, a complex mixture and is different from batch to batch (H7). It is easily eluted in 40 % acetic acid (O4).

Besides the four mentioned individually, the following dyes seem to have some value for staining proteins: orange G, säure violet, and Biebrich scarlet. It is worth noting that the following series proved impracticable: brilliant cresyl blue, thymolphthalein, toluidine blue, aniline blue, alizarin, Congo red, crystal violet, vital carmine, indigo carmine, induline, vital blue, methyl violet, Jodgrün, Janus green, malachite green, methylene green, magenta, trypan blue, tartrazine, orange III, soluble blue, neutral red, Nile blue, Scharlach red, carmine, Bismarck brown, safranine, methylene blue, rhodamine, and azorubine (G18).

Lipidogram

The more important dyes are Sudan black B, introduced by Swahn in 1952 (D16, E2, S32, S33, T1) or its acetyl derivative (C2, L11), Sudan III (F1, F2, G8), Sudan IV (R5), oil red O (D16, S16, S24) or a mixture of Ciba blue BZ2 and Sudan black B (L12, W22). Of these, Sudan black B (Ceresschwarz B) seems to give the best results. A low blank is very difficult to achieve without loss of dye. This led to trials of a different technique. McDonald and Bermes (M11) tried to get around the difficulty by staining the lipids before the run, thus avoiding having to stain and rinse the strip (H3, M10). As these dyes are not water soluble and are electrically neutral this procedure is interesting, although the adding of a saturated alcoholic solution of Sudan black B to the aqueous protein solution before the run may meet with objection.

One volume of dye solution is mixed with 10 volumes of serum at room temperature. Thirty minutes are allowed for reaction, after which the alcohol is evaporated under a nitrogen or air stream and the precipitated excess dye is centrifuged off. Parallel experiments with this and the usual methods show no difference; the dye is dissolved in the lipid fraction and does not influence migration as do dyes conjugated with a protein (S24).

3.2.1.2. *Color Reactions on the Strip.* Besides the use of dyes there are possibilities of developing colored protein zones by a chemical reaction on the strip itself.

For proteins, a reaction on the peptide groups is proposed in which after ethanol-acetone fixation on the strip, the strip is exposed to ClO_2. Development is done in a solution of benzidine in 10 % acetic acid. Unfortunately the color is not stable (R2, R3, R8).

For glyco- and mucoproteins the usual reagent is periodic acid-Schiff reagent (abbreviated PAS). This method was initiated by Koïw and Grönwall (K14) in 1952. Quantities of protein about four times as large as usual are applied to the paper (G19, K15, K16, S15).

The usual clearing solutions (K13) are difficult to use because they accentuate the shadow of the unstained proteins and thus diminish contrast; but the use of ethyl alcohol proposed by Sonnet (S22) eliminates this difficulty.

The same reaction can be done on mucoproteins. These are first precipitated from the protein solution, dialyzed, and concentrated. Then they are separated electrophoretically at pH 4.5 and stained on the paper either with Schiff reagent or with amido black (S22).

A new staining method for glycoproteins is the use of a boiling acetic acid solution of diphenylamine. The color is specific, there is no background, and a photographic recording of the stained strip is used for quantitative work (D12).

3.2.2. *Photometry of Fractions*

After staining, the amounts of dye taken up by the various fractions may be determined either by scanning the strip directly or by photometry of the eluted fractions. In the first method the paper acts as a source of error which is difficult to eliminate effectively, whether the paper is rendered translucent and the transmitted light is measured or whether it is untreated and the quantity of reflected light is measured. In the second method the cutting up of the paper to separate the fractions must be followed by perfect elution without influence on the color of the dye, but subsequent photometry is easy.

3.2.2.1. *Apparatus and Technique.* Many factors are common to all scanning methods whether they use transmitted or reflected light.

A *suitable filter* is needed to secure maximum light absorption and adherence to Beer's law (Fig. 27). In Table 8 a survey of a few dyes with the appropriate wavelength is given.

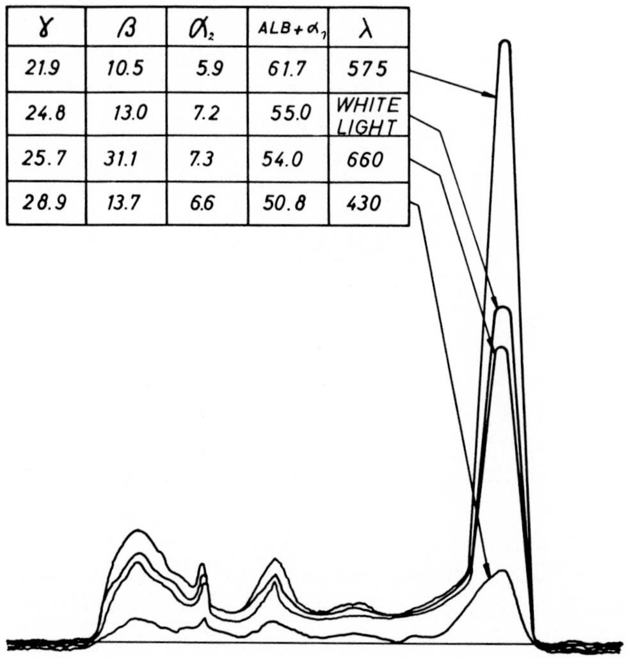

γ	β	α_2	ALB + α_1	λ
21.9	10.5	5.9	61.7	575
24.8	13.0	7.2	55.0	WHITE LIGHT
25.7	31.1	7.3	54.0	660
28.9	13.7	6.6	50.8	430

FIG. 27. Influence of wavelength on relative protein percentages for bromophenol blue-stained strip (D15).

A *slit* adjustable for both length and width is desirable, but improvisation with black tape can compensate for the absence of proper adjusting mechanisms. A slit width of about 1 mm is needed to provide enough light for the classic barrier-layer photocell, but as narrow a slit as possible should be used to minimize the degree of uneveness in the area covered. Toward the edges of the paper strip the protein (and therefore the dye) concentration of the zones decreases (variably because of different diffusion rates) to zero. The length of the slit should be such that only the central, protein-filled part of the paper strip is scanned (Fig. 28). The light leaving the paper should pass through an exit slit to avoid the effect of scatter on the surface of the barrier-layer photocell (Fig. 29). This scatter can be greatly reduced by the proper location of both

TABLE 8
WAVELENGTHS USED IN PHOTOMETRY OF PROTEIN FRACTIONS

Author	Dye	Wavelength mμ	
		Transparency	Elution
Grassmann *et al.* (G15)	Amido black	640	620
	Azocarmine	525	515
	Bromophenol blue	600	595
Loeffler *et al.* (L12)	Amido black		595
	Azocarmine		570
	Bromophenol blue		595
	Sudan III[a]		595
Bracco *et al.*	Unstained	280	
Pezold *et al.* (P16)	Azocarmine	500–550	
	Bromophenol blue	550–600	
Caspani *et al.* (C1a)	Lissamine green	600	

[a] Lipid stain.

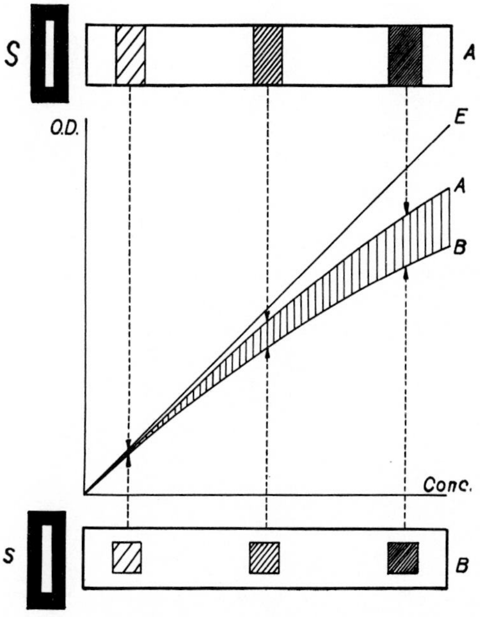

FIG. 28. Influence of slit/spot relation on translucency scanning. After elution the OD's follow Beer's law (Curve *E*). If the spot *A* fills the slit *S* completely and evenly, curve *A* is obtained, but with a too small spot *B*, a far greater error is made. The error of *A* and of *B* shows a logarithmic increase with concentration (P6).

entrance and exit slits (C9) and by the use of translucency fluid (see below).

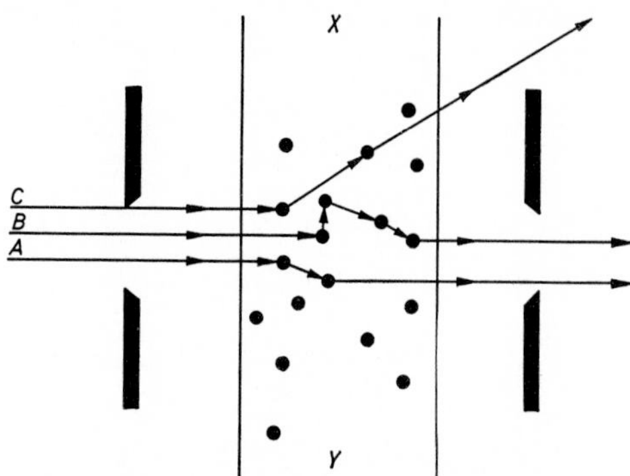

FIG. 29. Beam ABC reaches the photocell only partially after traveling through the substrate *XY*. Best results are obtained with narrow slits at both sides of the paper (C9).

The light source. In many cases a rather primitive light source, such as a long automobile lamp is used, but by chance this happens to compensate partially for the uneven distribution of protein on paper (Fig. 30), as the central and darker zone is better illuminated than the lateral zones (P17, S9; Fig. 31). If even illumination of the slit is obtained, this may cause an error analogous to that of a too wide slit whenever there is irregular protein distribution on the strip.

The photocell must be in a stable condition—have reached maximum sensitivity after a warming-up period—but not be suffering from the fatigue which results from excessive continuous exposure to light. Compensating mechanisms, based on the use of two photocells with a split light beam, improve reliability and are generally used in automatic scanning devices.

The power supply must be constant, whether it consists of batteries or is derived via a voltage stabilizer from the mains; failure here is the chief cause of erratic readings.

To evaluate the separate fractions two methods are available. Vertical dividing lines can be dropped from each trough of the curve to the base line, but this ignores the overlapping of the fractions (which varies since not all the fractions are equally widely separated). However, this method gives easily reproducible results, and since its inherent inaccuracy is not out of proportion to the other inaccuracies of the quantitation, it is commonly used (indeed, it is used for all automatic scanners). The alternative (M9) is to construct, within the framework of the complete protein curve and the base line, a series of Gaussian distribution curves corresponding to the various fractions (i.e. peaks). This is by no means easy, especially as "tailing" irregularities in the run, heterogeneity inside

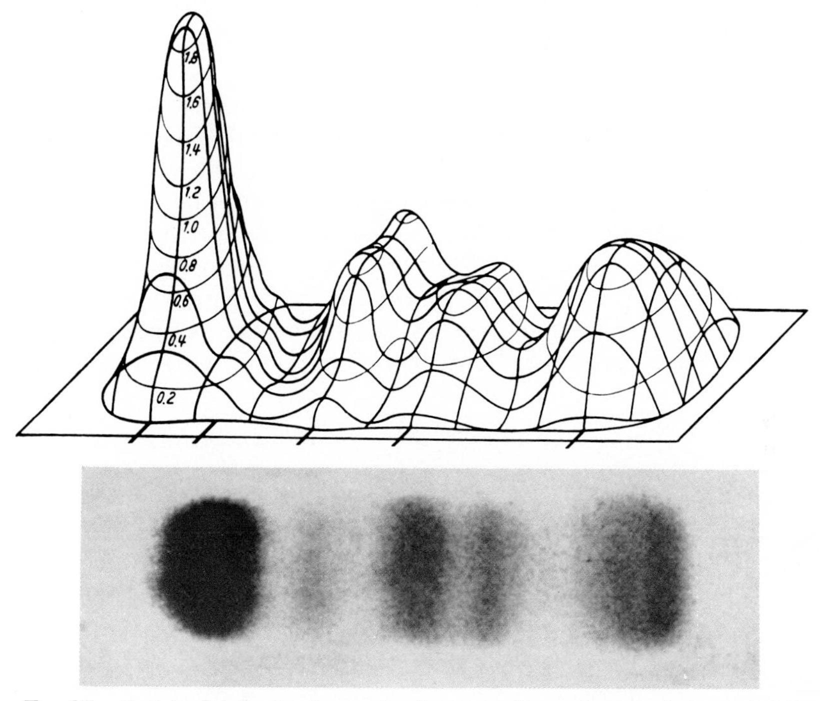

FIG. 30. Protein distribution in a strip after zone electrophoresis of serum (P16).

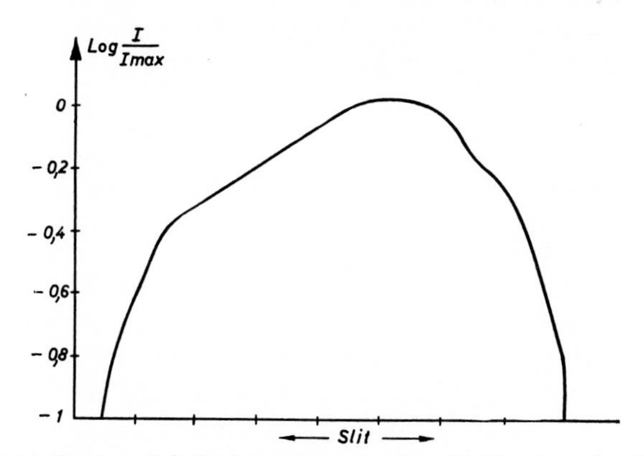

FIG. 31. Distribution of light intensity over the slit illuminated with a commercial bulb (P16). Each abscissa division corresponds to 3 mm.

the fraction, and the effect of the electroendosmotic flow result in skew distributions for many fractions (M13). In practice the method is little used.

Integration of the total area under the pattern is done by planimetry (G17), with an adding machine, or by (P6) cutting the areas and weighing the paper pieces.

Automatic scanners have an interesting feature when equipped with a double beam of the following type. From a single light source two similar beams are obtained, one scanning the strip and the other passing through a movable gray wedge. Each emergent beam falls on a separate photocell, the two cells being in opposition. At each position of the reading beam the wedge is moved by rack and pinion to a position of "no potential" and a pen or similar device records the position on the graph paper. A variant uses only one receiving photocell on which the two beams are compared after "chopping" (Fig. 32). A great advantage of

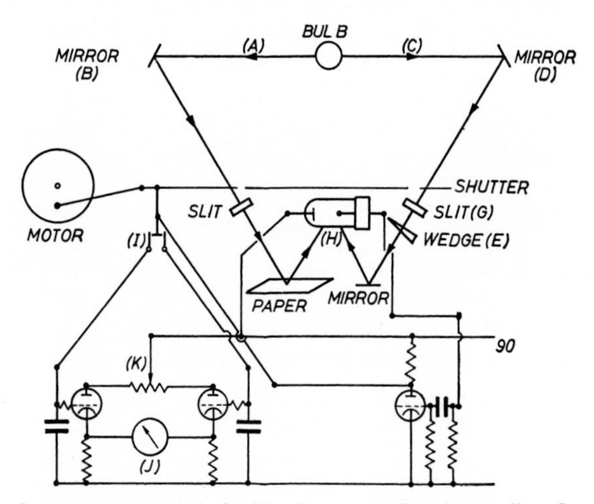

FIG. 32. Schematic layout of double beam single photocell reflectometer. The shutter cuts off alternatively beam *A* or *C* and the photocell *H* receives alternatively the reflected rays from paper or mirror. Via the electric circuit, (*IJK*), wedge *E* compensates the differences between *A* and *C* (L3).

automatic scanners is their independence of voltage fluctuations and light-source changes. If two photosensitive elements are used, their ageing must be parallel, but if a single photocell is used, ageing has no influence on the readings. Apart from stability and reproducibility of readings (C5), there is also the possibility of correcting the compensating wedge according to the paper error (*vide infra*). Corrections are of course limited to the systematic errors of a given type of paper and dye, and application of the individual protein factor is impossible.

Another advantage of automatic devices, and a very important one when many strips have to be evaluated, is the possibility of adapting an integrator to the reading mechanism. It supposes the dropping of theoretical vertical lines for isolation of fractions.

Very often the integrator writes a saw-tooth tracing. It is easy to count the number of teeth corresponding to each peak because the saw-tooth line runs parallel to the conventional concentration curve, and the ease of counting is increased by having, for zones of high optical density, large teeth each equivalent to ten small teeth. Other integrators write a continuous curve from which the cumulative values of the optical density can be read.

Difficulties of scanning on paper. The problem of photometry on paper may be considered to arise partly from the irregular distribution of protein on paper and partly from the structure of the paper itself.

Apart from tailing and irregular diffusion produced by electroendosmosis, heterogeneity of the protein fractions, and similar causes, irregularities of protein distribution throughout the "spot" corresponding to a single protein fraction may depend upon various factors. Thus if the paper is heat dried after the electrophoretic separation, the protein is concentrated on the surface, whereas paper stained without drying tends to have the protein evenly distributed throughout its thickness (Fig. 33). This may not make much difference if scanning is done by transmitted light but will seriously affect the results obtained by the use of reflected light. Distribution is, however, always irregular to some extent because of localized differences in evaporation, diffusion, adsorption, etc. To minimize this, the narrowest possible slit (giving the greatest possible number of readings per "spot") should be used. It is evident that elution of all the protein (or dye) in the spot is the only way of abolishing completely errors due to irregular distribution.

The structure of the paper itself affects the behavior with respect to *Beer's law* which establishes a relation between extinction E and concentration C of transparent solutions, according to the expression $E = kC$. When extinction measured on paper is plotted against concentration, not a linear but a hyperbolic curve is obtained (C10, C11). The cause of the error is the sievelike structure of the paper, since only the threads are covered with stained proteins while the meshes remain completely permeable to light (Fig. 34). This light falling directly on the photosensitive layer of the cell gives, on the microscopic scale, the same error as found for uneven distribution of stained spots on a clear background. The effect of this unabsorbed quantity of direct light is to produce a smaller total absorption of light in heavily stained zones than would be expected by comparison with the total absorption in lightly stained zones.

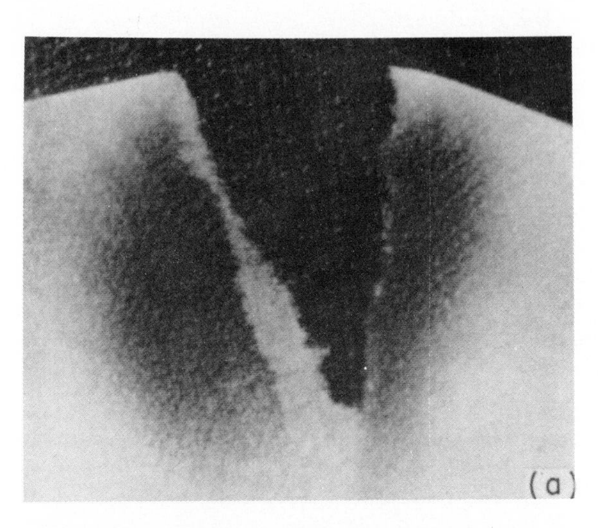

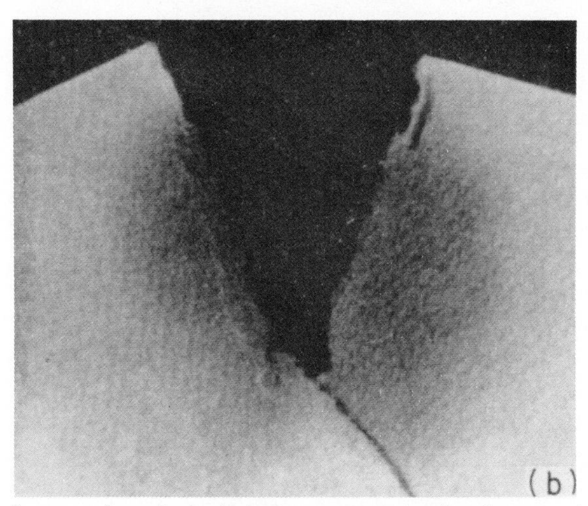

Fig. 33. Influence of method of drying on protein distribution. (a) By rapid drying in a hot air stream the protein collects on the surface of the paper and the inner layers remain colorless after staining. (b) Staining after slow drying in room air gives regular distribution through full thickness of paper (E1). Magnification: × 5.

This error may be called the "paper error" and differs according to the type of paper. A correction for the paper error has been sought in various ways. Thus Rees and Laurence (R1a) scanned a strip carrying a series of standard concentrations of serum. The protein solutions were applied in streaks between siliconed barriers and were thus ideal for

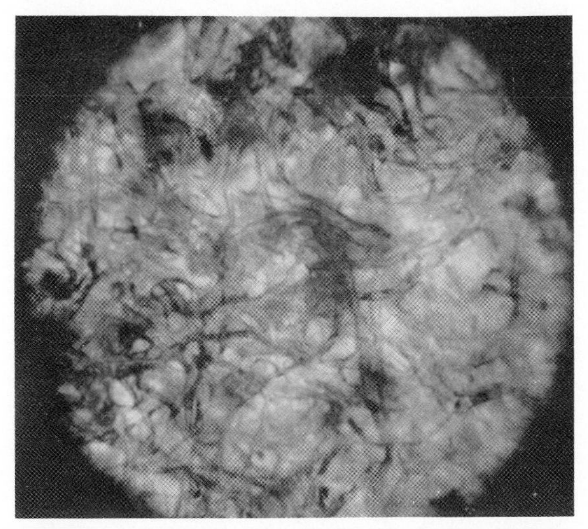

FIG. 34. Microphotographic view of a stained strip (E1). Magnification: × 100.

correct scanning. The values seem to correspond to an exponential equation of the form $E = kC^b$ where b is an empirical factor which can be calculated from the experimental figures (J6). This factor b permits correction of the photometric density scale which can be altered to give a linear optical density (OD) to concentration relationship as far as the particular paper and dye used are concerned. This procedure may be applied to any photometer with either manual or automatic scanning (W25). In automatic scanners the compensating wedge itself may be corrected and a linear relationship established (O4, S13). It must be emphasized again that this correction is valid only for the dye and paper and the experimental conditions used for making the calibrating curve. Instead of using serum, it is possible to use some synthetic chemical such as polyethylenimine (Polymine), as was proposed by Wunderly (W18, W19), for the correction is intended only for the paper error as a function of absorption of a given dye, so that protein itself can be omitted. The advantage of using a synthetic chemical is reproducibility.

The base line is also affected by structure, and papers with a cloudy structure are unsatisfactory for spectrophotometry because the blanks will not be identical over the whole surface. In practice all papers present some irregularity of their blank value, although the paper industry tries to produce uniform papers for chromatographic and electrophoretic purposes. The best possible base line is drawn as the mean of several blank readings taken in front of and behind the stained zones (S20).

Apart from this sporadic error, there is a systematic error due to adsorption of proteins on the paper, which causes the base line to rise toward the starting line. The general adsorption produces impurity of the fractions and favors their optical density against that of the albumin fraction, which has the longest migration distance. A corrected base line has been proposed (G15) as shown in Fig. 35, but the exact value of

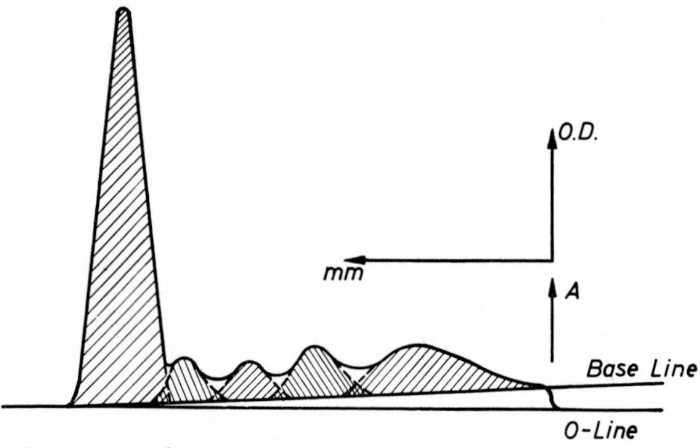

FIG. 35. Correction of Base Line. Base line according to Grassmann, giving correction for adsorption of albumin (G15).

the correction differs from paper to paper. The device is not used in practice because it is only a rough approximation; in any case automatic scanners cannot apply this correction.

Translucency scanning brings its specific difficulties due to the high optical density of paper and its rough optical structure. It calls for a strong light source in order that a sufficient quantity of light may strike the photocell. As the quantity of light absorbed by the stained spots is rather small in comparison with the density of the paper itself, small differences in the optical density of the background have a great influence on the correct reading of a fraction. Fortunately the use of translucency liquids reduces the optical density of paper (Fig. 36) and also has a favorable influence on the mesh of the paper network. A light ray falling on untreated paper leaves it at an entirely different place and the optical density read at one spot is in fact a resultant of those of the adjacent points (Fig. 29). Translucency liquids tend to cut down the paper optical density and also fill the holes in the network, replacing the air which has a refractive index different from that of the paper fibers. A translucency liquid must have several qualities: its refractive index must be nearly identical with that of the paper fibers (approximately 1.50–1.55), it must have a low optical density, and it should fill the paper interstices easily and completely.

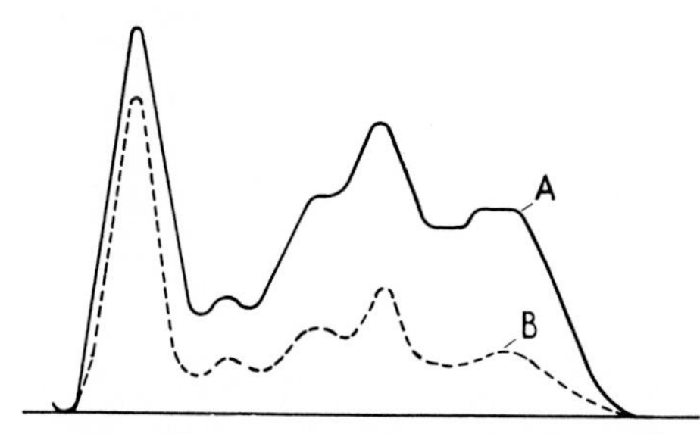

FIG. 36. Influence of translucency agents on OD: Curve *A*, without translucency agents; Curve *B*, after dipping into translucency agent. Curves drawn with "Extinction Schreiber Zeiss."

Results of Planimetry in Percentage of Total Protein

	Albumin	Globulins			
		α_1	α_2	β	γ
Curve *A*: not oiled	31.0	5.2	13.9	27.9	22.9
Curve *B*: oiled	45.0	7.4	13.3	19.6	14.7

The use of reduced pressure helps to eliminate rapidly the enclosed air bubbles (G15), otherwise immersion for several hours is required before the strip is completely penetrated by the translucency liquid. The transparent or translucent strips are then squeezed between two glass plates for subsequent manipulation. This should be done inside the solution to avoid trapping air bubbles.

A translucency liquid in current use is a mixture in equal amounts, of α-bromonaphthalene and paraffin (G14), but many other solutions have been proposed, including methyl salicylate (L3), 2 % sodium carbonate dissolved in equal parts of water and methanol (P6), and Soalex NB (K13). Anisole (P19) is also recommended because reading can be done within 20 minutes, air bubbles seem easy to eliminate, and anisole itself evaporates spontaneously afterward. Sonnet used ethanol for glycoproteins (S22). Some types of varnish have proved useful (D7), but as they dry rather slowly we ourselves have abandoned their use.

After scanning, the solution may be washed out with a solvent to avoid spotting other documents, and the dry strips may be stored in a polythene bag for future reference.

To avoid the difficulties of translucency and the irregularities in the structure of the paper, *reflection photometry* has been proposed (K17, L4, R6, R7, S14, T3). Originally a linear protein/dye relationship was claimed for reflectance measurements (L4) but soon the same error as for translucency readings was found (Fig. 37). A theoretical account

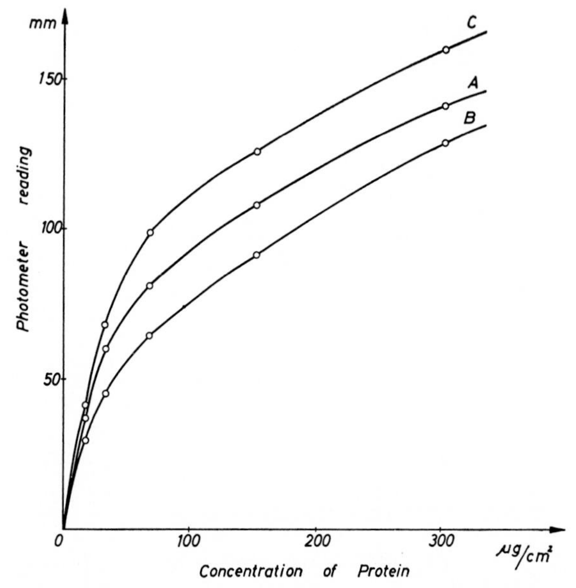

FIG. 37. Relationship between photometer reading and protein concentration. The protein was contained in circular areas approximately 16 mm in diameter and the photometer readings are those corresponding to the central portion of each area. Dyes: Curve *A*, light green; Curve *B*, bromocresol green; Curve C, naphthalene black (O4).

was given by Grüttner (G24), and a correction of the scanner was made by Owen (O4) with, of course, applicability only to a given dye and paper. A serious theoretical objection against reflection is the uneven protein distribution on the surface and inside the paper (C11). With chemical fixation this error is rather small (F1, O4), but with fixation in a hot air stream the exposed surface contains nearly all the proteins (E1; Fig. 33), and the error is indeed very serious. This irregular protein distribution seems an unfavorable aspect of reflectance scanning which was not present in the photometry of a translucent strip.

To relate the area under a peak to actual protein concentration many *correction factors* may be applied, e.g., for the protein error, for deviation from Beer's law, and for the base line.

Each correction may be applied separately. It must be noted that the application of a protein factor is essentially independent of the application of a paper factor, which accounts for deviation from Beer's law. This last correction is a systematic error due to reading on paper and can be fitted either into the scale of the photometer or into the wedge of a double-beam automatic reader. The protein factor may then be applied afterward to particular fractions. In the case of planimetric integration a corrected base line can be drawn, but this is rarely done.

A second possibility is to divide the fractions into the albumin on one side and the globulins on the other and to apply one single correcting factor to the albumin. This factor is a compound of the protein and of the paper factor. Protein and paper errors are most important for the large and concentrated albumin fractions. Fortunately they partly compensate each other, the paper error tending to lower the albumin value and the protein error tending to raise it. We use 1.5 in our own experimental setup and multiply the amido black-albumin surface by this figure.

So many and such contradictory factors have however been proposed that a logical review is not possible. It may be wise not to use any factor at all and to report results as they are measured (S6), for instance, as amido black values. This procedure has the advantage of being easy and also of admitting that paper electrophoretic fractions are a world of their own, full of problems which makes it impossible to compare them to free boundary electrophoretic fractions which are measured as concentration gradients by means of differences in refraction.

3.2.2.2. *Elution.* After staining, the strip is cut up into fractions. The cutting up corresponds to the drawing of theoretical vertical lines between the fractions, but with overlapping of fractions this job is not easy to perform. Each fraction is dropped into a numbered test tube for elution and photometry.

Procedures

For naphthalene black there is a perfect eluant, composed of equal parts ethanol and 0.5 *N* sodium hydroxide. The alkaline pH and the ethanol are required to overcome adsorption of protein to paper, but as alkaline solutions of the dye are very unstable, probably due to the presence of traces of heavy metal ions, the addition of 500 mg of Complexon (EDTA) per liter is required (D3; Fig. 26). In this condition an alkaline solution of amido black is stable for indefinite periods of time.

TABLE 9A

RELATIVE VALUES OF PROTEINOGRAMS MEASURED BY TRANSLUCENCY, REFLECTOMETRY, AND ELUTION

Author and reference	Number of sera	Number of runs per serum	Dye	Correction factor	Albumin %	Albumin σ[a]	Albumin σ%[b]
Translucency							
Grassmann and Hannig (G14)	1	10	Am Bl[c]	1.3		1.20	
Sommerfelt (S17)	1	10	Am Bl	1.3		1.60	
Sonnet and Rhodain (S23)	1	10	Am Bl	1.3	52.90	2.34	4.4
Drevon and Donakian (D11)	1	29	Am Bl	?	59.5	4.5	7.7
Knapp and Sieler (K10)	1	24	Am Bl	1.3		1.70	
Bodart *et al.* (B9)	1	19	Am Bl	1.3		1.38	
Schwartzkopff *et al.* (S8)	1	10	Am Bl	1	46.7	2.0	4.3
van Kampen and Zondag (V3)	1	6	Azo K[d]	?		1.4	
Grassmann and Hannig (G14)	25	3	Am Bl	1.3		1.2	2.0
Sonnet (S21)	13	1	Am Bl	1.3		3.1	5
Riva and Martini (R4)	30	7	Am Bl	1.3			6.5
Scheurlen (S3)	x	1	Am Bl	1.3		4.0	
van Kampen and Zondag (V3)	x	6	Azo C	?		1.9	
Reflectometry							
Owen (O4)	17	2	Light green	?	58.0	3.7	6.4
Elution							
Sonnet and Rhodain (S23)	1	10	Am Bl	1.3	48.9	1.53	3.1
de Jong (D3)	1	12	Am Bl	1.40		0.40	
Schwartzkopff *et al.* (S8)	1	10	Am Bl	1	66.6	3.3	5.0
Körver (K19)			Azo C	1	57.8	1.0	1.6
Plückthun and Götting (P20)	1	12	Azo C	1	59.9	1.20	2.0
Verschure (V6)	1	?	Azo C			2.3	
Owen (O4)	1	12	Light green		66.9	1.75	2.6
de Jong (D3)	45	1	Am Bl	1.46	61.1	1.5	2.5
Grass (G11)	26	1	Br Ph[e]	1.6	66.6	2.78	4.2
Antonini and Riva (A5)	15	1	Am Bl	1.3	55.7	2.02	3.6
Sonnet and Rhodain (S23)	13	1	Am Bl	1.3	61.7	1.98	3.2
Layani *et al.* (L8)	8	1	Am Bl	1.3	60	2.9	4.8
Plückthun and Götting (P20)	15	1	Azo K	1	59	2.45	4.2
de Jong (D3)	49	2	Am Bl	1.46		0.49	
de Jong (D3)	19	2	Am Bl	1.46		0.40	
de Jong (D3)	23	2	Am Bl	1.46		0.43	
Owen (O4)	36	2	Light green		55.0	2.3	4.2

[a] σ = Standard error in absolute values.

[b] σ% = Standard error expressed in per cent of the given value. The values missing in the table could not be found in the original sources.

TABLE 9A (*continued*)

α_1-Globulin			α_2-Globulin			β-Globulin			γ-Globulin			
%	σ[a]	σ%[b]	%	σ[a]	σ%[b]	%	σ[a]	σ%[b]	%	σ[a]	σ%[b]	Remarks
	0.60			0.70			0.60			1.0		
	0.40			0.70			0.50			1.50		
6.53	0.82	12.0	9.57	1.06	11.0	11.7	0.72	6.0	19.28	1.25	6.0	
5.1	1.3	25.0	6.75	1.5	22.0	12.3	2.0	16.2	15.8	2.2	13.9	
	0.80			1.0			1.10			1.80		
	0.05			0.07			0.11			0.19		
5.8	0.7	12.0	9.8	0.9	9.0	13.2	0.7	5.0	24.8	2.0	8.0	
	0.4			0.3			0.7			1.2		The same serum 6
	0.6	14.5		0.7	8.5		0.7	6.3		1.2	7.8	times on several days
	0.7	14.0		1.4	18.0		1.1	11.0		1.9	13.0	
		27.0			19.0			16.0			21.0	
	0.98			1.12			1.92			2.52		
	0.7			1.0			0.9			1.5		
7.0	1.0	14.3	8.0	1.2	15.0	9.0	1.4	15.6	17.0	1.9	11.2	Modified instrument
6.0	0.44	7.2	11.5	0.54	4.7	10.9	1.13	10.3	22.7	1.36	6.0	Path. serum
	0.20			0.34			0.31			0.37		The same serum on
3.3	0.6	18.0	5.7	0.7	13.0	8.5	1.0	12.0	15.6	1.6	10.5	several days
4.5	0.30	6.6	8.2	0.7	8.5	9.6	0.75	7.0	19.9	1.05	5.0	
4.1	0.48	11.7	8.9	0.62	7.0	10.1	0.84	8.3	17.0	1.45	8.5	
	1.0						0.8			1.4		$\alpha_1 + \alpha_2 = 1$ fraction
4.4	0.52	11.8	6.9	0.55	8.0	9.2	0.78	8.5	12.8	1.21	9.5	
4.2	0.3	7.1	8.1	0.6	7.4	10.1	0.8	7.9	16.5	1.3	7.9	
10.1	1.31	13.0				8.7	1.07	12.3	14.6	1.55	10.6	$\alpha_1 + \alpha_2 = 1$ fraction
4.0	0.84	21.0	7.9	1.5	19.0	12.6	1.45	11.5	19.5	1.97	10.1	
4.3	1.08	25.1	7.5	1.19	15.8	10.2	0.71	7.0	16.3	2.41	14.7	
12.0	0.8	6.7				15.0	0.9	6.0	13.0	0.7	5.4	$\alpha_1 + \alpha_2 = 1$ fraction
4.2	0.67	15.9	8.0	1.14	14.3	10.6	1.85	17.5	18.2	2.54	14.0	
	0.27			0.17			0.21			0.30		
	0.13			0.22			0.23			0.25		
	0.26			0.25			0.27			0.36		Path. serum
9.0	1.4	15.5	10.0	1.1	11.0	10.0	1.4	14.0	16.0	1.6	10.0	

[c] Am Bl = Amido black.
[d] Azo C = Azocarmine.
[e] Br Ph = Bromophenol blue.

TABLE 9B
COMPARISON OF METHODS

Author and reference	Method	Number of sera	Number of runs per serum	Dye	Correction factor	Albumin %	Albumin σ[a]	Albumin σ%[b]
Comparison of Elution and Translucency								
Schwartzkopff *et al.* (S8)	El.	1	10	Am Bl[c]	1	66.6	3.3	5.0
	Tr.	1	10	Am Bl	1	46.7	2.0	4.3
Pezold and Kofes (P16)	El.	1	3	Am Bl		57.0	4.0	7.0
	Tr.	1	3	Am Bl		64.0	2.0	3.1
Dittmer (D10)	El.			Am Bl			1.2	
	Tr.			Am Bl	1,4–1,6		1.6	
Comparison of Elution and Reflectometry								
Owen (O4)	El.	1	12	Light green		66.9	1.75	2.6
	Refl.	1	12	Light green		66.6	2.53	3.8

[a] σ = Standard error in absolute values.
[b] σ% = Standard error expressed in per cent of the given value. The values missing in the table could not be found in the original sources.
[c] Am Bl = Amido black.

For bromophenol blue elution is very easy. It is, on the other hand, a cause of error during the rinsing procedure where dye is easily lost. For this dye and also for bromocresol green (F5), azocarmine (K19, V2), and lissamine green (C1a), alkaline eluates are the most suitable. For lissamine green 40 % (v/v) acetic acid has also been used (O4).

A new method consists in staining by means of naphthylaminetrisulfonic acid, followed by fluorimetry of the eluate (W2).

Evaluation and Correction

If the blank or background value of the paper is low, and this is the result of correct rinsing, photometry of the eluate is done in ideal circumstances, and a spectrophotometer increases the accuracy of the readings.

Photometry of the eluate in itself is much easier than on the paper strip, but it presents the serious difficulty of cutting up the strip into the separate fractions, and thus losing the possibility of rechecking.

As correction factor, only the protein factor has to be considered, although trailing of albumin is also a theoretical error related to migration, which could be taken into account.

TABLE 9B (*continued*)

α_1-Globulin			α_2-Globulin			β-Globulin			γ-Globulin			
%	σ^a	$\sigma\%^b$	%	σ^a	$\sigma\%^b$	%	σ^a	$\sigma\%^b$	%	σ^a	$\sigma\%^b$	Remarks
3.3	0.6	18.0	5.7	0.7	13.0	8.5	1.0	12.0	15.6	1.6	10.5	
5.8	0.7	12.0	9.8	0.9	9.0	13.2	0.7	5.0	24.8	2.0	8.0	
3.3	2.0	60.6	8.7	2.0	23.0	10.0	3.0	30.0	21.0	7.0	33.3	
3.3	0.6	6.9	8.7	0.6	6.9	8.7	1.2	13.8	15.0	1.0	6.6	
	0.5			0.5			0.7			1.2		
	0.6			0.8			1.0			1.4		Correction fact 1.4 1.6
4.4	0.52	11.48	6.9	0.55	8.0	9.2	0.78	8.5	12.8	1.21	9.5	
3.9	0.67	17.2	6.2	0.77	12.4	7.6	0.88	11.6	15.6	2.11	13.5	

4. Results

4.1. Expression of Results

A complete insight into the protein composition of a biological fluid demands relative and absolute evaluation of the fractions. After quantitation only relative values are obtained. If a total protein determination is performed on the mixture, absolute values are easily calculated.

Mobility measurements of fractions would of course be another important goal. But at this point it will already be clear that mobility measurements on paper are theoretically impracticable. However, the relative migration distances are easily observed on the pattern.

Some authors (P10) express their results taking the migration distances of their own setups into account (Fig. 38).

4.2. Reproducibility

The reproducibility of electrophoretic work is lower than most experimenters seem to wish. At the moment, migration length and even fractionation (e.g. β_1- and β_2-globulins) differ, according to the technique used and even from day to day in the same laboratory (P6).

If experimental conditions were standardized, the migration distance might become of practical importance in clinical work.

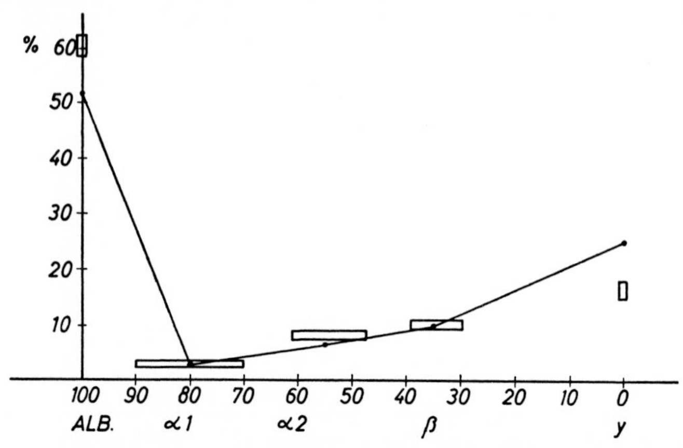

Fig. 38. The results for each fraction are noted on the graph by plotting the percentage of total protein of each fraction as ordinates against their migration distance as abscissas, expressed as percentage of the migration distance of the albumin fraction. The areas covered by the results of normal sera are indicated as rectangles.

In Tables 9A and 9B a few normal values are given, and from consideration of these figures it seems obvious that the following points should be stressed.

Each fraction is expressed in percentage of the total serum protein together with its standard deviation or σ. The fractions constituting a large percentage of the total protein have a low standard deviation and are determined with fair accuracy. For small fractions, however, there is a false impression of accuracy, and an example will show us the importance of this remark. A given fraction representing 3 % of the total protein has a standard deviation of ± 1 %, which is at first sight small, indeed perhaps even lower than the variation for a larger fraction. This represents, however, a deviation of 33 % for the given fraction, and for this reason each σ in the table is also calculated in percentage of the given fraction.

Another point is the great difference in so-called normal values given by different authors (V4). After a careful survey of the many physical problems involved in electrophoretic separation on a stabilizing medium, this seems easy to understand (G16, H7, K9). As we have shown, it is even difficult to establish an exact correction factor.

Elution gives the best and most consistent values (D3) if a careful technique is observed because this method rules out the serious difficulties inherent in photometry on paper.

TABLE 10
RELATIVE VALUES OF NORMAL GLUCIDOGRAMS (S22)

Authors and ref.	Albumin	Globulins				Number of cases
		α_1	α_2	β	γ	
Raynaud *et al.* (1954) (R1)	3.57	10.70	28.60	28.60	28.60	Not given
Biserte and Guérin (1954) (B8a)	4.3	20.5	33.8	25.8	16	Not given
Andreani (1955) (A4a)	15.2	7.5	29.4	30.5	17.4	Not given
Hirsch-Marie (1955) (H13a)	10.89	10.89	26.1	26.1	26.1	Not given
Romani (1954) (R4a)	10.89	10.89	26.1	26.1	26.1	Not given
Björnesjö (1955) (B8b)	13.1 ± 4.1	17.9 ± 3.8	27.9 ± 4.7	23.8 ± 3.7	17.2 ± 4.5	30
Sonnet, J. (1956)	12.6 ± 3.3	20.4 ± 5.3	26.5 ± 3.7	23 ± 3.5	17.5 ± 3.7	15

TABLE 11
RELATIVE VALUES OF NORMAL LIPIDOGRAMS

Authors	Ref.	Number of sera	Dye	α-fractions		β-fractions		O-fractions		α/β
				%	σ	%	σ	%	σ	
Biserte, G., *et al.*, 1954	(C1)	20	Sudan black (cold saturation)	22		52		25		
Joyeuse, R., 1954	(H1)	30	Sudan black	25		74				1/3
Franken, F., 1956	(C1)		Sudan black	20		80				1/4
Haag, W., 1956	(H1)	42	Sudan black	27.3		72.7				1/3
Nijs, A., 1956	(C1)	16	Sudan black B	23.8	4.6	70.3	6.3			1/3
Ricci, G., *et al.*, 1956	(C1)		Sudan black							1/3
Schettler, G., 1956	(C1)		Sudan black	20	10	80	10			1/4
Wieme, R., 1956	(C1)		Sudan black							1/5
Wunderly, C., 1956	(C1)		Sudan black	25–30		45–50		10–25		
Sohar, E., *et al.*, 1957	(S16)	10	Oil red O	35.3	6.7	52.4	7.2	12.3	3.8	

In Table 10 (after Sonnet) a few comparative figures for glucidograms are given. The only point of agreement is the value of the glycoproteins in the α_2 position. The recent figures of Hjörnson and Sonnet are concordant, however, and more likely to be correct.

For lipids, authors are far from agreement and it is difficult to compare the values reported. A selection of published figures is, however, given in Table 11. Comparison between electrophoretic and ultracentrifuge fractions seems an interesting field (P12, P13, P14, P15). Two-dimensional techniques are modifying our concepts on lipoproteins and give consistent results, as will be seen at the end of Part II.

References

A1. Abdel-Wahab, E. M., Rees, V. H., and Laurence, D. J. R., Evaluation of the albumin-globulin ratio of blood plasma or serum by paper electrophoresis. *In* "Paper Electrophoresis," Ciba Foundation Symposium (G. E. W. Wolstenholme and E. C. P. Millar, eds.), pp. 30-39. Churchill, London, 1956.

A2. Ackerman, P. G., Toro, G., and Kountz, W. B., Zone electrophoresis in the study of serum lipoproteins. I. Method and preliminary results. *J. Lab. Clin. Med.,* **44**, 517 (1954).

A3. Aly, F. W., Untersuchungen über die electrophoretisch isolierte Vorfraktion aus Liquor cerebrospinalis. *Biochem. Z.* **325**, 505 (1954).

A4. Anderson, A. B., The direct determination of nitrogen in protein fractions separated by electrophoresis in filter paper. *Biochem. J.* **52**, Proc. x (1952).

A4a. Andreani, D., and Pagni, G., Polisaccaridi protidici del siero nel diabete mellito e nelle complicazioni degenerative. *Boll. soc. med.-chir. Pisa,* **23**, 186 (1955).

A5. Antonini, F., and Riva, G., *Rec. progr. med.* **15**, 243, 258 (1953).

A6. Antweiler, H. J., Quantitative Mikroelektrophorese. *Kolloid-Z.* **115**, 130 (1949).

A7. Ardry, R., Action des détergents sur le sérum humain en présence d'éther. *Ann. biol. clin.* (*Paris*) **15**, 417 (1956).

A8. Audubert, R., and de Mende, S., *In* "Les principes de l'électrophorèse," pp. 88, 101, 113-116. Presses Universitaires de France, Paris, 1957.

B1. Baerts, A. M., Claes, J. H., and Joossens, J. V., Le dosage du cholestérol dans les lipoprotéines alpha et bêta. *4de Colloq. Sint Jans Hospitaal, Brugge* pp. 58-63. De Tempel, Brugge (1956).

B2. Balle-Helaers, E., Dosage polarographique des fractions d'électrophorèse. "Protides of the Biological Fluids, Proceedings of the Fifth Colloquium, Bruges," pp. 51-57. Elsevier, Amsterdam, 1957.

B3. Barac, G., Interaction de la thyroxine avec les protéines plasmatiques et son élimination par le rein. "Protides of the Biological Fluids, Proceedings of the Fifth Colloquium, Bruges," pp. 99-104. Elsevier, Amsterdam, 1957.

B4. Baum, and Dunkelman, Ultraviolet radiation of the high pressure Xenon Arc. *J. Opt. Soc. Am.* **40**, 782 (1950).

B5. Beaven, G. H., and Durrum, E. L., Discussion on "Protein-dye interaction considered in relation to the estimation of protein in paper electrophoresis" by Franglen, G. *In* "Paper Electrophoresis," Ciba Foundation Symposium (G. E. W. Wolstenholme and E. C. P. Millar, eds.), pp. 181-182. Churchill, London, 1956.

B6. Benson, F. A., Series-Parallel valve voltage stabilizer. *Electronic Eng.* **24**, 118-119 (1952).

B7. Berkes, I., and Karas, V., Fibrinogenbestimmung mittels der Papierelektrophorese. *Biochem. Z.* **324**, 499 (1953).

B8. Biserte, G., Dispositif d'électrophorèse sur papier. *Biochim. et Biophys. Acta* **4**, 416 (1950).

B8a. Biserte, G., and Guérin, F., Que peut-on attendre de l'étude comparative des protéines et des complexes lipoprotéiques et glucoprotéiques au moyen de l'électrophorèse sur papier. "Compte rendu 2e Colloque. Les protides des liquides biologiques, Bruges 1954." Ed. Arscia, Bruxelles, 1954.

B8b. Björnesjö, K. B., Analysis of protein-bound serum polysaccharides with anthrone reagent. *Scand. J. Clin. Lab. Invest.* **7**, 147 (1955).

B9. Bodart, P., Osinski, P., and Stein, F., Electrophorèse des proteines sériques dans l'hémochromatose. *Ann. endocrinol.* (*Belg.*) (1953).

B10. Böhle, E., and Fischer, H., Eine colorimetrische Routinemethode zur Bestimmung kleiner Eiweissmengen in Körperflüssigkeiten und Geweben. *Klin. Wochschr.* **31**, 798 (1953).

B11. Bon, W. F., Een nieuwe methode voor het aantonen van eiwit in papier bij de papierelectrophorese. *Chem. Weekblad* **50**, 131 (1954).

B12. Booij, J., The electrophoretic protein pattern of the cerebrospinal fluid. *4de Colloq. Sint Jans Hospitaal, Brugge* pp. 92-102. De Tempel, Brugge (1956).

B13. Boyd, G. S., The estimation of serum lipoproteins—a micromethod based on zone electrophoresis and cholesterol estimations. *Biochem. J.* **58**, 680 (1954).

B14. Bücher, T., Matzelt, D., and Pette, D., Papierelektrophorese von Liquor cerebrospinalis. *Klin. Wochschr.* **30**, 325 (1952).

B15. Bush, I. E., Chromatography of steroids on Alumina-impregnated filter paper. *Nature* **166**, 445 (1950).

C1. Caroli, J., and Charbonnier, A., Enquête sur l'électrophorèse des lipoprotéines. *4de Colloq. Sint Jans Hospitaal, Brugge* p. 245. De Tempel, Brugge (1956).

C1a. Caspani, R., and Magistretti, M., Nuova metodica di colorazione dei tracciati elettroforetici su carta da filtro. *Plasma* (*Milan*) **2**, 33 (1954).

C2. Casselman, W. G. B., Acetylated Sudan black B as a reagent for Lipids. *Biochim. et Biophys. Acta* **14**, 450 (1954).

C3. Consden, R., and Stanier, W. M., Ionophoresis of sugars on paper and some applications to the analysis of protein polysaccharide complexes. *Nature* **169**, 783 (1952).

C4. Consden, R., and Stanier, W. M., A simple paper electrophoresis apparatus. *Nature* **170**, 1069 (1952).

C5. Cooper, G. R., and Mandel, E. E., Paper electrophoresis with automatic scanning and recording. *J. Lab. Clin. Med.* **44**, 636 (1954).

C6. Cremer, H. D., and Tiselius, A., Elektrophorese von Eiweiss in Filtrierpapier. *Biochem. Z.* **320**, 273 (1950).

C7. Crestfield, A. M., and Allen, F. W., Improved apparatus for zone electrophoresis. *Anal. Chem.* **27**, 422 (1955).

C8. Crestfield, A. M., and Allen, F. W., Resolution of ribonucleotides by zone electrophoresis. *Anal. Chem.* **27**, 424 (1955).

C9. Crook, E. M., Analysis of separated materials, I. *In* "Paper Electrophoresis," Ciba Foundation Symposium (G. E. W. Wolstenholme and E. C. P. Millar, eds.), pp. 132-143. Churchill, London, 1956.

C10. Crook, E. M., Harris, H., and Warren, F. L., Continuous direct photometry of proteins separated by electrophoresis in filter paper. *Biochem. J.* **51**, Proc. xxvi (1952).

C11. Crook, E. M., Harris, H., Hassan, F., and Warren, F. L., Continuous direct photometry of dyed materials in filter paper, with special reference to the estimation of proteins separated by electrophoresis. *Biochem. J.* **56**, 434 (1954).

D1. Dangerfield, W. G., and Smith, E. B., An investigation of serum lipids and lipoproteins by paper electrophoresis. *J. Clin. Pathol.* **8**, 132 (1955).

D2. Datta, S. P., Overell, B. G., and Stack-Dunne, M., Chromatography on Alumina-impregnated filter paper. *Nature* **164**, 673 (1949).

D3. de Jong, E. B. M., Eine genaue Methode zur papierelektrophoretischen Auswertung von Proteinen. *Rec. trav. chim.* **74**, 1290 (1955).

D4. Demoulin-Brahy, L., and Franckx-Toussaint, Y., Etude électrophorétique des protéines dans épanchements pleuraux tuberculeux. *3de Colloq. Sint Jans Hospitaal, Brugge* pp. 57-63 (1955).

D5. Dettker, A., and Anduren, H., A new time-saving apparatus for paper electrophoresis. *Scand. J. Clin. & Lab. Invest.* **6**, 74 (1954).

D6. de Wael, J., The combined influence of evaporation and diffusion on the separation of serum proteins by paper electrophoresis. *In* "Paper Electrophoresis" Ciba Foundation Symposium (G. E. W. Wolstenholme and E. C. P. Millar, eds.), pp. 105-110. Churchill, London, 1956.

D7. de Wael, J., Discussion on "Some practical points regarding the reading of paper electrophoretic strip," by Sommerfelt, S. C. *In* "Paper Electrophoresis" Ciba Foundation Symposium (G. E. W. Wolstenholme and E. C. P. Millar, eds.), p. 155. Churchill, London, 1956.

D8. de Wael, J., General discussion by Beaven, G. H., Dent, C. E., Durrum, E. L., Franglen, G., Grassmann, W., Jencks, W. P., Laurell, H., McDonald, H. J., Martin, N. H., Svensson, H., de Wael, J., White, J. C., *In* "Paper Electrophoresis" Ciba Foundation Symposium (G. E. W. Wolstenholme and E. C. P. Millar, eds.), p. 191. Churchill, London, 1956.

D9. Discombe, G., Jones, R. F., and Winstanley, D. P., Estimation of γ-globulin. *J. Clin. Pathol.* **7**, 106 (1954).

D10. Dittmer, A., *In* "Papierelektrophorese." Fischer, Jena, 1956.

D11. Drevon, B., and Donakian, R., Précision obtenue dans l'électrophorèse sur papier des protéines sériques. *Bull soc. chim. biol.* **37**, 605-611 (1955).

D12. Drevon, B., and Donakian, R., Sur une nouvelle méthode d'étude des glycoprotéines séparées par électrophorèse sur papier: réaction à la diphénylamine. *Bull. soc. chim. biol.* **37**, 1321 (1955).

D13. Durrum, E. L., A microelectrophoretic and microionophoretic technique. *J. Am. Chem. Soc.* **72**, 2943 (1950).

D14. Durrum, E. L., Paper electrophoresis. *Science* **113**, 66 (1951).

D15. Durrum, E. L., Paper Electrophoresis. *In* "A Manual of Paper Chromatography and Paper Electrophoresis" (R. J. Block, E. L. Durrum, and G. Zweig, eds.), 2nd ed., p. 508. Academic Press, New York, 1958.

D16. Durrum, E. L., Milton, H., Paul, L., and Smith, E. R. B., Lipid detection in paper-electrophoresis. *Science* **116**, 428 (1952).

D17. Durrum, E. L., and Svensson, H., General Discussion. *In* "Paper Electrophoresis," Ciba Foundation Symposium (G. E. W. Wolstenholme and E. C. P. Millar, eds.), pp. 119, 215. Churchill, London, 1956.

D18. Dustin, J. P., Electrophorèse sur papier avec évaporation controlée. *1ste Colloq. Sint Jans Hospitaal, Brugge* pp. 33-42. Union Clinique Belge S.A. Bruxelles (1953).

D19. Dustin, J. P., Wodon, C., Medard, O., and Bigwood, E. J., Isolement et électrophorèse sur papier de l'hormone gonadotrophique de l'urine humaine. *Ann. endocrinol. (Paris)* **13**, 687 (1952).

E1. Eberhardt, H., Untersuchungen über die Bindung von Farbstoffen an Proteine auf Filtrierpapier und ihre Bedeutung für die Auswertung von Papierelektropherogrammen. *Röntgen- u. Lab.-praxis* 9, pp. 249-261; and pp. 283-295 (1956).

E2. Eggstein, M., Lipoidelektrophorese. *Ärztl. Lab.* **6**, 205 (1956).

E3. Elliott, F., De ionisatie der proteinen. *Mededel. Vlaam. Chem. Ver.* **2**, 79 (1955).

E4. Esser, H., Heinzler, F., Kazmeier, F., and Scholtan, W., Quantitative Serumeiweissfraktionierung mit der Elektrophorese in Filtrierpapier. *Münch. med. Wochschr.* **93**, 985 (1951).

E5. Esser, H., Heinzler, F., Kazmeier, F., and Scholtan, W., *Arzneimittel-Forsch.* **6**, 156 (1952).

E6. Ewerbeck, H., Die elektrophoretische Darstellung normalen menschlichen Liquors. *Klin. Wochschr.* **28**, 692 (1950).

F1. Fasoli, A., Electrophoresis of serum lipoproteins on filter paper. *Lancet* **262**, 106 (1952).

F2. Fasoli, A., Electrophoresis of serum lipoproteins on filter paper. *Acta Med. Scand.* **145**, 233 (1953).

F3. Flynn, F. V., and de Mayo, P., Micro-electrophoresis of proteins on filter paper. *Lancet* **261**, 235 (1951).

F4. François, J., Wieme, R. J., Rabaey, M., and Neetens, A., L'électrophorèse sur papier des protéines hydrosolubles du cristallin. *Experientia* **10**, 79 (1954).

F5. Franglen, G. T., Protein-dye interactions considered in relation to the estimation of protein in paper electrophoresis. *In* "Paper Electrophoresis," Ciba Foundation Symposium (G. E. W. Wolstenholme and E. C. P. Millar, eds.), pp. 172-177. Churchill, London, 1956.

F6. Franglen, G. T., The dye uptake of native and modified serumproteins. "Protides of the Biological Fluids. Proceedings of the Fifth Colloquium Bruges," pp. 63-67. Elsevier, Amsterdam, 1957.

F7. Franglen, G. T., Martin, N. H., and Treherne, J. D., An apparatus for paper electrophoresis. *J. Clin. Pathol.* **8**, 144 (1955).

G1. Gabrieli, E., Goulian, D., Kindersly, T., and Collet, R., *J. Clin. Invest.* **33**, 136 (1954).

G2. Geinitz, W., and Hobitz, H., Ueber die Eiweisse der Perikardflüssigkeit des Menschen. "Protides of the Biological Fluids. Proceedings of the Fifth Colloquium, Bruges," pp. 84-88. Elsevier, Amsterdam, 1957.

G3. Girard, M., and Rousselet, Fr., Extracteur continu pour décoloration de fond des protéinogrammes. *Ann. biol. clin.* **15**, 239 (1957).

G4. Goa, J., A simplified method for the separation and quantitative determination of serum proteins by paper electrophoresis. *Scand. J. Clin. & Lab. Invest.* **3**, 236 (1951).

G5. Goa, J., On protein-bound carbohydrates in human serum. *Scand. J. Clin. Lab. Invest.* **7**, Suppl. 22 (1955).

G6. Goldberg, C., Identification of human hemoglobins. *Clin. Chem.* **3**, 1 (1957).

G7. Gordon, A. H., Gross, J., O'Connor, D., and Pitt-Rivers, R., Nature of the circulating thyroid hormone-plasma protein complex. *Nature* **169**, 19 (1952).

G8. Gottfried, S. P., Pope, R. H., Friedman, N. H., and Di Mauro, S., A simple method for the quantitative determination of α and β lipoproteins in serum by Paper electrophoresis. *J. Lab. Clin. Med.* **4**, 651 (1954).

G9. Grabar, P., and Williams, C. A., Méthode permettant l'étude conjuguée des propriétés électrophorétiques et immunochimiques d'un mélange de protéine. Application au sérum sanguin. *Biochim. et Biophys. Acta* **10**, 193 (1953).

G10. Graf, E., and List, P. H., Ueber einen Umkehreffekt bei der Papierelektrophorese im Blutseren. *Naturwiss.* **40**, 273 (1953).

G11. Gras, J., *Rev. españ. fisiol.* **8**, 59 (1952).

G12. Grassmann, W., General discussion. *In* "Paper Electrophoresis," Ciba Foundation Symposium (G. E. W. Wolstenholme and E. C. P. Millar, eds.), pp. 193, 196. Churchill, London, 1956.

G13. Grassmann, W., and Hannig, K., Ein einfaches Verfahren zur Analyse der Serumproteine und anderer Proteingemische. *Naturwiss.* **37**, 496 (1950).

G14. Grassmann, W., and Hannig, K., Ein quantitatives Verfahren zur Analyse der Serumproteine durch Papierelektrophorese. *Z. physiol. Chem.* **290**, 1 (1952).

G15. Grassmann, W., and Hannig, K., Beiträge zur Methodik der papierelektrophoretischen Serumanalyse. *Klin. Wochschr.* **32**, 838 (1954).

G16. Grassmann, W., and Hannig, K., Stellungnahme zur Arbeit V. W. Fuchs und A. Flach. *Klin. Wochschr.* **33**, 907 (1955).

G17. Grassmann, W., Hannig, K., and Knedel, M., Ueber ein Verfahren zur elektrophoretischen Bestimmung der Serumproteine auf Filtrierpapier. *Deut. med. Wochschr.* **11**, 333 (1951).

G18. Griffiths, L. L., The electrophoresis of serum and other body fluids in filter paper. *J. Lab. Clin. Med.* **41**, 188 (1953).

G19. Grönwall, A., On paper electrophoresis in the clinical laboratory. *Scand. J. Clin. & Lab. Invest.* **4**, 270 (1952).

G20. Gross, D., Paper electrophoresis of sugars at high potential gradient. *Nature* **172**, 908 (1953).

G21. Gross, D., Paper electrophoresis of the oligosaccharides synthesized from sucrose by yeast invertase. *Nature* **173**, 487 (1954).

G22. Gross, D., Paper electrophoresis of amino-acids and oligopeptides at very high potential gradients. *Nature* **176**, 72 (1955).

G23. Grüne, A., Papiers à filtrer pour la chromatographie sur papier et l'électrophorèse sur papier. Schleicher & Schüll, Dassel (1953).

G24. Grüttner, R., Zur Auswertung von Papierelektrophoresen und Chromatogrammen durch Messung reflektierter Strahlung. *Klin. Wochschr.* **32**, 263 (1954).

G25. Guillien, R., and Hermann, G., Ueber die Adsorption von Bromphenolblau an zellulosegebundene Serumeiweisse. *Naturwiss.* **42**, 72 (1955).

G26. Guillot, M., Divers aspects physico-chimiques de l'électrophorèse et en particulier de l'electrophorèse de zones. *Ann. biol. clin.* (*Paris*) **15**, 8 (1956).

H1. Haag, W., Contribution à la technique de l'électrophorèse des lipoprotéines. *Extr. des Ann. biol. clin.* (*Paris*) **14**, 1 (1956).

H1a. Haag, W., Précisions sur la technique de l'électrophorèse des lipoprotéines. *4de Colloq. Sint Jans Hospitaal, Brugge* p. 144. De Tempel, Brugge (1956).

H2. Harders, C. L., Beschouwingen naar aanleiding van klinische ervaringen met Papierelectrophorese. *Chem. Weekblad* **49**, 246-247 (1953).

H3. Harms, H., Neuere Verfahren zur Anfärbung von Lipoproteiden in Papierpherogrammen. *Lab. Blätter* **2**, 37 (1956).

H4. Heide, K., and Biel, H., Versuche zur Senkung der mittleren Streubreite der Papierelektrophorese. *Ärztl. Lab.* **2**, 360 (1956).

H5. Heilmeyer, L., Clotten, R., Sano, I., Sturm, A., Jr., and Lipp, A., Analyse des Reststickstoffes mit Hilfe des Hochspannungspherogrammes. *Klin. Wochschr.* **32**, 831 (1954).

H6. Henning, N., Kinzlmeier, H., and Demling, L., Ueber die elektrophoretisch darstellbaren Proteine normaler und pathologischer Magensäfte. *Münch. med. Wochschr.* **95** (15), 423 (1953).

H7. Henry, R. J., Golub, O. J., and Sobel, C., Some of the variables involved in the fractionation of serum proteins by paper electrophoresis. *Clin. Chem.* **3**, 49 (1957).

H8. Heremans, J., Résultats de l'électrophorèse des colloides urinaires. "Protides of the Biological Fluids. Proceedings of the Fifth Colloquium, Bruges," pp. 34-39. Elsevier, Amsterdam, 1957.

H9. Hermann, G., Electrophorèse sur papier de liquides pleuraux et péritonéaux en comparaison avec le séroelectrophérogramme. *4de Colloq. Sint Jans Hospitaal, Brugge* pp. 230-237. De Tempel, Brugge (1956).

H10. Hess, K., *Melliand Textilber.* **29** (1934).

H11. Hill, W., Analysis of voltage regulator operation. *Proc. IRE* **33**, 38-45 (1945).

H12. Hill, W., Analysis of current stabilizer circuit. *Proc. IRE* **33**, 785-792 (1945).

H13. Hirs, C. H. W., Moore, S., and Stein, W. H., Isolation of amino acids by chromatography on ion exchange columns; use of volatile buffers. *J. Biol. Chem.* **195**, 669 (1952).

H13a. Hirsch-Marie, H., Sur la mise en évidence des polysaccharides sériques par l'électrophorése sur papier (glucidogramme). *Ann. biol. clin.* (*Paris*) **13**, 593 (1955).

H14. Holiday, E. R., and Johnson, E. A., Location of paper chromatogram spots of pyrimidine derivatives in U.V. light. *Nature* **163**, 216 (1949).

H15. Holt, C. v., Voigt, K. D., and Gaede, K., Papierelektrophorese von Eiweisskörpern bei erhöhter Spannung. *Biochem. Z.* **323**, 345 (1952).

H16. Homolka, J., *Radiometer Polarographics* **1**, 110 (1952).

H17. Horst, W., and Schumacher, H. H., Papierelektrophoretische Untersuchungen von Schilddrüsenextracten und Serum nach in vitro-Zusatz von Radiojodid, Radiomangan, und Radiocobalt sowie nach in vivo-Gabe von Radiojodid. *Klin. Wochschr.* **32**, 361 (1954).

H18. Horst, W., and Schumacher, H. H., Transport und Bindung im Serum untersucht mit Papierelektrophorese und radioaktiven Indicatoren. *Klin. Wochschr.* **32**, 961 (1954).

H19. Huisman, T. H. J., Abnormal human hemoglobins. "Protides of the Biological Fluids. Proceedings of the Fifth Colloquium, Bruges," pp. 184-194. Elsevier, Amsterdam, 1957.

J1. Jaenicke, L., Die Papierelektrophorese von Zuckern und Zuckerderivaten. *Naturwiss.* **39**, 86 (1952).

J2. Janssen, L. W., Movement of uncharged particles and molecules in the electrical field. *Rec. trav. chim.* T. **65**, 564 (1946).

J3. Jencks, W. P., cited by Durrum, E. L., *In* "A Manual of Paper Chromatography and Paper Electrophoresis" (R. J. Block, E. L. Durrum, and G. Zweig, eds.), 2nd ed., p. 554. Academic Press, New York, 1958.

J4. Jencks, W. P., Jetton, M. R., and Durrum, E. L., Paper electrophoresis as a quantitative Method (serum proteins). *Biochem. J.* **60**, 205 (1955).

J5. Jermyn, M. A., and Thomas, R., Reduction of liquid flow in paper electrophoresis. *Nature* **172**, 728 (1953).

J6. Joossens, J. V., and Claes, J. H., Quantitative study of dye-protein relationship by scanning and elution methods in paper electrophoresis. Paper presented at *First Intern. Congr. Clin. Chem., New York* (1956).

J7. Josephson, B., St. Eriks Sjukhus, Stockholm. Personal communication (1957).

K1. Kallee, E., Ueber J-signiertes Insulin. I. Mitteilung (Nachweis). *Z. Naturforsch.* **7b**, 661-663 (1952).

K2. Karcher, D., Lowenthal, A., and van Sande, M., Discussion sur l'origine des protéines du liquide céphalorachidien. "Protides of the Biological Fluids. Proceedings of the Fifth Colloquium, Bruges," pp. 76-82. Elsevier, Amsterdam, 1957.

K3. Keuning, F. J., Proeven over de quantitatieve bepaling van serum-eiwit fracties in de papierelectrophorese met kleurstof elutie. *Chem. Weekblad* **41**, 702 (1954).

K4. Kickhöfen, B., High voltage paper electrophoresis. *In* "Paper Electrophoresis," Ciba Foundation Symposium (G. E. W. Wolstenholme and E. C. P. Millar, eds.), pp. 206-209. Churchill, London, 1956.

K5. Kickhöfen, B., and Westphal, O., Papierelektrophorese bei hohen Spannungen zur Trennung von Peptiden. *Z. Naturforsch.* **7b**, 655 (1952).

K6. Kickhöfen, B., and Westphal, O., Ueber eine einfache Kombination von Papierelektrophorese und Papierchromatographie. *Z. Naturforsch.* **7b**, 659 (1952).

K7. Kimbel, K. H., Proteinnachweis auf Elektrophoresestreifen durch U.V.-Absorption. *Naturwiss.* **40**, 200 (1953).

K8. Knapp, A., Die Papierelektrophorese des Liquor cerebrospinalis. *Ärzt. Lab.* **5**, 189 (1955).

K9. Knapp, A., Ueber Fehlerquellen bei der Auswertung der Papierelektrophorese durch Transparenzmessung. *Arztl. Lab.* **2**, 53 (1956).

K10. Knapp, A., and Sieler, H., Zur Methodik der Papierelektrophorese. *Z. ges. inn. Med. u. ihre Grenzgebiete* **8**, 741 (1953).

K11. Knedel, M., Quantitative Glykoproteidbestimmungen in isolierten Serumeiweissfraktionen. *Verhandl. deut. Ges. inn. Med.* (Kongr.) **61**, 277 (1955).

K12. Knedel, M., Ueber eine erbliche Proteinanomalie. "Protides of the Biological Fluids. Proceedings of the Fifth Colloquium, Bruges," pp. 72-75. Elsevier, Amsterdam, 1957.

K13. Koïw, E., Paper Electrophoresis. *In* "Paper Electrophoresis," Ciba Foundation Symposium (G. E. W. Wolstenholme and E. C. P. Millar, eds.), pp. 79-85. Churchill, London, 1956.

K14. Koïw, E., and Grönwall, A., Staining of protein-bound carbohydrates after electrophoresis of serum on filter paper. *Scand. J. Clin. & Lab. Invest.* **2**, 244 (1952).

K15. Koïw, E., Wallenius, G., and Grönwall, A., L'utilisation clinique de l'électrophorèse sur papier filtre. Comparaison avec l'électrophorèse à tube en U selon la méthode de Tiselius. *Bull. soc. chim. biol.* **33**, 1940 (1951).

K16. Koïw, E., Wallenius, G., and Grönwall, A., *Scand. J. Clin. & Lab. Invest.* **4**, 47 (1952).

K17. Kolmar, D., and Wendler, J., *Deut. Gesundheitswesen* **3**, 1519 (1953).

K18. König, P., *Actas e trabalhos 3 congr. sud-am. quim., Rio de Janeiro e São Paulo* **2**, 334 (1937).

K19. Körver, G., Elektrophorese im Filtrierpapier; ein einfaches Verfahren zur Bestimmung der Serumeiweisskörper. *Klin. Wochschr.* **28**, 693 (1950).

K20. Kraft, D., and Biel, H., Einfluss der Albuminadsorption auf die papierelektrophoretisch ermittelte Zusammensetzung des Normalserums. *Ärzt. Lab.* **11**, 397 (1956).

K21. Kunkel, H. G., *In* "Methods of Biochemical Analysis" (D. Glick, ed.), Vol. 1, pp. 141-170. Interscience, New York, 1954.

K22. Kunkel, H. G., and Tiselius, A., Electrophoresis of proteins on filter paper. *J. Gen. Physiol.* **35**, 89 (1951).

L1. Lang, N., Die Reaktions- oder Ueberwanderungs-Elektrophorese mit markierten Eiweisskörpern in der Immun-Biologie. *In* "Radioaktive Isotope in Klinik und Forschung" (K. Fellinger and H. Vetter, eds.), Vol. II, pp. 95-103. Urban & Schwarzenberg, München and Berlin, 1957.

L2. Langan, T. A., Durrum, E. L., and Jencks, W. P., Paper electrophoresis as a quantitative method: measurement of alpha and beta lipoprotein cholesterol. *J. Clin. Invest.* **34**, 1427 (1955).

L3. Latner, A. L., Braithwaite, F., and Nunn, R. A., A photoelectric scanning device for electrophoresis of proteins on filter paper. *Biochem. J.* **51**, Proc. x (1952).

L4. Latner, A. L., Molyneux, L., and Rose, J. D., A semiautomatic recording densitometer for use after paper-strip electrophoresis. *J. Lab. Clin. Med.* **43**, 157 (1954).

L5. Laurell, C. B., and Laurell, S., Electrophoresis of plasmaproteins. *Lancet* ii, 40 (1955).

L6. Laurell, C. B., Laurell, S., and Skoog, N., Buffer composition in Paper Electrophoresis. *Clin. Chem.* **2**, 99 (1956).

L7. Laurence, D. J. R., Discussion on: Protein-dye interactions considered in relation to the estimation of protein in paper electrophoresis, by Franglen, G. *In* "Paper Electrophoresis," Ciba Foundation Symposium (G. E. W. Wolstenholme and E. C. P. Millar, eds.), p. 177. Churchill, London, 1956.

L8. Layani, F., Bengui, A., and de Mende, S., *Semaine hôp.* **28**, 80 (1952).

L9. Levin, B., and Oberholzer, V. G., Paper electrophoresis of serum proteins with micro Kjeldahl nitrogen analysis of the protein fractions: A comparison with free electrophoresis and salt fractionation methods. *Am. J. Clin. Pathol.* **23**, 205 (1953).

L10. Liesegang, R. E., *In* "Kolloidchemische Technologie." Dresden and Leipzig, 1927.

L11. Lillie, R. D., and Burtner, H. J., Stable sudanophilia of human neutrophil leucocytes in relation to peroxidase and oxidase. *J. Histochem.* **1**, 8 (1953).

L12. Loeffler, W., and Wunderly, C., The staining of proteins and lipoids after electrophoresis on filter paper. *J. Clin. Pathol.* **4**, 282 (1953).

L13. Longsworth, L. G., and Jacobson, C. F., An electrophoretic study of the binding of salt ions by lactoglobin and bovine serum albumin. *J. Phys. & Colloid Chem.* **53**, 126 (1949).

L14. Loos, R., Aperçu des méthodes employées dans l'électrophorèse sur papier. *1ste Colloq. Sint Jans Hospitaal, Brugge* pp. 21-27. Union Chimique Belge S.A., Bruxelles. (1953).

M1. Mach, W., and Geffert, R., Eine neue Methode der Schnellelektrophorese durch mittel- oder hochfrequente Impulse *Arzneimittel-Forschung* **3**, 534 (1953).

M2. Macheboeuf, M., Dubert, J. M., and Rebeyrotte, P., Microélectrophorèse sur papier avec évaporation continue du solvant (Electrorhéophorèse). II. Etude théorique. Application à la mesure des mobilités électrophorétiques. *Bull. soc. chim. biol.* **35**, 346 (1953).

M3. Mackay, I. R., Volwiler, W., and Goldsworth, P. D., Paper electrophoresis of serum proteins; Photometric quantitation and comparison with free electrophoresis. *J. Clin. Invest.* **33**, 855 (1954).

M4. Markham, R., Jacobs, J., and Fletcher, F., Zone electrophoresis of serum and urine at pH 4.5 and its application to the isolation and investigation of mucoproteins. *J. Lab. Clin. Med.* **35**, 559 (1956).

M5. Martin, N. H., Analysis of separated materials, II. *In* "Paper Electrophoresis," Ciba Symposium (G. E. W. Wolstenholme and E. C. P. Millar, eds.), pp. 160-171. Churchill, London, 1956.

M6. Martin, N. H., and Franglen, G. T., Use and limitations of filter-paper electrophoresis. *J. Clin. Pathol.* **7**, 87 (1954).

M7. Maurer, W., and Reichenbach, L., Ueber die Bindung des organischen Jods im Serum an einzelne Serum-Eiweissfraktionen. *Naturwiss.* **39**, 261 (1952).

M8. McDonald, H. J., Ionography: a new frontier in electrophoresis. *J. Chem. Educ.* **29**, 428 (1952).

M9. McDonald, H. J., Area under peaks: dropping perpendiculars versus extending curves to baseline. *In* "Paper Electrophoresis," Ciba Foundation Symposium (G. E. W. Wolstenholme and E. C. P. Millar, eds.), pp. 149-150. Churchill, London, 1956.

M10. McDonald, H. J., A new approach to the staining of lipoproteins. *In* "Paper Electrophoresis," Ciba Foundation Symposium (G. E. W. Wolstenholme and E. C. P. Millar, eds.), pp. 183-186. Churchill, London, 1956.

M11. McDonald, H. J., and Bermes, E. W., A new procedure for staining lipoproteins in ionographic separations. *Biochim. et Biophys. Acta* **17**, 290 (1955).

M12. McDonald, H. J., and Spitzer, R. H., Polyvinylpyrrolidine: the electromigration characteristics of the blood plasma expander. *Circulation Research* **1**, 396 (1953).

M13. McDonald, H. J., Durrum, E. L., Warren, F. L., and Crook, E. M., Discussion on some practical points regarding the reading of the paper electrophoresis strip. *In* "Paper Electrophoresis," Ciba Foundation Symposium (G. E. W. Wolstenholme and E. C. P. Millar, eds.), p. 155. Churchill, London, 1956.

M14. McDonald, H. J., Marbach, E. P., and Urbin, M. C., *Clin. Chemist* **5**, 17 (1953).

M15. McDonald, H. J., Lappe, R. J., Marbach, E. P., Spitzer, R. H., and Urbin, M. C., "Ionography: Electrophoresis in Stabilized Media." Year Book Publishers, Chicago, 1953.

M16. McKinley, W. P., Oliver, W. F., and Comma, R. H., The determination of serum proteinfractions on filter paper electropherograms by the biuret reaction. *Can. J. Biochem. and Physiol.* **32**, 189 (1954).

M17. Mellon, M. G., General principles of absorptimetric measurements. *In* "Analytical Absorption Spectroscopy" (M. G. Mellon, ed.). Wiley, New York, 1950.

M18. Merklen, F. P., and Masseyeff, R., L'adsorption d'albumine, cause d'erreur dans l'appréciation quantitative des protéines sériques dans l'électrophorèse sur papier-filtre. (1954).

M19. Merlevede, E., Pottiez, F., and Verhelle, O., Etude qualitative des acides aminés libres dans le suc gastrique, à l'état normal et à l'état pathologique par une méthode combinée de chromatographie et d'ionophorèse sur papier. Les acides aminés dans les hydrolysats de tissus normaux et néoplasiques de l'estomac. *Arch. intern. pharmacodynamie* **100**, 265 (1955).

M20. Merten, R., Schram, G., Grassmann, W., and Hannig, K., Untersuchungen zur Reindarstellung des Magenkathepsins. *Z. physiol. Chem.* **289**, 173 (1952).

M21. Meulemans, O., Een eenvoudig electrophorese-apparaat. *3de Colloq. Sint Jans Hospitaal, Brugge* pp. 191-196. De Tempel, Brugge. (1955).

M22. Meulemans, O., Relatieve loopsnelheidsmetingen van serumeiwitfracties met behulp van papier-electro-rheo-phorese. *Maandschr. kindergeneesk.* **12**, 488 (1955).

M23. Michaelis, L., Der Acetat Veronal Puffer. *Biochem. Z.* **234**, 139 (1931).

M24. Michl, H., Ueber Papierionophorese bei Spannungsgefällen von 50 Volt/cm. *Monatsh. Chem.* **82**, 489 (1951).

M25. Michl, H., Ueber Papierionophorese bei Spannungsgefällen von 50 V/cm. *Monatsh. Chem.* **83**, 737 (1952).

M26. Motulsky, A. G., Paul, M. H., and Durrum, E. L., Paper electrophoresis of abnormal hemoglobin and its clinical applications. A simple semi quantitative method for the study of the hereditary hemoglobinopathies. *Blood* **9**, 897 (1954).

N1. Newton, G. G. F., and Abraham, E. P., Degradation, structure and some derivatives of Cephalosporin N., *Biochem. J.* **58**, 103 (1954).

N1a. Neyt, J., Verhelle, O., and Maertens, R., Powersupply for constant voltage or constant current. Exhibit presented at the 5th Colloquium Sint Jans Hospitaal, Brugge (1957).

N2. Nikkilä, E., Studies on the lipoprotein relationships in normal and pathological sera and the effect of heparin on serum lipoproteins. *Scand. J. Clin. & Lab. Invest.* **5**, Suppl. 8, 9-100 (1953).

O1. Oosterhuis, H. K., Studies on paper electrophoresis. *J. Lab. Clin. Med.* **44**, 280 (1954).

O2. Oosterhuis, H. K., Prins, G., and Verleur, H., A new apparatus for automatic coloration and extraction on the filter paper strips in electrophoresis on filter paper. *Rec. trav. chim.* **73**, 963 (1954).

O3. Owen, J. A., Electrophoresis of plasma proteins. *Lancet* **i**, 868 (1955).

O4. Owen, J. A., Determination of serum-protein fractions by zone electrophoresis on paper and direct reflection photometry. *Analyst* **81**, 26 (1956).

O5. Owen, J. A., and McKenzie, B., Application of paper electrophoresis to separation of blood-clotting factors. *J. Appl. Physiol.* **6**, 696 (1954).

O6. Owen, J. A., Rider, W. D., and Stewart, C. P., The value of paper electrophoresis of serum and urinary proteins in the diagnosis of myelomatosis. *4de Colloq. Sint Jans Hospitaal, Brugge* pp. 103-111. De Tempel, Brugge. (1956).

P1. Partridge, S. M., Aniline hydrogen phthalate as a spraying reagent for chromatography of sugars. *Nature* **164**, 443 (1949).

P2. Patchett, G. N., "Automatic Voltage Regulators and Stabilizers." Pitman, London, 1954.

P3. Pearson, S., Stern, S., and McGavack, T., A rapid procedure for the determination of serum cholesterol. *J. Clin. Endocrinol.* **12**, 1245 (1952).

P4. Peeters, H., and Anthierens, L., Eenvoudige apparatuur voor collecterende papierelectrophorese. *3de Colloq. Sint Jans Hospitaal, Brugge* pp. 197-204. De Tempel, Brugge. (1955).

P5. Peeters, H., and Vuylsteke, P., Natriumconcentratie veranderingen tijdens papierelectrophorese. *3de Colloq. Sint Jans Hospitaal, Brugge* pp. 231-236. De Tempel, Brugge. (1955).

P6. Peeters, H., and Vuylsteke, P., Methodology of paper electrophoresis; Results of a survey with comments and conclusions. *3de Colloq. Sint Jans Hospitaal, Brugge* pp. 239-281. De Tempel, Brugge. (1955).

P7. Peeters, H., and Vuylsteke, P., Factors influencing paper electrophoresis of proteins. Exhibit presented at the 3rd Intern. Congr. Biochem., Brussels (1955).

P8. Peeters, H., and Vuylsteke, P., Oorzaken van beperkte migratielengte in de electrophorese. *4de Colloq. Sint Jans Hospitaal, Brugge* pp. 63-68. De Tempel, Brugge. (1956).

P9. Peeters, H., and Vuylsteke, P., Ionophoresis of the buffersalts during zone- and bidimensional electrophoresis. Paper presented at *1st Intern. Congr. Clin. Chem. New York* (1956); *J. Chromatography* **2**, 308 (1959).

P10. Peeters, H., and Vuylsteke, P., Principles and results of paper electrophoresis of proteins. Exhibit presented at 1st Intern. Congr. Clin. Chem., New York (1956).

P11. Perilhou, P., and Cayzac, J., Une alimentation stabilisée pour 5000 V., 50 mA. *Philips Rev. Tech.* **13**, 302 (1952).

P12. Pezold, F. A., Isolierung des α-Lipoproteins aus Humanserum in der Ultrazentrifuge und sein Identitätsnachweis mittels Papierelektrophorese. *Naturwiss.* **43**, 280 (1956).

P13. Pezold, F. A., Zur Kenntnis der elektrophoretisch trennbaren Lipoproteidfraktionen in Blutserum. *Klin. Wochschr.* **35**, 475 (1957).

P14. Pezold, F. A., Vergleichende Untersuchungen der Lipoproteine in Humanseren und Ultrazentrifugaten mittels der Zonen-elektrophorese in Filtrierpapier und Agargel. "Protides of the Biological Fluids. Proceedings of the Fifth Colloquium, Bruges," pp. 40-44. Elsevier, Amsterdam, 1957.

P15. Pezold, F. A., De Lalla, O., and Gofman, J., Experimentelle Untersuchungen über die Zuordnung der papierelektrophoretisch bestimmbaren Lipoproteidgruppen zu den mittels präparativer Ultrazentrifugierung trennbaren Fraktionen. *Clin. Chim. Acta.* **2**, 43 (1957).

P16. Pezold, F. A., and Kofes, A., Beitrag zur Photometrie gefärbter Filtrierpapierstreifen mittels des Transparenzverfahrens. *Klin. Wochschr.* **32**, 504 (1954).

P17. Pezold, F. A., and Kofes, A., Schlusswort zu der vorstehenden Entgegnung von K. Sellier. *Klin. Wochschr.* **33**, 769 (1955).

P18. Pezold, F. A., and Peiser, U., Untersuchungen über die Affinität papierelektrophoretisch aufgetrennter und denaturierter Serumproteinfraktionen gegenüber den Eiweissfarbstoffen Amidoschwarz, Azocarmin und Bromphenolblau. *Klin. Wochschr.* **31**, 982 (1953).

P19. Pieper, J., and Molinski, H., Zur Methodik der Papierelektrophorese nach Grassmann und Hannig. *Klin. Wochschr.* **32**, 985 (1954).

P20. Plückthun, H., and Götting, H., *Klin. Wochschr.* **29**, 415 (1951).

P21. Pučar, Z., Beitrag zur Methode der electrophoretischen Untersuchung der Serumeiweisskörper auf Filtrierpapier. *Arkiv Kemi* **25**, 205 (1953).

P22. Pučar, Z., Ueber die Färbung der Papier-Elektropherogramme mit Amidoschwarz. *Z. physiol. Chem.* **296**, 62 (1954).

P23. Putzeys, P., and Bouckaert, L., The ionization of proteins, I. *Mededeel. Koninkl. Vlaam. Acad. Wetenschap. Belg.* **4**, 3 (1942).

R1. Raynaud, R., d'Eshougues, J. R., and Pasquet, P., Technique de l'électrophorèse sur papier pour l'étude des glucides et des lipides liés aux protéines sériques. *Alg. Méd.* **59**, 263 (1954).

R1a. Rees, V. H., and Laurence, D. J. R., The correspondence with Beer's Law for the optical density of stained protein patterns on filter paper as a function of surface protein concentration. *Clin. Chem.* **1**, 329 (1955).

R2. Reindel, F., and Hoppe, W., Ueber eine neue Färbenmethode zum Nachweis von Aminosäuren, Peptiden und Eiweisskörpern auf Papierchromatogrammen und Elektropherogrammen. *Naturwiss.* **40**, 221 (1953).

R3. Reindel, F., and Hoppe, W., Ueber eine Färbenmethode zum Anfärben von Aminosäuren, Peptiden und Proteinen auf Papierchromatogrammen und Papierelektropherogrammen. *Chem. Ber.* **87**, 1102 (1954).

R4. Riva, G., and Martini, V., *Schweiz. med. Wochschr.* **4**, 73 (1953).

R4a. Romani, J. D., La recherche et l'évaluation des glycoprotéines du sérum à l'aide de l'électrophorèse sur papier. *Presse Méd.* **62**, 1578 (1954).

R5. Rosenberg, I. N., Serum lipids studied by electrophoresis on paper. *Proc. Soc. Exptl. Biol. Med.* **80**, 751 (1952).

R6. Röttger, H., Vergleichsmessungen an Papierelektrophoresestreifen mit vier verschiedenartigen optischen Geräten. *Experimentia* **9**, 150 (1953).

R7. Röttger, H., Zur Reflektionsmessung von Papierpherogrammen. *Ärztl. Lab.* **2**, 58 (1956).

R8. Rydon, H. N., and Smith, P. W. G., A new method for the detection of peptides and similar compounds on paper chromatograms. *Nature* **169**, 922 (1952).

R9. Ryle, A. P., Sanger, F., Smith, F. L., and Kitai, R., The disulfide bonds of insulin. *Biochem. J.* **60**, 541 (1955).

S1. Sanger, F., Thompson, E. O. P., and Kitai, R., The amide groups of insulin. *Biochem. J.* **59**, 509 (1955).

S2. Schaeffer, A., *Textilpraxis* 307 (1947).

S3. Scheurlen, P. G., Über Serumeiweissveränderungen beim Diabetes Mellitus. *Klin. Wochschr.* **33**, 198 (1955).

S4. Schneider, R. G., and Galveston, T., Incidence of Hemoglobin C trait in 505 normal negroes: a family with homozygous hemoglobin C and sickle cell trait union. *J. Lab. Clin. Med.* **44**, 133 (1954).

S5. Schneider, G., and Wallenius, G., Electrophoretic studies on cerebrospinal fluid proteins. *Scand. J. Clin. & Lab. Invest.* **3**, 145 (1951).

S6. Schulte, M. S., Reproduceerbaarheid van de papierelectrophoretische eiwitbepaling. *Chem. Weekblad* **41**, 704 (1954).

S7. Schulz, D. M., and Holdcraft, M. A., Quantitative Paperstrip electrophoresis. *Lab. Invest.* **4**, 262 (1955).

S8. Schwartzkopff, W. Remmer, H., and Hübner, E. *Ärztl. Wochschr.* **53**, 1929 (1954).

S9. Sellier, K., Beitrag zur Photometrie gefärbter Filtrierpapierstreifen mittels des Transparenzverfahren. (Zur Arbeit gleichen Titels von F. A. Pezold und A. Kofes. *Klin. Wochschr.* 1954. 504) *Klin. Wochschr.* **33** (1955).

S10. Semm, K., and Fried, R., Einfache Apparatur zur Fluoreszenzmessung auf Filtrierpapierstreifen. *Naturwiss.* **39**, 326 (1952).

S11. Simon, N., Radioactive gold in filter paper electrophoresis patterns of plasma. *Science* **119**, 95 (1954).

S12. Sluyterman, J. A., Electrophoretic behaviour in filter paper and molecular weight of insulin. *Biochim. et Biophys. Acta* **17**, 169 (1955).

S13. Smith, R. L., The estimation of dye in stained paper electrophoresis strips using the EEL-scanner. *Clin. Chem. Acta* **2**, 122 (1956).

S14. Smith, J. D., and Markham, R., Structure of ribonucleic acid. *Nature* **168**, 406 (1951).

S15. Sohar, E., Bossak, F., and Adlersberg, D., Species differences in serum glycoproteins of man, dog and rabbit by paper electrophoresis. *Proc. Soc. Exptl. Biol. Med.* **90**, 305 (1955).

S16. Sohar, E., Bossak, F., Adlersberg, D., Correlative studies of serum proteins lipoproteins and glycoproteins in inborn disorders of lipid metabolism. *J. Lab. Clin. Med.* **5**, 716 (1957).

S17. Sommerfelt, C., Reproductibility with paper electrophoresis of serum proteins. *Scand. J. Clin. & Lab. Invest.* **4**, 307 (1952).

S18. Sommerfelt, C., Comments regarding the evaluation of paper electrophoresis of proteins. *Scand. J. Clin. & Lab. Invest.* **5**, 105 (1953).

S19. Sommerfelt, C., Paper electrophoresis of isolated plasma protein fractions. *Scand. J. Clin. & Lab. Invest.* **5**, 299 (1953).

S20. Sommerfelt, C., Some practical points regarding the reading of the paper electrophoretic strip. *In* "Paper Electrophoresis," Ciba Foundation Symposium (G. E. W. Wolstenholme and E. C. P. Millar, eds.), pp. 151-154, 177. Churchill, London, 1956.

S21. Sonnet, J., Etude des protéines sériques par l'électrophorèse sur papier. Reproductibilité des résultats. *1ste Colloq. Sint Jans Hospitaal, Brugge* pp. 63-72. Union Chimique Belge. S. A., Bruxelles (1953).

S22. Sonnet, J., "Les glycoprotéines sériques à l'état normal et pathologique." Edit. Arscia, Bruxelles, 1956.

S23. Sonnet, J., and Rhodain, J., Etudes des protéines sériques par l'électrophorèse sur papier. I. Technique et résultats normaux. *Rev. belge pathol. et méd. exptl.* **22**, 226 (1952).

S24. Spitzer, R. H., and McDonald, H. J., Equilibrium dialysis and ionographic studies of proteins and polyvinylpyrolidane interactions with bromophenol blue. *Federation Proc.* **14**, 285 (1955).

S25. Standaert, L., L'électrophorèse des protéinuries. *2de Colloq. Sint Jans Hospitaal, Brugge* pp. 119-129. Editions Arscia, Bruxelles. (1954).

S26. Steigerwald, H., and Schütte, F., Präparative elektrophoretische Darstellung einer blutzuckersteigernden Substanz aus der Bauchspeicheldrüse. *Klin. Wochschr.* **31**, 385 (1953).

S27. Sterz, H., and Klementschitz, W., Eine Methode zur raschen Trennung der Eiweissfraktionen des menschlichen Serums. *Wien. klin. Wochschr.* **64**, 103 (1952).

S28. Svensson, H., Physicochemical aspects and their relationship to the design of apparatus. *In* "Paper Electrophoresis," Ciba Foundation Symposium (G. E. W. Wolstenholme and E. C. P. Millar, eds.), pp. 86-104. Churchill, London, 1956.

S29. Svensson, H., General Discussion. *In* "Paper Electrophoresis," Ciba Foundation Symposium (G. E. W. Wolstenholme and E. C. P. Millar, eds.), p. 192. Churchill, London, 1956.

S30. Svensson, H., A discussion on the meaning of equivalent weights and transport (transference) numbers for amphoteric electrolytes, especially protolytes. *Sci. Tools,* **3**, 30 (1956).

S31. Svensson, H., and Olhagen, B., Electrophoresis of serum and plasma in free solution without preceding dialysis. *Sci. Tools,* **1**, 2 (1954).

S32. Swahn, B., A method for localization and determination of serum lipids after electrophoretical separation on filter paper. *Scand. J. Clin. & Lab. Invest.* **4**, 98 (1952).

S33. Swahn, B., Studies on blood lipids. *Scand. J. Clin. & Lab. Invest.* **5**, Suppl. 9, 1 (1952).

T1. Tigaud, J., Technique des lipidogrammes. *J. méd. Lyon* n° **826** (1954).

T2. Tiselius, A., Elektrophorese und Adsorptionsanalyse als Hilfsmittel zur Untersuchung hochmolekularer Stoffe und ihrer Zerfallsprodukte. *Naturwiss.* **37**, 25 (1950).

T3. Turba, F., "Chromatographische Methoden in der Proteinchemie." Springer, Berlin-Göttingen-Heidelberg, 1954.

T4. Turba, F., and Enenkel, H. J., Elektrophorese von Proteinen in Filterpapier. *Naturwiss.* **37**, 93 (1950).

T5. Turba, F., Pelzer, H., and Schuster, H., Trennung von Nukleinsäureabkömmlingen durch Ionenaustausch, Papierchromatographie und Hochspannungsionophorese in Filtrierpapier. *Z. physiol. Chem.* **296**, 97 (1954).

T6. Tompsett, S. L., and Tennant, W. S., A method for determining esterified fatty acid with zone electrophoresis of serum proteins. *Am. J. Clin. Pathol.* vol. 26, **10**, 1226 (1956).

V1. Valmet, E., and Svensson, H., *Sci. Tools,* **1**, 3 (1954).

V2. van Gool, J., Het kleurbindingsvermogen van albumine en γ-globuline voor azokarmijn B. *Chem. Weekblad* **41**, 704 (1954).

V3. van Kampen, E., and Zondag, H., Quantitative papierelectrophorese der serumeiwitten. Eijn bedrage tot standaardisatie. *Chem. Weekblad* **51**, 535 (1955).

V4. Vera, J., Roche, M., A note on the distribution of the serum protein fractions in apparently normal persons in Caracas. *J. Lab. Clin. Med.* **47**, 418 (1956).

V5. Verhelle, O., and Merlevede, E., Méthode combinée de chromatographie et d'ionophorèse sur papier pour la séparation des acides aminés. *2de Colloq. Sint Jans Hospitaal, Brugge* pp. 181-197. Editions Arscia, Bruxelles. (1954).

V6. Verschure, J. C. M., and Boom, J., De algemene klinische toepasbaarheid van de papier electrophorese. *Ned. Tijdschr. Geneesk.* **98**, 2607 (1954).

V7. von Frijtag, C. A. J., and Reinhold, J. G., Application of zone electrophoresis to analysis of serum proteins. Technique for horizontal strip method and evaluation of its precision and accuracy. *Anal. Chem.* **27**, 1090 (1955).

V8. von Holt, C., Voigt, K. D., and Gaede, K., see reference (H15).

W1. Wahab, E., Adjutantis, G., and Laurence, D. J. R., Borate buffer as a medium for the ionophoresis of human serum proteins. *Biochem. J.* **60**, Proc. xiii (1955).

W2. Weber, G., Cited by Laurence, D. J. R., in General Discussion. *In* "Paper Electrophoresis," Ciba Foundation Symposium (G. E. W. Wolstenholme and E. C. P. Millar, eds.), p. 216. Churchill, London, 1956.

W3. Weismann-Netter, R., and Hirsch-Marie, H., Aperçu sur la formule protéique du liquide synovial. *4de Colloq. Sint Jans Hospitaal, Brugge* pp. 71-76. De Tempel, Brugge. (1956).

W4. Werner, G., and Westphal, O., Stofftrennungen durch Hochspannungs-Papierelektrophorese. *Angew. Chem.* **67**, 251 (1955).

W5. Wieland, T., and Fischer, E., Ueber Elektrophorese auf Filtrierpapier. Trennung von Aminosäuren und ihren Kupferkomplexen. *Naturwiss.* **35**, 29 (1948).

W6. Wieland, T., and Fischer, E., Die "Retentionsanalyse," eine quantitative Ultramikromethode. Ultramikrobestimmung von Aminosäuren. *Naturwiss.* **35**, 29 (1948).

W7. Wieland, T., and Wirth, L., Ueber Retentionsanalyse. III. Retentionsanalyse von Papierelektropherogrammen natürlicher Proteingemische. *Angew. Chem.* **62**, 473 (1950).

W8. Wieland, T., and Wirth, L., Ueber Retentionsanalyse. IV. Quantitative Bestimmung von Aminosäuren mit der verbesserten Retentionstechnik. *Angew. Chem.* **63**, 171 (1951).

W9. Wieme, R. J., and Rabaey, M., Les protéines du cristallin et leur rapport avec la cataracte. *2de Colloq. Sint Jans Hospitaal, Brugge* pp. 29-44. Edition Arscia, Bruxelles. (1954).

W10. Wieme, R. J., and Rabaey, M., Acquisitions nouvelles dans le fractionnement des protéines du cristallin. *3de Colloq. Sint Jans Hospitaal, Brugge* pp. 157-162. De Tempel, Brugge. (1955).

W11. Williams, F. G., Jr., Pickels, E. G., and Durrum, E. L., Improved hanging-strip paperelectrophoresis technique. *Science* **3154**, 829 (1955).

W12. Witmer, R., Elektrophorese pathologischen menschlichen Kammerwassers. *Experientia* **7**, 347 (1951).

W13. Wolff, R., and Magnin, P., Remarques sur l'emploi de l'Amidoschwarz B comme révélateur des fractions de protides obtenues par électrophorèse sur papier. *Bull. soc. chim. biol.* **36**, 925 (1954).

W14. Wolff, H. P., Lang, N., and Knedel, M., Untersuchungen mit Cu^{64} über die Bindung des Kupfers an Serumeiweisskörper. *Z. ges. exptl. Med.* **125**, 359 (1955).

W15. Wood, S. E., and Strain, H. H., Electro-osmosis in paper electrochromatography with electrodes on the paper. *Anal. Chem.* **26**, 1869 (1954).

W16. Woods, E. F., and Gillespie, J. M., *Australian J. Biol. Sci.* **6**, 130 (1953).

W17. Wunderly, C., Ueber methodische Verbesserungen der Papierelektrophorese. *Plasma (Milan)* **2**, 143 (1954).

W18. Wunderly, C., Die Kontrolle des Färbung nach Papierelektrophorese mit einer Bezugssubstanz. *4de Colloq. Sint Jans Hospitaal, Brugge* pp. 29-34. De Tempel, Brugge. (1956).

W19. Wunderly, C., Erfahrungen mit dem Polymin-standard als Kontrolle für die Proteinfärbung nach Papierelektrophorese. *Klin. Wochschr.* **34**, 1123 (1956).

W20. Wunderly, C., Ueber Fortschritte mit Papierelektrophorese. *Chimia (Switz.)*, **10**, 1 (1956).

W21. Wunderly, C., *In* "Electrophorèse sur papier." Vigot Frères, Paris, 1956.

W22. Wunderly, C., and Pezold, F. A., Die Lösung cancerogener Kohlenwasserstoffe im Blutserum. *Naturwiss.* **39**, 493 (1952).

W23. Wunderly, C., Hässig, A., Lottenbach, F., Die Differenzierung des Serumalbumins vom Albumin in Liquor, Pleuraexsudat und der Oedemflüssigkeit. *Klin. Wochschr.* **31**, 49 (1953).

W24. Wunderly, C., Steiger, R., and Böhringer, H. R., Neue mehrfache Untersuchungen am gleichen Kammerwasser menschlicher Augen. *Experienta* **10**, 432 (1954).

W25. Wurm, M., and Epstein, F. H., Quantitative electrophoresis of serum proteins on paper. *Clin. Chem.* **2**, 303 (1956).

Part II. Two-Dimensional Methods

5. Introduction

The review of two-dimensional methods starts with the description of the principle and the advantages and main problems of the technique. Under a second heading the apparatus will be described, and many theoretical points will be discussed along with the development of the instrument because in two-dimensional techniques the relation between theory and practice is far more intricate than in zone electrophoresis. After this critical survey of the apparatus, a description of the run according to two different working conditions will close the review.

6. General Problems

6.1. Principles

In two-dimensional electrophoresis the charged particle migrates in a field of two forces which act perpendicularly to one another. The first force F creates a vertical hydrodynamic field. A flow of liquid runs by gravity down a vertical curtainlike supporting medium to which we shall refer as the *substrate*. The liquid is a buffer solution which through its pH and ionic strength determines the mobility of the particle.

The second force F^1 is electrical and creates a horizontal electrical field across the supporting curtain. The force derived from the liquid flow F is identical for all particles, while the electrical force F^1 differs according to the charge on the particles. Thus the resultant R of the two forces F and F^1 is different according to the charge on a given particle, and migration occurs according to R_1, R_2, R_3, . . R_n (Fig. 39). When a mixture of particles is applied at a given point on the substrate, each of the different components of the sample at once takes its own direction which it follows down to the bottom of the substrate, where it appears as a single fraction at a point which is determined by the direction and magnitude of the resultant force concerned. Insofar as only these two forces are considered, there is no theoretical possibility of finding particles of a different electrophoretic mobility inside a given fraction as long as the experimental conditions do not vary.

If the sample is *continuously applied* at a uniform rate, this allows for a continuous collection of particles of given mobility in test tubes located at the bottom of the substrate. This is called "collecting electrophoresis" because the collection of isolated fractions is the aim of the technique. The method was originated by Haugaard and Kroner in 1948 (H2), Grassmann in 1949 (G1, G2), and Svensson and Brattsten in 1949 (S6).

If the sample is applied *as a spot,* it splits up into a series of spots which lie farther apart as the time interval grows longer. This method with discontinuous application of the sample is called "star electrophoresis" because of the appearance of the stained substrate (P4).

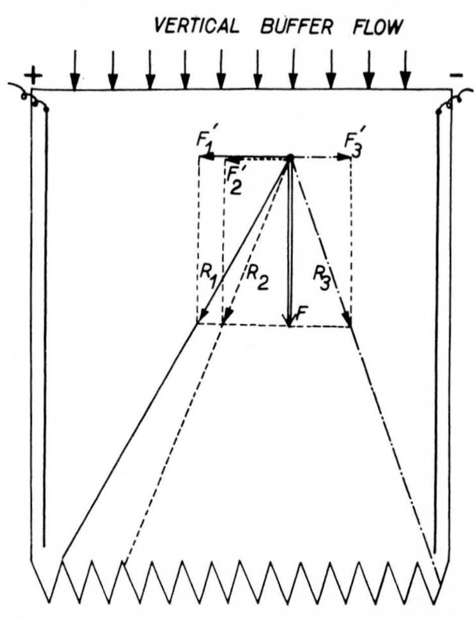

FIG. 39. Principle of migration in two-dimensional electrophoresis. Each particle migrates according to the resultant vector (R_1, R_2, R_3) of the horizontal electrophoretic velocity (F_1', F_2', F_3') and the vertical velocity of the buffer (F).

An important advantage of two-dimensional methods is the solitary migration of each fraction on its own path. In contrast with unidimensional or zone electrophoresis where each fraction migrates its own distance but over a common path, no contamination of fractions is possible (Fig. 40). If the substrate is sufficiently long and wide and a high intensity electrical field is applied, it becomes possible to separate fractions reasonably completely from each other. If the physical forces involved are deployed skillfully, the electrical field can be driven up so that rapid and wide separations are obtained. This point is especially important in star electrophoresis, whereas in collecting electrophoresis mastery of the physical forces provides an excellent continuous preparative tool.

At this stage we must take note of an important theoretical point which is usually overlooked. The migration of a particle on a supporting medium is partially counteracted through adsorption, and in the case of the two-dimensional field both driving forces will be proportionally affected (P8; Fig. 41). This applies both to collecting electrophoresis and to star electrophoresis.

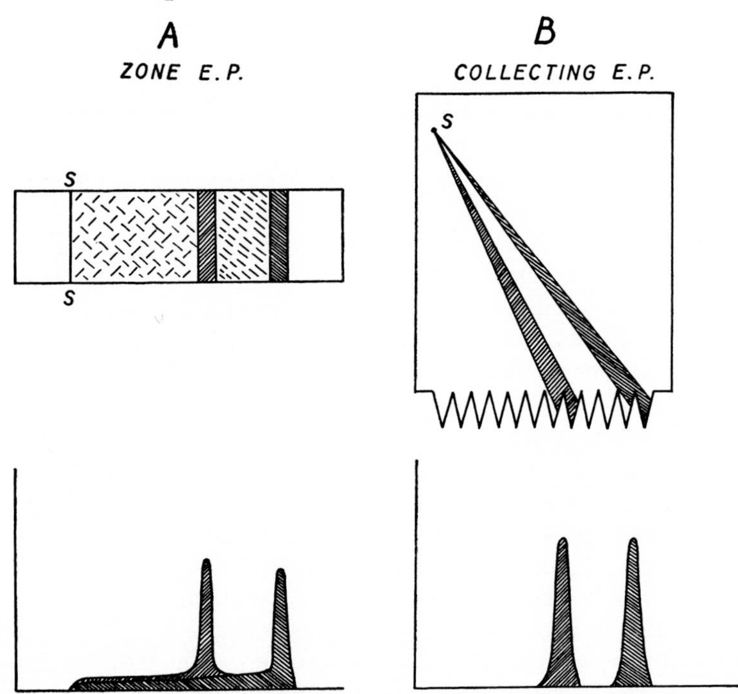

FIG. 40. Influence of adsorption in zone and two-dimensional electrophoresis. (*A*) In zone electrophoresis adsorption causes contamination of the fractions. (*B*) In two-dimensional electrophoresis adsorption has no influence on the purity of the fractions.

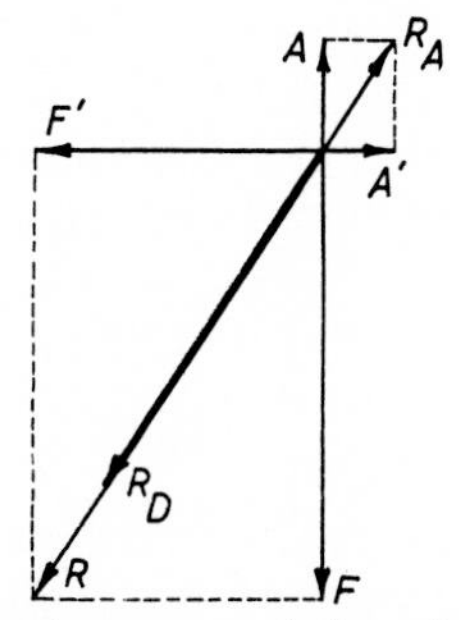

FIG. 41. The vector R must be corrected for adsorption (R_A) on the substrate which acts as a resistive force (A' and A) to the driving forces of electrical field (F') and buffer flow (F). R_D is the resulting driving force.

During collecting electrophoresis the sample is fed onto the substrate continuously and the degree of adsorption causes only a difference in the time needed for a fraction to reach the bottom of the paper. Insofar as two fractions have the same electrophoretic mobility, but a different degree of adsorption on the curtain, they will be collected together at the same place underneath the substrate, and the so-called pure fraction is a mixture of two components. This means that in general for collecting electrophoresis two fractions can only be separated if they have different electrophoretic mobilities. On the contrary, in zone electrophoresis two substances having the same mobility, but a different adsorption, can be separated.

In star electrophoresis with application of the sample as a spot, two components with identical mobility but different adsorption will soon appear as two isolated spots along the same vector leaving the application zone. This means that only fractions having identical mobility together with identical adsorption cannot be separated. It follows from these theoretical considerations that star electrophoresis gives rise to more and better individualized fractions than zone or collecting electrophoresis. A general survey of the possibilities of separation is given in Fig. 42.

Some authors, who consider zone electrophoresis as electrochromatography, seem to attach much importance to adsorption on the substrate, which can indeed be called the chromatographic effect of the paper. Hence two-dimensional techniques may be called two-dimensional electrochromatography, but from the above theoretical considerations it will be clear that collecting electrophoresis is not a chromatographic method and that star electrophoresis is. For this reason a terminology based on the general outlook and the scope of a technique—zone, collecting, or star electrophoresis—is simpler than a terminology which inadequately reflects intricate theoretical principles.

A last remark must be made about the use of the word "continuous electrophoresis," covering collecting two-dimensional techniques in which both sample and background buffer feeding are continuous. As there is a two-dimensional technique with discontinuous application of the sample but a continuous feeding of buffer, the use of the term continuous is difficult, and again the terms "collecting" and "star" electrophoresis are preferred.

6.2. Problems

The interweaving of electrical field and buffer flow gives rise to several problems. Rectilinear migration is theoretically the result of constant forces at all points of the substrate curtain at all times. Varying the intensity of the field during the run causes the fractions to rejoin or to spread like a fan with curved ribs during the experiment.

The measurement of the buffer flow by itself can be done by applying eosine spots which must run vertically downward over the same distance during a chromatographic experiment when no current is applied (Fig. 43, A). The isolated measurement of the electrical field can be made by determining isopotential lines which should run parallel at equal intervals from top to bottom of the substrate (Fig. 43, B). To measure evaporation in the presence of current, an alternating current generating the same quantity of heat as the corresponding DC is applied and the eosine spots are followed. When direct current is applied, there is moreover an ionophoretic movement of the buffer salts together with an electroendosmotic effect. In this case the buffer flow can be measured by applying spots of glucose or some other electrically neutral substance (Fig. 43, C).

6.2.1. *Problems of the Vertical Field (Hydrodynamics)*

The vertical field must be considered on its own because it exists even in the absence of an electrical field and involves several anomalies in connection with the capillary movement of the buffer.

In the vertical sense there is, as a result of evaporation, a slower buffer speed toward the lower edge of the paper. In the horizontal sense evaporation has a tendency to gather the buffer toward the center of the paper.

As soon as the electrophoretic chamber is saturated with water vapor and under good working conditions, the horizontal evaporation flow practically disappears and only the vertical evaporation flow remains to compensate for condensation of water on the walls of the chamber.

As a logical consequence of evaporation there is concentration of buffer salts in the dryer zones, and this increases the intensity of the electrical field which in turn aggravates the phenomenon. This stresses the great importance of an equilibration period of several hours, with current applied, before considering the hydrodynamic flow on the curtain as being stabilized.

In addition to the loss of buffer water through evaporation, there is the danger either of sucking dry or of flooding the paper by abnormal feeding of a given zone as a result of the electrode structure. In any case the influence of differences in humidity extends over a large part of the paper. In Fig. 44 a composite picture of evaporation and flooding in different setups is shown.

It is clearly important to minimize evaporation but also to avoid

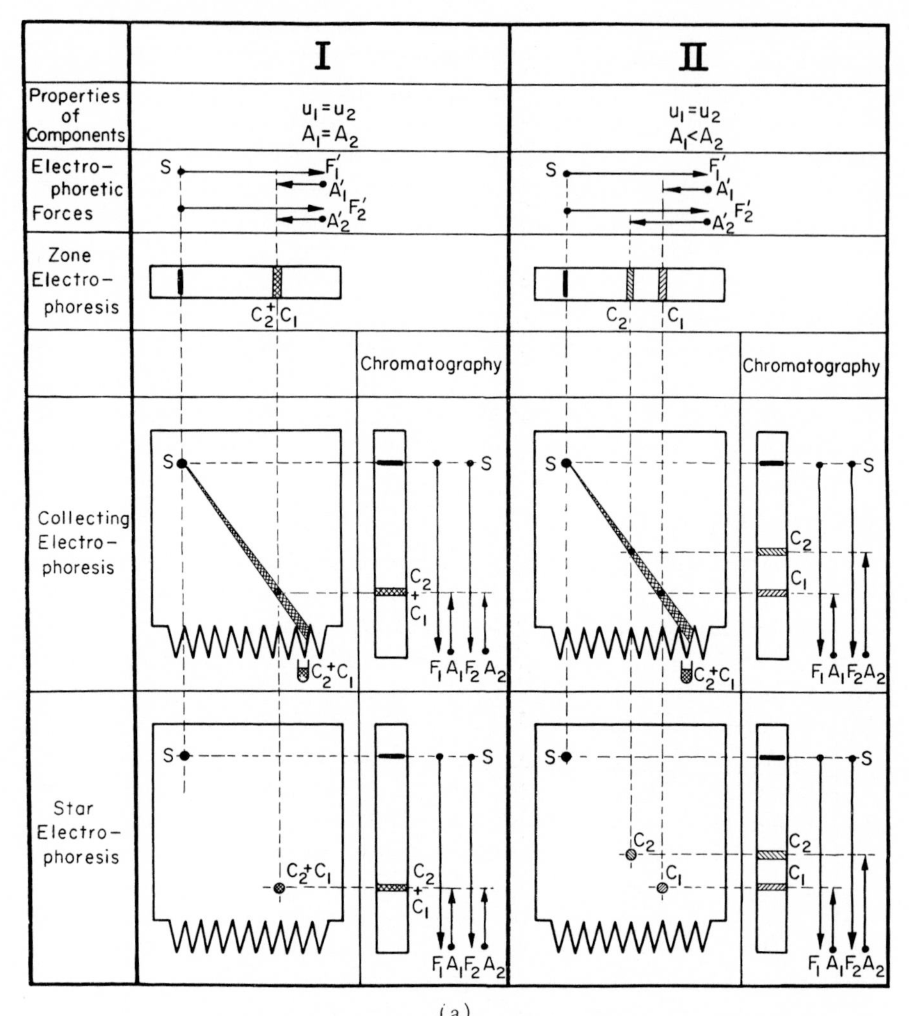

(a)

Fig. 42a and b. A few typical examples of the separation of a mixture of two components are given for zone, collecting, and star electrophoresis. It will be noted that with a given mixture the results may be different from one technique to the other. In this figure evaporation, electroendosmosis, and other factors are left out of the picture. In the figure A_1 and A_2 denote adsorption; C_1, component 1; C_2, component 2; F'_1 and F'_2, electrophoretic force; F_1 and F_2, chromatographic force; S, starting point; and u_1 and u_2, mobility.

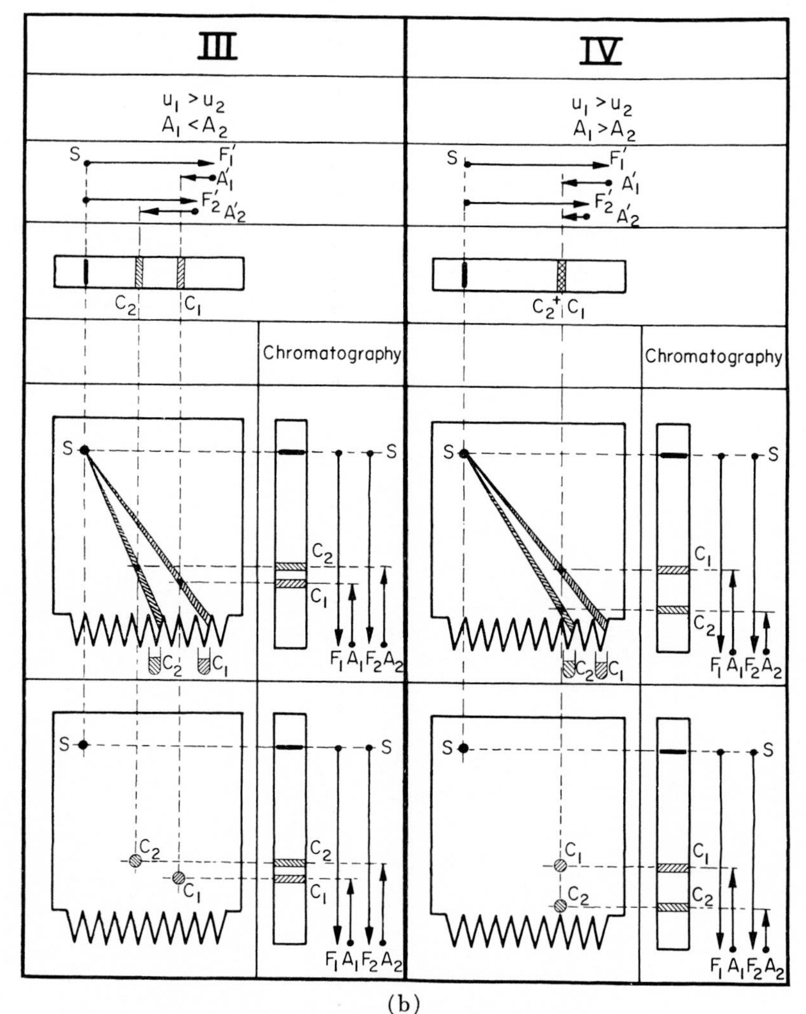

(b)

FIG. 42b. For legend see opposite page.

flooding of the curtain in order to obtain correct application of the vertical forces for runs in electrophoresis.

6.2.2. *Problems of the Electrical Field*

The application of electrodes to a curtain is a difficult problem. To obtain a homogeneous field and at the same time to eliminate products of electrolysis requires paradoxically contradictory conditions. If the field is applied outside, at the bottom of the curtain, the field is completely distorted; if the electrodes are applied alongside the paper, the field is regular, but important changes in pH are the result of electrolysis occurring nearly on the curtain itself. To avoid this, the electrolysis products must be washed away, but then the hydrodynamic problem of buffer flow alongside the edges, with danger of flooding, is apparent. In the description of the apparatus the solution to this problem will be given.

Electroendosmosis (Fig. 43c) is based on the same theoretical con-

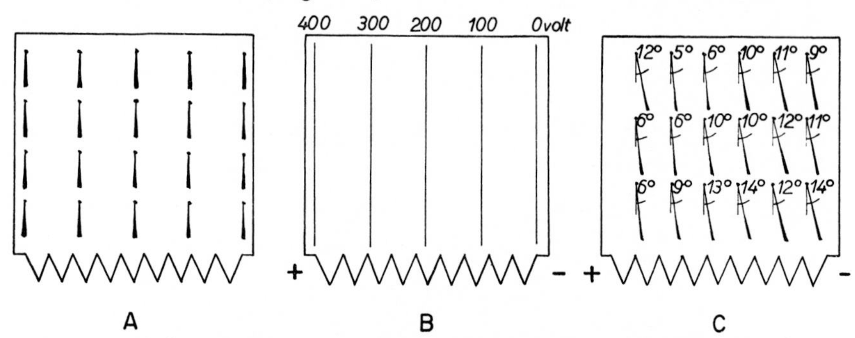

FIG. 43. (A) Uniformity of background buffer flow. Eosin spots applied on a curtain through which no current is passing reflect the buffer flow (P5). (B) Uniformity of electrical field. Potentials are measured by means of an electronic voltmeter and represent an ideal electrical field (P5). (C) Electroendosmosis. Glucose spots are applied to a curtain through which current is passing, and the curtain is developed after the run by means of infrared (400 volts across 60 cm, Whatman 3 MM paper, Veronal buffer, pH 8.6, and ionic strength, 0.02) (P2). Electroendosmosis deflects the vertical buffer flow to an angle of about 5°–14° toward the cathode.

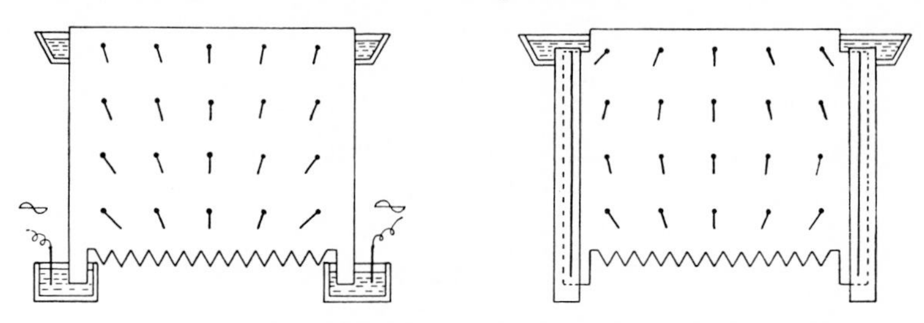

FIG. 44. (A) Influence of evaporation on background buffer feed. Alternating current at 400 volts is applied, and eosin spots are allowed to migrate during 30 minutes. Convergence of spots toward the curtain as a result of strong evaporation is a consequence of the lack of stabilizing period. (B) Influence of lateral electrodes with feeding wicks on the background buffer feed of the curtain. Eosin spots are sucked toward the electrodes in the upper part and pushed away from the electrodes in the lower parts of the curtain (P3).

siderations as in zone electrophoresis (see Section 1.1.2). Endosmosis is small, but not completely uniform over the entire field. This force deviates the originally vertical buffer flow to an angle of approximately 6–12 degrees toward the cathode, at least for alkaline barbital buffer.

A main problem in two-dimensional methods is the fate of the buffer salts and this needs careful consideration. The clue to the cause of many erroneous runs and the way to easy and reproducible working conditions are found in a study of the ionic pattern during curtain electrophoresis.

After electrophoresis of a sodium barbiturate buffer the sodium was determined. The same result was obtained consistently as given in Fig. 45 which summarizes the mean of the figures of four similar experiments

FIG. 45. Shifts in sodium concentration during two-dimensional electrophoresis with a Veronal-sodium Veronalate buffer, pH 8.6; ionic strength, 0.02; 300 volts. There is an increase in sodium from the top toward the bottom of the curtain. There is also a para-anodic zone of sodium increase. *B* represents the original sodium concentration of the buffer (P1a).

(P2). Considering the pattern in the vertical sense, the influence of evaporation from the top toward the bottom is striking. In the horizontal sense a very typical para-anodic sodium peak is observed. This has also been found in zone electrophoresis and is due to the large difference in mobility between the two ions of the buffer (P8).

The origin of the sodium peak was controlled at first with a potassium barbiturate buffer which showed the same ionic distribution as the sodium buffer. Next, a sodium and a potassium barbiturate buffer were used simultaneously, each being fed in turn into one of the two electrode compartments or on the curtain itself, while the other buffer was used for the two remaining compartments (P6). The results of these experiments confirmed again the existence of a constant para-anodic concentration zone.

As in zone electrophoresis, no migration in the opposite sense occurs, and in the para-anodic zone, where cation concentration is found, a concomitant increase of anion was demonstrated. As a logical consequence there is a rise in conductivity and in ionic strength in the same zone. This also means that the electrical field is not homogeneous.

The results of an experiment with simultaneous use of a potassium buffer in one electrode compartment, a lithium buffer on the curtain, and sodium barbiturate buffer in the other electrode compartment are shown in Fig. 46 (P7). The pattern of the different ions can be repeated

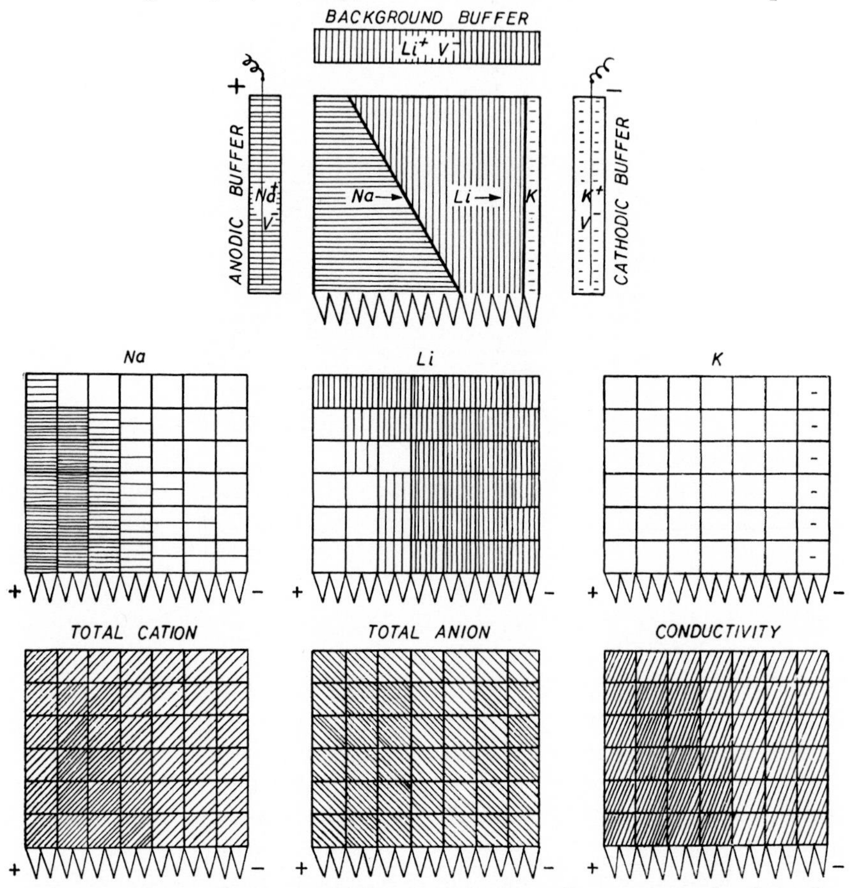

Fig. 46. Ionic pattern in two-dimensional electrophoresis: cascade electrodes, 6 volts/cm, Veronal-Veronalate buffer, $\mu = 0.022$ and pH 8.6, 4 hours. The background buffer flow is fed with lithium buffer, the positive cascade electrode with a sodium buffer, and the negative cascade electrode with a potassium buffer. After the run, sodium, lithium, potassium, Veronal, and conductivity are determined over the entire field. Sodium and lithium migrate toward the cathode. Potassium does not leave the cathode. The total number of cations increases from top to bottom and there is also a para-anodic zone of salt concentration. Veronal and conductivity follow the same outline (P7).

changing over to another cation in any one of the three parts of the system. Although in a normal experiment, these ion concentration changes are not large enough to cause modification of the original pH, there is, nevertheless, a definite ionic pattern, which causes slight differences in mobility of the migrating particles of the sample. Only when these "physiological" changes become too large, do pH changes occur sufficient to ruin the experiment (P7a).

6.2.3. *Synthetic View*

In practice, the rectilinear migration path which can often be developed on the substrate after a good run is not the expression of completely constant conditions at all points of the curtain, as one would expect at first sight. In fact the buffer flow is slower and thus the electrical field acts longer in the lower segments of the curtain. But in that same lower region, there is, for example for the albumin, the presence of the para-anodic buffer peak, which causes lower mobility due to higher ionic strength and a smaller electrical field. It means that rectilinear migration results by chance from a coincidence of several factors.

On many stained curtains, fractions may be seen separating suddenly at a given point, out of a smoothly running composite fraction at the exact spot where a modification in ionic strength and conductivity create favorable circumstances for this separation.

7. General Conditions

7.1. Apparatus

Several important structural characteristics of the chamber, such as the electrodes, the buffer feeding, and the means of controlling evaporation and heat dissipation are considered separately before an adequate apparatus is presented as a conclusion.

7.1.1. *Electrodes*

Two completely different types of electrodes have been suggested: "bottom electrodes" lying at the bottom and outside the curtain and "lateral electrodes" lying alongside the paper along the full height of the curtain.

7.1.1.1. *Bottom Electrodes.* A typical arrangement of the original type of bottom electrodes is due to Durrum (D4, D5). This consists of an electrode vessel in which the bottom corners of the paper are dipping. If an adequate baffle system is added, this arrangement has the advantage of preserving the pH of the buffer on the substrate, as contamination by electrolysis products is almost entirely excluded, but it has the serious disadvantage of producing a curved electrical field (Fig. 47).

A rather high field strength must be applied because only the lower part of the curtain is effective for separation, the higher zones being submitted to a very weak field. On the other hand, the current has to pass along a rather narrow strip of paper connecting the buffer vessel to the curtain itself. There is also some rather selective drying of the lateral

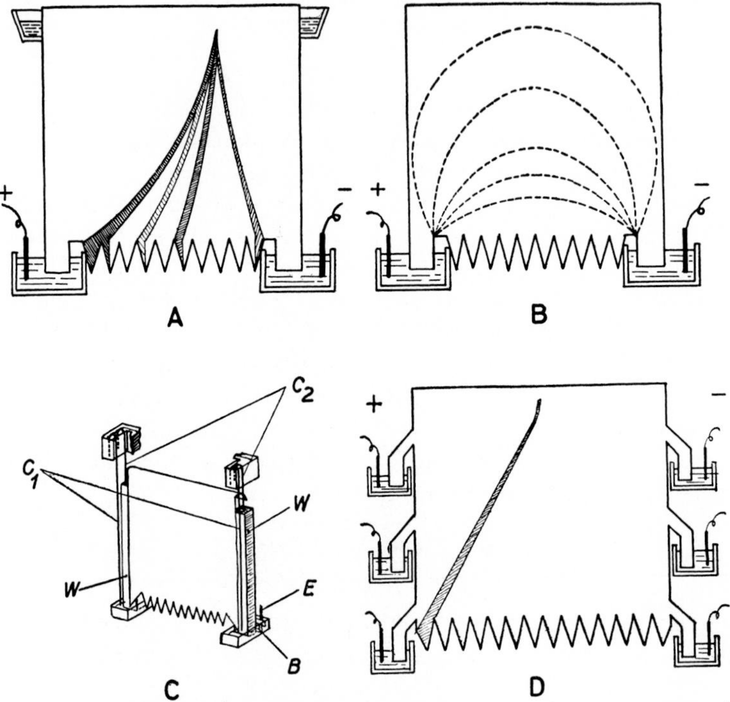

FIG. 47. (A) Apparatus with bottom electrodes and curvilinear separation (D4). (B) Curved electrical field with bottom electrodes (L1). (C) Apparatus with bottom electrodes and lateral wicks (*W*) 4 x 4 cm thick. Electrode *E* behind baffle system *B*. Capillaries C_2 feed the lateral wicks. The capillaries C_1 carry the overflow from the top vessels and rinse the electrodes at the same time (B1). (D) Correction of the electrical field by the introduction of supplementary electrodes. Migration is more rectilinear (D3).

edges, and to combat this, extra wetting by supplementary layers has been proposed (D4).

Lateral wicks have been introduced (B1) in order to avoid a curved field without affecting the original arrangement of the electrodes, very thick conductive wicks being placed along both edges of the paper. To prevent these wicks from sucking the buffer out of the substrate curtain, they are specially fed with buffer solution in such a way that they neither flood nor drain the central curtain. As a result, the electrical field is more uniform and the products of electrolysis are washed away easily and completely, but an accessory and accurately adjustable buffer supply to the wicks is essential to achieve these objects. At best, however, the actual surface used for separation reaches only 30 by 30 cm, and the shape of the substrate cannot be varied.

Supplementary electrodes were introduced by Dicastro and San Marco (D3) who proposed the use of several buffer electrode vessels distributed over the height of the paper. This system retains the advantage of electrodes outside the curtain and favors the uniformity of the field (Fig. 47, D).

In practice, however, bottom electrodes demand a complicated technical arrangement which does not allow for versatility of shape or of dimensions of the curtain.

7.1.1.2. *Lateral Electrodes.* The use of lateral electrodes is far more general, in spite of many difficulties, and gives a more regular electrical field (Fig. 48). The electrode wire lies alongside the edge of the paper and over its full height, the electrode and the curtain being pressed together between two press-plates. The electrical field is of course much more uniform than with bottom electrodes, but there still remains a distortion in the upper part of the field due to a shunt caused by the central buffer vessel (G4).

A second problem is the elimination of electrolysis products, which are

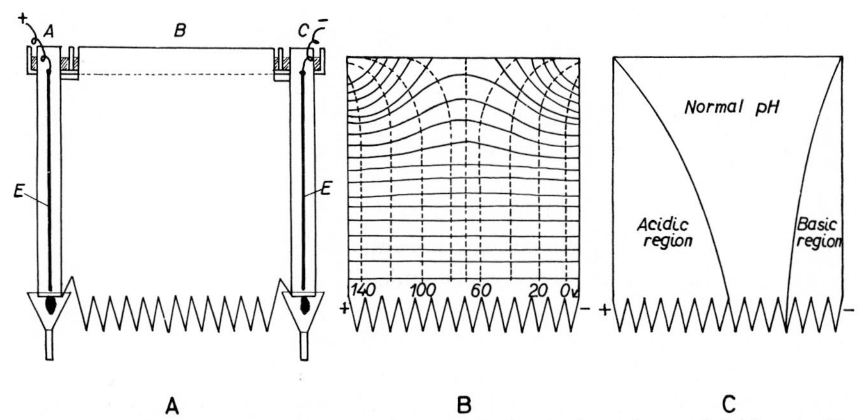

FIG. 48. Influence of electrodes and central buffer feed on electrical field and pH. (A) Apparatus with lateral wicks covering the electrodes starting in buffer vessels *A* and *C* separated from central buffer feed *B*. (B) The electrical field is curved as the result of the shunt over buffer feed *B* (G4). (C) Insufficient rinsing of the electrodes causes modification of pH on the curtain (L1).

usually washed away by pleated filter paper or by a layer of cotton wool in order to obtain a flow of buffer along the electrode wire (G1, S3). Others do not apply even this intermediary layer but leave the curtain in direct contact with the electrodes (S2). In any case, however, when the potential is raised, the elimination of electrolysis products is insufficient and pH changes occur alongside anode and cathode. There is also a limit to the use of a buffer flow between the press-plates because a hydrostatic pressure builds up in the lower parts with a tendency to flood, while paper or cotton wool have a tendency to suck the buffer solution out of the curtain in the higher zones.

Membrane electrodes have the electrode wires immersed in buffer solution and enclosed within a semipermeable membrane which can be irrigated efficiently without the risk of flooding the paper. Their advantage is that they allow a free buffer flow along the electrode wire. By using a pump instead of Mariott bottles to effect the circulation of buffer, the flow from both electrodes can be mixed and recirculated with a simple and efficient renewal of the original qualities of the buffer. With this type of electrode, however, it is impossible to adjust the buffer flow along the edges of the curtain (H3, J1).

An ideal system would include the rinsing realized in membrane electrodes, avoid the difficulty of the membrane, and present the possibility of a perfect adjustment of the hydrostatic flow along the edges of the paper. This means that the electrode would be rinsed and the paper curtain fed at the same time.

A first attempt at such a "free electrode" is the construction of a hollow electrode carrier where a slit or a series of holes gives contact with the paper (P3, R1). With a long and high electrode, however, there is a serious hydrostatic disturbance, and buffer runs from the curtain into the electrode at the top segments and vice versa in the lower segments (Fig. 44).

To maintain a regular hydrostatic pressure, cascade electrodes were built (H3). The electrode carrier is shown in Fig. 49 and consists of a series of small slanted buffer vessels, which lie one under the other, along the edge of the curtain. Each buffer vessel has a window *W* against which the curtain substrate *S* is pressed by means of the press-plate *P* and the screw *H*. As the buffer vessel is higher at the "back" than at the "front"—calling "front" the window which is pressed against the curtain—the buffer solution overflows into the underlying buffer vessel and the succession of overflows forms a series of cascades. The electrode wire runs in a zigzag through the series of buffer vessels. The buffer is fed into the uppermost vessel, and through overflow the second and consequently the following compartments are filled. The buffer is collected underneath the last cascade of the two electrodes, mixed, and recirculated by a pump.

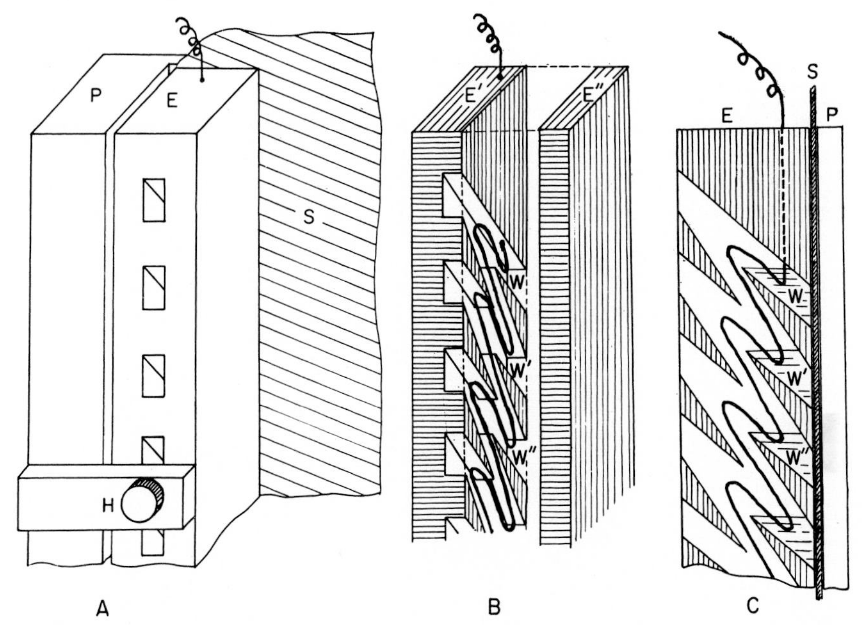

FIG. 49. Cascade electrodes. S, substrate; *P*, press-plate; *H*, screw for maintaining electrode carrier *E* against press-plate *P*; W, W′, W″, windows of the successive buffer vessels.

(A) Working position. (B) Vertical section through *E* to show its internal structure. (C) Side view of *E* to show the buffer vessels with their contact window *W*.

As the substrate is pressed against the "front" window of each buffer vessel, the paper is fed with buffer, while the free overflow at the "back" of each compartment to the one below avoids the building up of hydrostatic pressure even if the electrode is very high. Each "front" window bears the same pressure of a few millimeters of buffer, and the buffer can be fed at any desired rate without danger of flooding the curtain. That the cascade electrodes provide correct hydrodynamic qualities can be proved by the perfectly perpendicular chromatography of eosin spots on a curtain field of 60×60 cm (Fig. 50).

As the electrode itself lies near and practically on the curtain, there is no loss of the applied potential, and because the successive windows form a series of contact zones lying very close to one another, the field is also uniform (Fig. 51). The slight deviation of the field from strict uniformity in the upper zone is not due to the electrode system but to the background buffer feeding and is discussed under the next heading.

It is an important supplementary advantage that substrates of any size can be used without necessitating any modification, either in the qualities of the curtain or in the structure of the apparatus.

7.1.2. *Development of Background Buffer Feeding*

The importance of buffer feeding lies simultaneously in the hydrodynamic aspect of the question and in its effect on the electrical field. Lateral feeding has already been considered in connection with the

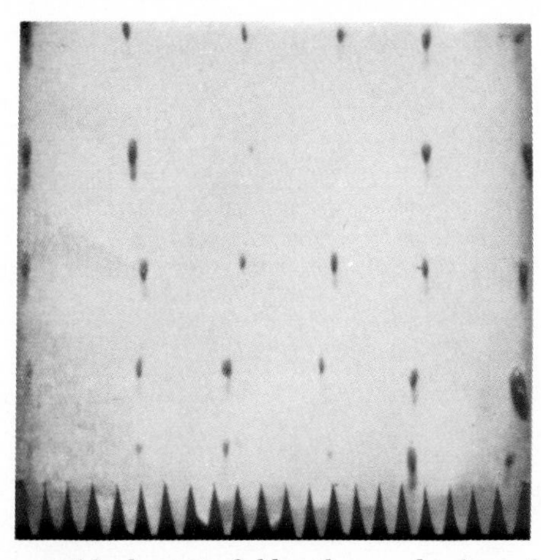

FIG. 50. Correctness of hydrostatic field with cascade electrodes proved by means of eosin spots without application of current (P5).

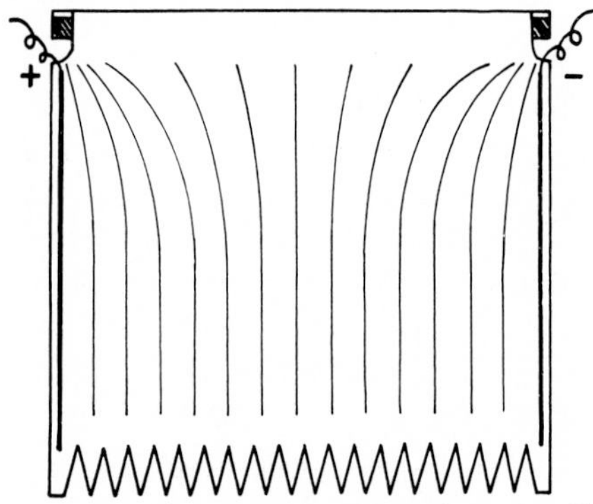

FIG. 51. Electrical field with cascade electrodes. Parallel isopotential lines, except in the upper segment near the central buffer feed.

electrodes, and the interweaving of hydrodynamic and electrical qualities was stressed. The background feeding of the curtain is a problem which has been solved by almost all workers as in descending paper chromatography.

As the top strip of the paper is folded back and dips into a trough of buffer solution, there is a limit to the quantity of buffer fed to the curtain caused by the shape of the fold. There is also a shunt over the buffer trough which causes a distortion of the electrical field (Fig. 52a). Until

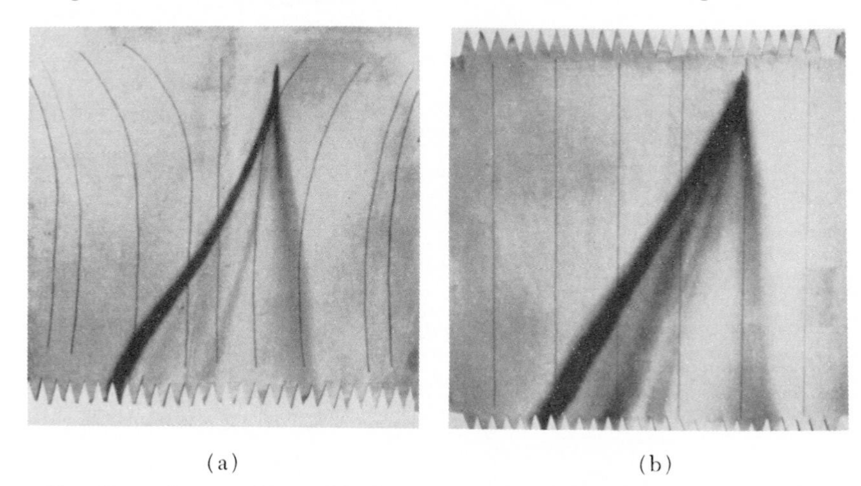

(a) (b)

FIG. 52. Influence of central buffer feed on electrical field and migration. Buffer, Veronal-sodium Veronalate, pH 8.6; μ 0.02; substrate Whatman 3 MM, dimensions 60 x 60 cm; field, 6.6 volts/cm; duration, 12 hours; sample, normal human serum; dye, amido black. Measurement of isopotential lines with electronic voltmeter. (a) Chromatographic feeding with central buffer vessel. Albumins and globulins are separated over a distance of 30 cm. (b) Trickle feeding of the curtain. Albumins and globulins are separated over a distance of 34 cm.

recently, no proposals had been made to solve these problems except to feed the paper by means of a supplementary curtain, which drips onto the folded upper edge of the actual separating curtain (B1). By varying the rate of feeding of the upper curtain, the curtain on which the electric field is applied receives more or less buffer, but this was not intended to correct the electrical field.

A new type of background electrolyte feeding has now been proposed (P5). A horizontal channel, which has a series of small holes, provides a drip of electrolyte which falls onto the folded edge of the curtain. The folded edge lies on a drain board, and while the paper is able to suck the amount of water it needs, the excess flows from the drain board into a drain. By varying the angle of the drain board, more or less buffer flows over the curtain itself (Fig. 53). The rate of trickle feeding is of no importance as long as it is greater than the minimum required to wet the curtain to the desired extent. In combination with cascade electrodes a single pumping device allows for the recirculation and feeding of the buffer without any problem of rate adjustment.

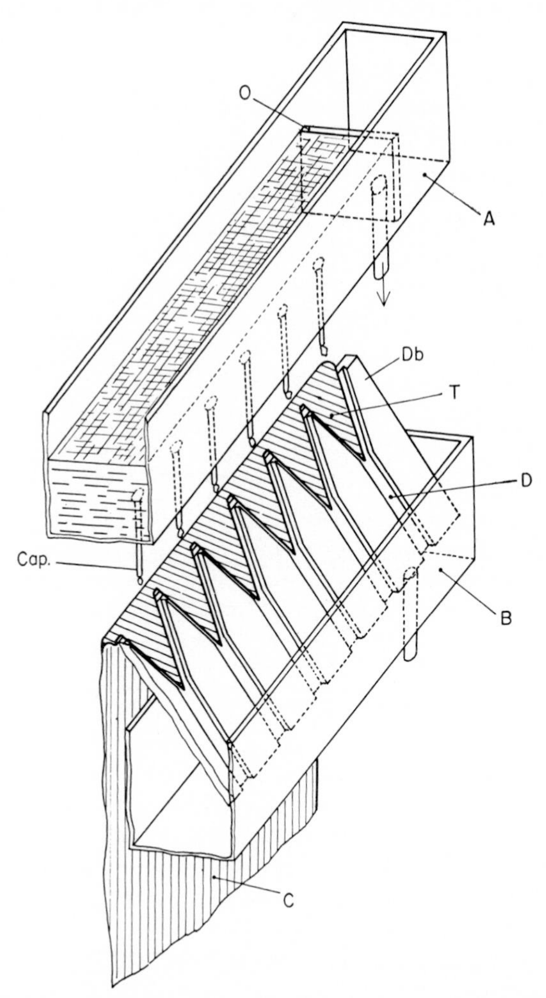

FIG. 53. Trickle feeding of the curtain. The gutter *A*, equipped with the overflow *O*, trickles on the teeth cut in the upper edge of the curtain *C* through the capillaries (*Cap*). Each tooth (*T*) of the curtain lies in a separate drain *D* of the drain board (*Db*). All drains are collected in the general drain *B*.

This setup also provides a solution to the problem of a perfectly uniform electrical field. With a system of trickle feeding, the upper edge of the paper may be cut into a series of teeth, analogous to the teeth at the lower edge of the substrate. These teeth provide efficient buffer feeding to the curtain. Moreover, since each tooth lies on the drain board, isolated from the others in its own separate gutter, any shunt is avoided; the field is perfectly uniform up to the top of the electrodes. The sample may be applied very high near the upper edge of the curtain without danger of insufficient separation (Fig. 52b).

A last advantage of this feeding device is that it can be constructed at any length desired and also that the same trickle feeder will take a narrow or a wide curtain without needing any adjustment.

7.1.3. *Evaporation*

Evaporation must be kept low in two-dimensional techniques, and in this respect the walls and the air seal of the chamber play an important part. The use of a cooling system results in an advantage under given conditions.

A small chamber is more rapidly saturated. This is a reason for keeping all accessory mechanisms, such as the actual fraction collector, out of the chamber.

As condensation on the walls calls for supplementary evaporation, a thin wall is advisable. Most chambers are built of glass or semirigid plastic. These materials are thick, have low heat conductivity and high specific heat, as easily observed from the profuse condensation which runs in heavy drops from the wall. A thin transparent plastic sheet (P1) is a far cheaper cover and easy to handle. Only a very fine mist condenses on the sheet, and the possibility of using a high current density without the need for any cooling is proof of good functional conditions.

A perfect seal is essential. This presupposes sealing the chamber after the curtain is hung and also arrangements for applying the sample, for controlling the in- and outflow of the buffer for background electrolyte and electrode feeding, and for collecting fractions without opening the chamber and thus without loss of vapor-saturated air. In rigid glass or Plexiglass (Perspex) chambers an entire wall is usually removed, either the top or the front section. Tightening with screws and using soft plastic for gaskets may seem perfect, but proves unsatisfactory in practical use chiefly because large plastic panels are not rigid and warp rapidly. These problems are nonexistent when a soft plastic sheet, e.g., polyvinyl, is used to cover the chamber and is allowed to hang into a water-filled moat in the base of the chamber. Several stoppered portholes may be provided permanently in the cover, but a sample may also be injected by a needle straight into the sample cuvette through the plastic and the puncture hole sealed by means of Scotch tape. The inflow and outflow of buffer present no problem if a siphon is used to deal with the loss of vapor-saturated air when eliminating excess buffer.

The main difficulty of the air seal seeems to be the fraction collector, but a capillary channel will hold a drop permanently and let it pass only on the pressure of the next one, so that a permanent seal is achieved. In older types of apparatus the fraction collector stood inside the electrophoretic chamber, but this is of course a serious practical disadvantage. A problem in rigid chambers is the change of pressure due to slight rises in temperature during the run, which pushes the sealing drops out of the fraction collector with loss of water-saturated air, but with a soft plastic sheet such pressure changes are compensated by the hood itself.

The importance of cooling has already been mentioned. Heat generated from the electric current may be eliminated by means of a liquid or gaseous refrigerant if spontaneous heat loss through air and walls is insufficient. Cooled liquid passes through a hollow plate against which the substrate curtain is hung (H3). A cylindrical backplate has also been proposed (S1). Silicone grease avoids contact between plate and curtain, but the manipulation of a paper applied to a rigid support is not easy. A stream of conditioned air has also been used (H1), but cooling systems are necessary only if very thick substrates are used and are not required for paper with a thickness of Whatman 3 MM, provided the walls of the chamber are thin.

7.1.4. *Summary*

Efficient equipment should include the features summarized in Table 12. A very important point is adaptability to substrates of any size or shape (Fig. 54). Cascade electrodes allow any height of paper, and if the distance between the two electrode carriers is adaptable, any width of paper can be used. Background buffer feeding, either chromatographic or trickle feeding, allows for any width of substrate. Unused channels of the collector may be stoppered.

A diagram summarizing these possibilities is given in Fig. 55.

7.2. Substrate

Rapid filter papers with high water capacity favor the separation of large quantities of material, but because of greater diffusion are less favorable for a neat separation of neighboring fractions. Thick papers call for high current, which generates a large quantity of heat. The

TABLE 12
FEATURES OF TWO-DIMENSIONAL APPARATUS

	Problem	Solution
Electrodes		
Hydrodynamics	Equal hydrostatic pressure at all heights of the electrode	"Cascade electrodes"
Electrical field	Homogeneous field	Electrode over full height
Electrolysis products	Washing away of electrolytes	"Cascade electrodes"
Background electrolyte feeding		
Hydrodynamics	Regulation of the flow	Trickle feeding
Electrical field	Shunt over the central buffer vessel	Trickle feeding
Evaporation		
	Small chamber	
	Perfect seal	
	walls	Thin and soft plastic
	sample	Introduced without breaking seal
	buffer supply	Siphon outlet
	fraction collector	Capillary outlets
	Efficient cooling	Thin walls (or conditioned air)

(a)

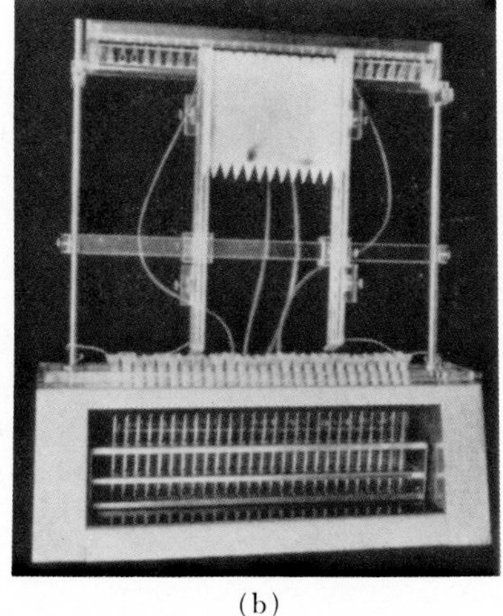

(b)

FIG. 54. Versatility of equipment. (a) Curtain 60 x 60 cm (with plastic covering sheet); (b) Curtain 20 x 30 cm (without plastic covering).

smaller surface and more rapid flow of Whatman 3 MM compensate for this effect, and for practical work this type is far better than Whatman No. 1. Homogeneity of the paper and its tensile strength are of course important. Adequate electrode carriers can be moved a little apart once the paper is wetted to compensate for dilation. Machine direction is important because paper hanging across the machine direction will have a tendency to show a series of horizontal swellings.

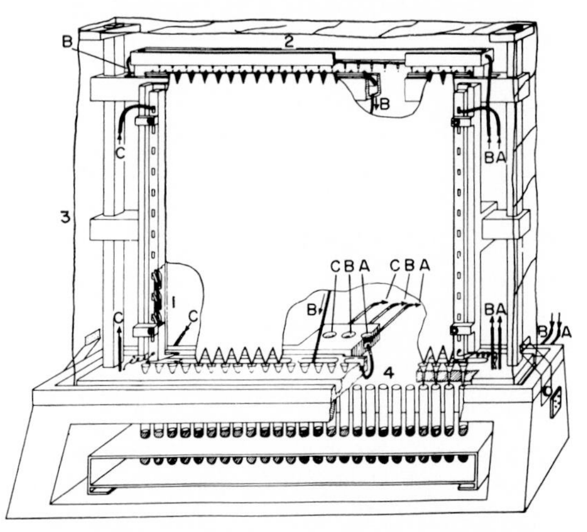

Fig. 55. General arrangement of two-dimensional apparatus. (1) Cascades; (2) trickle feeding (common pump system omitted); (3) cover (plastic sheet); (4) collector. A and C, electrode buffer circuit. B, trickle feeding buffer circuit.

New chambers have been constructed containing a compressed paper pad instead of paper, and these, if equipped with an efficient cooling system, allow the separation of large quantities of material (H1). Other substrates such as glass powder (G3) or cellulose powder (D2) have been proposed, but so far they have not had clinical application.

7.3. Buffer

There is a general tendency to use low ionic strength, especially for continuous separating work. On the contrary, the very rapid star electrophoresis calls for higher ionic strength, which allows a sharper definition of the fraction spots. As there is opposition between ionic strength and field strength, compromise is necessary, and practice will help to determine the correct composition of the buffer for specific conditions. Table 13 gives a summary of a few useful buffers and their characteristics.

7.4. Power Supply

An alternating ripple superimposed on DC must be avoided as it causes unnecessary evaporation. Stabilized DC is of utmost importance. The use of constant voltage is possible if isothermal conditions are realized, but in practice this does not happen and constant current compensates for changes in mobility due to changes in working temperature during the experimental run (see Section 1.1.4.1).

As the fields are often very wide, high voltage and low current are desirable; this calls for low conductivity of the buffer. A safety switch which automatically cuts off the current when the chamber is not closed should be included in the circuit.

8. The Run

8.1. Collecting Type

It is the aim of collecting electrophoresis to isolate fractions quantitatively, but resolution and processing rate are limited by the paper substrate. At the end of the run the substrate must be quickly dried if correct evaluation of the fractions remaining on the substrate itself is desired. They can be stained, eluted, and measured, as in strip electrophoresis.

The collected samples may be concentrated by dialysis (against dextran solution, for instance) or by ultrafiltration. For substances other than protein, freeze drying may be of interest, or the buffer may be removed by evaporation if the substances sought are sufficiently stable and the buffer components are volatile.

Serum proteins (B2, G6), collagen (G5), and insulin (G7) afford instructive examples of mixtures which have been studied with this method.

The sample for electrophoretic separation is often not dialyzed, although dialysis favors the separation at the contact point between sample feeder and substrate curtain through adjustment of pH and ionic strength. The sample can be applied by cutting out of the curtain a narrow tongue of paper which dips into the sample container and forms a wick.

In another arrangement a separate feeding strip can be applied against the intact substrate; this system is practical because it is easy and regular, but does not allow a large flow. A cotton wick allows faster feeding than a paper one, but there is danger of drying out. A glass or plastic capillary fed by gravity has been used, but tends to clog or to be blocked by an air bubble. Mechanical devices have been extensively used but the easiest method is still to take advantage of the spontaneous capillarity of a paper wick.

TABLE 13
BUFFERS USED FOR COLLECTING ELECTROPHORESIS AND STAR ELECTROPHORESIS

Kind of buffer	pH	μ	Components in 1 liter	Voltage volts/cm	Electrophoresis
For protein					
Hannig (Na)	8.6	0.022	4.666 g Na barbiturate 0.8 g Barbital	5–10	Collecting
Hannig (Na)	8.6	0.033	7 g Na barbiturate 1.2 g Barbital	15–20	Star
Barbital (K)	8.6	0.022	222 ml 0.1 *N* KOH 4.97 g Barbital	5–10	Collecting
Barbital (Li)	8.6	0.022	222 ml 0.1 *N* LiOH 4.97 g Barbital	5–10	Collecting
Barbital (Na)	9.0	0.022	4.666 g Na barbiturate 0.5 g Barbital	5–10	Collecting
Barbital (Na)	8.2	0.022	4.666 g Na barbiturate 1.53 g Barbital	5–10	Collecting
Borate	8.6	0.054	540 ml 0.1 *N* NaOH 12 g Boric acid	5–10	Collecting
Borate	8.6	0.036	360 ml 0.1 *N* NaOH 8 g Boric acid	5–10	Collecting
Borate	9.2	0.054	540 ml 0.1 *N* NaOH 6 g Boric acid	5–10	Collecting

TABLE 13 (*concluded*)

Kind of buffer	pH	μ	Components per liter	Voltage volts/cm	Electrophoresis
For protein (*continued*)					
Borate	9.2	0.036	360 ml 0.1 *N* NaOH 4 g Boric acid	5–10	Collecting
Tris	8.6	0.08	24.228 g Tris[a] 14.72 g Barbital	10–15	Star
Tris	8.6	0.04	12.114 g Tris[a] 7.36 g Barbital	10–15	Star
For insulin					
Lactate	3.25	0.09	31 ml Lactic acid 5.6 ml 40% NaOH	5–10	Collecting
Salicylate	3	0.005	1.380 g Salicylic acid 50 ml 0.1 *N* NaOH	5–10	Collecting
Acetate	4.62	0.05	6.8 g Sodium acetate 3 ml 100% Acetic acid	5–10	Collecting

[a] Tris (hydroxymethyl) aminomethane.

It is easy to hang the dry paper, cut so that the machine direction is vertical, in the chamber and to wet the paper with the buffer which fills the pump system. About 3 liters of buffer are required for a 48 hr-run on a 60 × 60 cm-substrate of Whatman 3 MM paper.

Before the sample is fed, the chamber is closed and current is applied for at least 3 hours, in case evaporation should cause disturbance of the pattern. After this equilibration period, the sample is applied without opening the chamber and constant working conditions for a field up to 12 volts/cm will be maintained for any length of time. When constant current is used, there is an initial voltage drop due simultaneously to the heating up of the substrate, to evaporation, and to ionophoresis. In a good run the experiment will start with a field of 14 volts/cm and fall to 10–12 volts/cm, where it remains constant.

About 2 ml of serum can be separated in 24 hours, and during the same period a total amount of about 300 ml of buffer is collected in all the different test tubes for a curtain 60 cm wide, i.e., about 5 ml per cm of length of substrate.

Table 14 which summarizes anomalies, their origin, and correction may help the experimental worker to find a clue to his problems.

8.2. Star Type

A rapid two-dimensional separation of a microspot of protein solution under high field intensity is a new application of the two-dimensional principle. The various fractions emerge as a series of rays out of the application zone and hence the designation "star electrophoresis" which was introduced by us in connection with work on human serum proteins (P4). A similar discontinuous procedure was already mentioned by Strain in 1951 (S4, S5) for the separation of metal ions and recently used by Pučar (P8) and applied by Kéler *et al.* (K1) to human serum.

This method makes use of the same apparatus as that proposed for collecting electrophoresis, and both the cascade electrodes and the trickle feeding for the background buffer flow are important to obtain good results. The general working conditions are also identical to those of the two-dimensional collecting method, but with the following modifications.

The sample of 0.01–0.02 ml is applied as a spot either by means of a micropipette or, more easily, by means of a small thin paper disk impregnated with the protein solution. The application zone is slightly dried beforehand to avoid diffusion and to favor rapid uptake of the sample. To obtain each fraction as an isolated spot and with little trailing along the migration path, the immediate uptake of the entire microspot by the application zone of the substrate is important. If the application

TABLE 14
ANOMALIES AND CORRECTIONS IN COLLECTING ELECTROPHORESIS

Anomaly	Origin	Correction
Broad fraction	Diffusion due to paper structure	Machine direction: vertical
No separation	Evaporation flow in the horizontal sense	Good dissipation of heat (thin wall and cooling system); reduction of heat (lower μ, thinner paper)
	Flooding of paper from the electrodes	Regulation of electrode flow or cascade electrodes
Increasing separation toward the bottom	Bottom electrodes	Feeding by supplementary wicks
	Buffer shunt	Trickle feeding of background
	Evaporation	Good dissipation of heat (thin wall and cooling system); reduction of heat (lower μ, thinner paper)
Malformation	Irregular buffer flow	
	Through folded substrate	Machine direction: vertical
	Through irregular buffer feeding from electrodes	Regulation of electrode flow or cascade electrodes
	Differences in μ over the paper	
	Through evaporation	Good dissipation of heat (thin wall and cooling system); reduction of heat (lower μ, thinner paper)
	Shunt over central buffer vessel (salt deposit in upper corner)	Trickle feeding of background
	Bad evacuation of electrolysis products	Washing of electrodes (cascade electrodes)
	Changes in pH over the paper	
	As a result of excessive changes in μ	Washing of electrodes (cascade electrodes)
	As a result of electrolysis products	Mixing of anodic and cathodic buffer flow
Initial lack of separation	Modification, due to the sample, of pH and μ of background buffer	Dialysis of sample or slower feeding

is not interrupted very soon, long streaks of protein run from the application zone toward the head of each fraction. The best images, and those which are most likely to give the best information, are those where the protein mixture is split up into a series of neatly isolated spots (Fig. 56).

FIG. 56. Star electrophoresis. Buffer, Veronal-sodium Veronalate, pH 8.6; μ 0.06; substrate, Whatman 3 MM, dimensions 22 x 30 cm; electrical field constant current 26 ma; starting point, 17 volts/cm; endpoint, 11 volts/cm; sample, 0.02 ml normal human serum; duration, 3 hours.

Differences in migration distance of the several fractions, due to differences in adsorption, related to a chromatographic effect, are always marked (see Section 6.1.).

To obtain well-isolated fractions the buffer needs to be a compromise between low conductivity and fair ionic strength. A Veronal-Veronalate buffer with an ionic strength of 0.033 or a Tris-Veronal buffer with an ionic strength of 0.08 is required in order to isolate clearly five fractions from normal human serum over a width of 10 to 12 cm after a 180-min run.

A field of 20 volts/cm is applied, and the use of constant current avoids salting-out on the substrate. When working with apparatus of efficient design, there is no need for a special cooling device. No equilibration period is needed because the rapidity of fractionation does not call for strictly stabilized working conditions.

As substrate, rather thick and rapid paper is required, such as Whatman 3 MM, because such paper compensates easily for evaporation due to heating, thanks to a relatively small surface combined with a good buffer speed.

On a field 20 cm high and 30 cm wide one single sample can be separated, but using larger fields several spots can be applied at once. A favorable shape is a very wide curtain, on which several spots can be applied one next to another, but only 20 cm high so as to maintain an efficient background flow (Fig. 57).

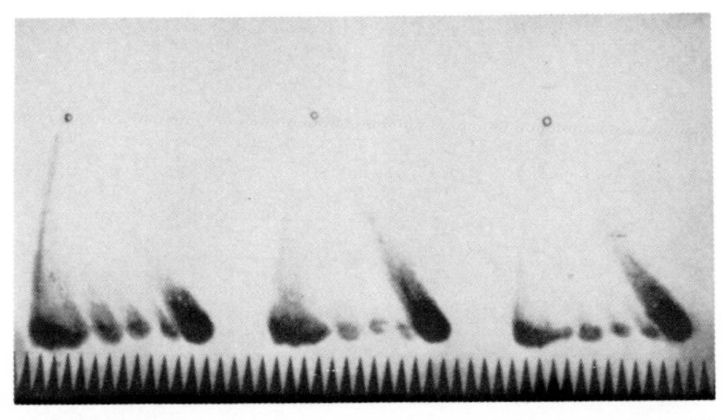

FIG. 57. Star electrophoresis: three simultaneous runs. Buffer, Veronal-sodium Veronalate, pH 8.6; $\mu = 0.06$; substrate, Whatman 3 MM, dimensions 22 x 58 cm; electrical field, constant current, 27 ma; starting point 14.5 volts/cm; endpoint, 13.6 volts/cm; sample, 0.02 ml normal human serum simultaneously applied at three different spots; duration, 3 hours.

After the run, rapid drying is essential in order to retain the clearness of the fractionation, especially if elution of the dyed spots is needed for quantitation. In this case standardized methods of fixation, staining, and rinsing are important (see Setion 3.2).

It is the main advantage of this rapid two-dimensional technique that it applies the important principle of a two-dimensional method to microsamples; there is no mixing of fractions running simultaneously on the same migration path as in zone electrophoresis, but on the contrary, uncontaminated fractions are separated within very short time. For clinical research the method has a bright future because of the theoretical purity of the spot, which is not obtainable with zone electrophoresis. Good results have already been gained in work on animal serum (D1).

Recently work on glyco- and lipoproteins was presented by Kéler-Bacôka, and several lipid patterns were described, which are different from those obtained in zone electrophoresis (K1). Our own experience with glyco- and lipoproteins runs parallel (Fig. 58), and this may prove the sound theoretical basis of the proposed technique, in a field where zone electrophoresis has given so many contradictory results (see Tables 10 and 11).

Many other applications of this method can be found. For rapid serial work, the evolution of a pattern of a protein-bound isotope can be followed, even if its half-life is very short (e.g. I^{132}). As only microsamples of fingertip blood are required, the scope of two-dimensional star techniques may be extended to work on children and infants, and also to surveys on larger groups, especially if wide curtains are used.

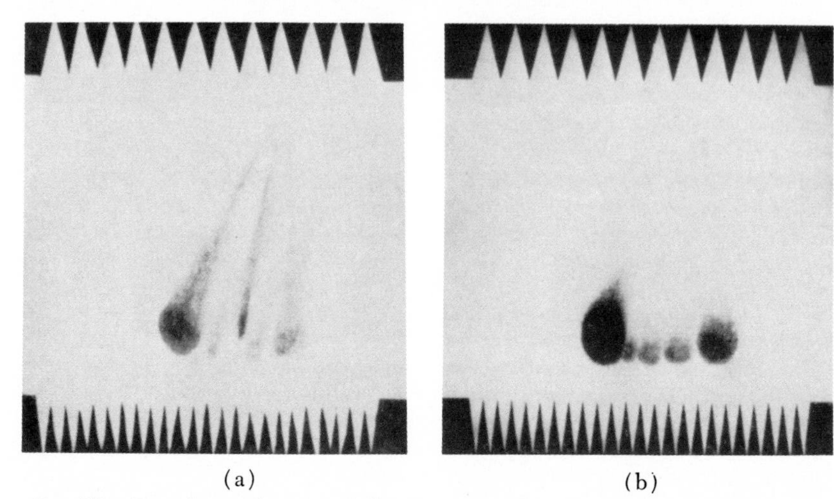

(a) (b)

FIG. 58. Star electrophoresis. Buffer, Veronal-sodium Veronalate, pH 8.6; μ 0.06; substrate, Whatman 3 MM, dimensions, 30 x 30 cm; electrical field, constant current 30 ma; starting point, 16.1 volts/cm; endpoint, 11.6 volts/cm; sample, 0.02 ml serum, concentrated 5 times; prestained with Sudan black; duration, 3 hours. (a) Prestained for lipids, but before staining for protides. (b) After supplementary staining with amido black.

9. General Conclusion

The two-dimensional techniques, both as a preparative tool for collecting fractions and perhaps especially as a separative tool in rapid star electrophoresis, will undoubtedly prove to be of great value in clinical laboratory work. The recent technical improvements in the apparatus allow for easy working conditions. These were previously lacking and delayed the clinical use of methods which are based on a different principle from that of zone electrophoresis. They yield important results in fields where the one-dimensional method has partly failed.

References

B1. Beckman Data Sheet. Spinco Division Form 4 - CP 6-56.

B2. Bodman, J., The measurement of serum inhibitor to chymotrypsin, its nature and clinical significance. *Rheumatism* **13**, 25 (1957).

D1. De Haene, Y., Invloed van de voeding en van de oogsteelhormonen op het normaal eiwitbeeld bij de chinese wolhandkrab. *4de Colloq. Sint Jans Hospitaal, Brugge* pp. 125-134. De Tempel, Brugge, 1956.

D2. Dicastro, G., Zone electrophoresis. *Bull. soc. chim. Belges* **65**, 97 (1956).

D3. Dicastro, G., and San Marco, M., Continuous electrophoresis on paper. *J. Chem. Soc.* p. 4157 (1954).

D4. Durrum, E. L., Continuous electrophoresis and Ionophoresis on Filter paper. *J. Am. Chem. Soc.* **73**, 4875 (1951).

D5. Durrum, E. L., *In* "A manual of Paper Chromatography and Paper Electrophoresis" (R. J. Block, E. L. Durrum, and G. Zweig, eds.), 2nd ed., pp. 489. Academic Press, New York, 1958.

G1. Grassmann, W., and Hannig, K., German Patent 805,339 (1951).

G2. Grassmann, W., Tagung der Physiologen und physiologischen Chemiker in Göttingen vom 31 Aug. 1949. *Angew. Chem.* **62**, 170 (1950).

G3. Grassmann, W., Neue Verfahren der Elektrophorese auf dem Eiweissgebiet. *Naturwiss.* **38**, 200 (1951).

G4. Grassmann, W., and Hannig, K., Trennung von Stoffgemischen auf Filtrierpapier durch Ablenkung im elektrischen Feld. *Z. physiol. Chem.* **292**, 32 (1953).

G5. Grassmann, W., Hannig, K., Endres, H., and Riedel, A., Zur Bindungsweise des Prolins und Hydroxyprolins. *Z. physiol. Chem.* **306**, 123 (1956).

G6. Grassmann, W., Hannig, K., and Plöckl, M., Eine Methode zur quantitativen Bestimmung der Aminosäurezusammensetzung von Eiweisshydrolysaten durch Kombination von Elektrophorese und Chromatographie. *Z. physiol. Chem.* **299**, 258 (1955).

G7. Grassmann, W., Strobel, R., Hannig, K., and Deffner-Plöckl, M., Zur Konstitution des Insulins. *Z. physiol. Chem.* **305**, 21 (1956).

H1. Hannig, K., Die präparative kontinuierliche Ablenkungselektrophorese. "Protides of the Biological Fluids. Proceedings of the Fifth Colloquium, Bruges," pp. 10-16. Elsevier, Amsterdam, 1957.

H2. Haugaard, G., and Kroner, T. D., U.S. Patent 2,555,487 (1948).

H3. Holdsworth, E. S., An apparatus for continuous electrophoresis on paper. *Biochem. J.* **59**, 340 (1955).

J1. Jonas, H., A Versatile Paper electrophoresis apparatus. *J. Lab. Clin. Med.* **1**, 135 (1957).

K1. Kéler-Bacôka, M., Pučar, Z., and Petek, M., Zweidimensionale Elektrochromatographie und Triasfärbung normaler und pathologischer humaner Sera. *Intern. Congr. Clin. Chem., Stockholm* (1957).

L1. Lederer, M., *In* "Introduction to Paper Electrophoresis and Related Methods." Elsevier, Amsterdam, 1955.

P1. Peeters, H., and Anthierens, L., Eenvoudige apparatuur voor Collecterende papier electrophorese. *3de Colloq. Sint Jans Hospitaal, Brugge* pp. 197-204. De Tempel, Brugge, 1955.

P1a. Peeters, H., and Vuylsteke, P., Natriumconcentratieveranderingen tijdens papierelectrophorese. *3de Colloquium, Sint Jans Hospitaal, Brugge* p. 231. De Tempel, Brugge, 1955.

P2. Peeters, H., and Vuylsteke, P., Oorzaken van beperkte migratielengte in de electrophorese. *4de Colloq. Sint Jans Hospitaal, Brugge* pp. 63-68. De Tempel, Brugge, 1956.

P3. Peeters, H., Vuylsteke, P., and Noë, R., Waterval-electroden in collecterende electrophorese. *4de Colloq. Sint Jans Hospitaal, Brugge* pp. 177-182. De Tempel, Brugge, 1956.

P4. Peeters, H., Vuylsteke, P., and Noë, R., Snelle bidimensionale "Sterelectrophorese." *4de Colloq. Sint Jans Hospitaal, Brugge* pp. 164-171 (1956).

P5. Peeters, H., Vuylsteke, P., and Noë, R., New type of bufferfeeding in two-dimensional electrophoresis. "Protides of the Biological Fluids. Proceedings of the Fifth Colloquium, Bruges," pp. 159-166. Elsevier, Amsterdam, 1957.

P6. Peeters, H., Ionophoresis of the buffersalts during zone- and two-dimensional electrophoresis. Paper presented at *1st Intern. Congr. Clin. Chem., New York* (1956).

P7. Peeters, H., Ionophoretic pattern during two-dimensional electrophoresis. Paper presented at *2nd Intern. Congr. Clin. Chem., Stockholm* (1957).

P7a. Peeters, H., Vuylsteke, P., and Noë, R., Ionophoretic pattern in two-dimensional electrophoresis. *J. Chromatography* **2**, 308 (1959).

P8. Pučar, Z., Beiträge zur Kenntnis der Papierelektrophorese in feuchter Kammer. IV. Kontinuierliche Elektrophorese und zweidimensionale Elektrochromatographie. *Croatica Chem. Acta* **28**, 195 (1956).

R1. Rondelet, J., and Lontie, R., Le fractionnement continu des albumines et globulines de l'orge. *Proc. Congr. Eur. Brewery Convention, Nice* (1953).

S1. Saroff, H. A., An easily assembled continuous electrophoresis apparatus. *Nature* **175**, 896 (1955).

S2. Sato, T., Norris, W., and Strain, H. H., Apparatus for continuous electrochromatography. *Anal. Chem.* **24**, 776 (1952).

S3. Selden, G. L., and Westphal, U., An apparatus for continuous paper electrophoresis. *J. Lab. Clin. Med.* **5**, 786 (1957).

S4. Strain, H. H., Chromatographic systems. *Anal. Chem.* **23**, 25 (1951).

S5. Strain, H. H., Qualitative separations by two-way and three-way electrochromatography. *Anal. Chem.* **24**, 356 (1952).

S6. Svensson, H., and Brattsten, I., An apparatus for continuous electrophoretic separation in flowing liquids. *Arkiv Kemi* **1**, 401 (1949).

Part III. New Developments

10. One-Dimensional Methods

Simultaneously with the results which were achieved by paper electrophoresis, the technical problems inherent in its use became apparent and stimulated a search for new substrates. Among others, starch and gelose were proposed.

Starch was introduced by Kunkel and Slater (K3) and used for preparative work, as the starch block can be cut up according to fractions after separation of rather large quantities of serum. The separations are very sharp, but in practice the cutting up is difficult, as in some cases the zones are numerous and narrow and are often curved or jagged. Others studied lipoproteins (B1) and lipids (A1) with an analogous technique.

After careful hydrolysis of starch a starch gel of suitable quality is obtained; this substrate was used by Smithies (S2, S3) on normal human serum. He obtained a fractionation into thirteen fractions. The size of the mesh of the gel seems to act as a sieve for the molecules of different size.

Extensive work was done with gelose as a supporting medium, especially in the field of immunoelectrophoresis (C1, G1, G2, G3, U2). It is important to note that 98 % of the substrate is composed of buffer solution and that adsorbtion of protein is very low. The equipment itself is of simple design, and at the same time accurate reproduction of the experimental conditions is easier to obtain than with paper strips. Photometry is far less liable to errors than on paper (U1, U3) due to a very low blank and the absence of a structure of the substrate.

On the other hand, the preparation of the substrate is far from easy and demands considerable technical skill. Another drawback is the strong electroendosmosis which amounts to half the main electrophoretic mobility and drives β- and γ-globulins toward the cathode. This means that the application zone lies in the middle of the run and disturbs photometric scanning. Microelectrophoresis, according to Scheidegger (S1), is among the promising developments in this field of electrophoresis. With this substrate it becomes possible to perform an electrophoresis on an ultramicroscale with as little as 0.1–1 μg protein, which needs about 3 minutes (W1).

Immunoelectrophoresis on agar gels has given splendid results in the differential diagnosis of a few important diseases (B2, B3, G3), but must form the subject of a later review since the technique is still in the course of development. Besides this immunological technique other reactions may also be done on the electrophoretic fractions, as is shown in enzymoelectrophoresis (W2). Here the enzymatic activity of the pattern is developed.

Attempts to get away from paper as a supporting medium, while using some chemically better defined form of cellulose, led to the use of ethanol-treated cellulose powder, packed in columns, and more recently to the introduction of cellulose acetate sheets.

To reduce adsorption of cellulose powder (T2) Flodin introduced ethanol-treated cellulose as a supporting medium for zone electrophoresis columns (F1). The main advantage is low adsorption, so that the column can be eluted and used over and over again. In the large models up to 2 liters of serum are separated, while microcolumns are under development which should give excellent results for clinical work (P1, T1).

Apart from the problems related to adsorption, so many factors interfere with photometry on paper, that the accuracy which is needed to give the electrophoretic separation of proteins its full value is fatally lost during the procedure.

Here again cellulose acetate provides an easy solution. Kohn introduced this substance recently (K1), and its two main advantages are the absence of adsorption and the possibility of rendering the strip perfectly translucent to all wavelengths between 250 and 1000 mμ.

This medium was derived from the filters used for the ultrafiltration of bacteria and is a pure chemical. For proteins and glycoproteins the existing methods for paper can be applied at once. Staining of lipoproteins after the run is difficult because of the rinsing solutions, which dissolve the substrate, but in our hands the prestained fractions of lipoproteins could be followed as easily or even better than on paper.

There is often a pre-albumin, as in work on gelose; α_1-globulin is well separated from the albumin, and β_1- and β_2-globulin are well isolated from one another.

Translucency is complete if the strip is soaked in cottonseed oil or decalin, and this is a great step forward toward accurate photometry.

A small disadvantage is the brittleness of the strip when dry, but on the other hand, it is very strong, once wet. There is also a lack of buffer flow and thus the danger of irregular wetting due to evaporation. As a whole, however, cellulose acetate is an important advance over paper. Once electrophoresis is performed the proteinogram can be transferred to an agar medium through diffusion by simple application of the strip on the upper surface of the gel. Thereafter immunoelectrophoresis is more easily and rapidly performed (K2).

Insofar as the splitting up of serum into a greater number of fractions may be expected to have clinical significance, a new buffer introduced by Aronsson and Grönwall (A2) gives nine fractions which are interpreted as a pre-albumin, an albumin, three α-globulins, three β-globulins and one γ-globulin. The buffer has a pH of 8.9 and is composed of 0.5 mole Tris, 0.021 mole EDTA and 0.075 mole boric acid per liter.

As photometric reading techniques using a light source passing through a slit give difficulties when the fractions have an irregular shape, a new type of photometer has been constructed by Wieme. A small light beam is moved over the entire area, and the density of this whole surface is integrated. This integration is made exponentially by means of an electronic system (W1).

It seems desirable to encourage other means of determining proteins than by photometry of dyed conjugates, and in this respect direct ultraviolet absorptiometry is apparently an important method (T1). For this purpose ethanol-treated cellulose powder and cellulose acetate strips are suitable, while buffers satisfactory both for good separation and for ultraviolet reading will be needed.

An interesting technique for reading very low protein concentrations was introduced by Foss (F2). After labeling a protein solution of very low concentration, such as is found in native cerebrospinal fluid, with I^{131}, it is run on paper strips saturated with nonradioactive protein beforehand, and the radioactivity of the separated fractions is registered directly or by means of radioautography, thus avoiding concentration previous to the run. In a similar way the immunological interaction of two proteins was followed. A rapid-moving labeled protein is applied behind a slower protein on the same path. At the end of the run the formed complexes can be found through radioactivity scanning (L1).

Accurate Kjeldahl determinations both on total serum and on serum fractions obtained by salting out show, according to Rein (R1), less spreading than clinical electrophoresis in its present form. This indicates that the accuracy and reproducibility of the electrophoretic results must be increased in order to obtain greater clinical significance.

11. Two-Dimensional Methods

The study of the ionic distribution of the buffer ions on the curtain which was discussed under the problems of the electrical field (Section 6.2.2, p. 97), led to the careful measurement of the field itself during the run. To this effect we scanned the curtain by means of a set of two platinum wires. The potential between these two contact points is registered while the set of wires slides horizontally along the curtain at a constant rate. The graph shows the variations in field strength and these are indirectly proportional to the conductivity.

The field strength of the general background has a typical configuration (Fig. 59), and confirms the typical salt concentration in the para-anodic zone, which was described earlier (Section 6.2.2, p. 99; P2a).

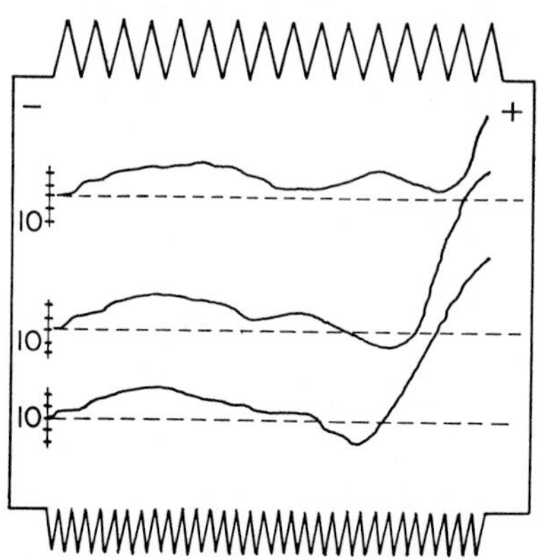

FIG. 59. Field strength of the general background. Curtain, Whatman 3 MM, 22 (height) × 30 cm (width); Buffer, Veronal/sodium Veronalate, μ 0.033, pH 8.6; Run, 3 hours, constant current, 30 ma; voltage 520–410 volt, no sample applied. The field was measured on 3 levels as indicated by the dashed lines (– – – – –) and the recording is reported onto the curtain itself. Ordinate: the field strength given in volts. Each subdivision corresponds to a variation of 1 volt. Abscissa; the localization on the curtain. Notice higher main value of the field in the upper region of the curtain.

The influence of the application of the undialyzed sample is demonstrated very efficiently as the salts and proteins of the serum greatly enhance conductivity and break the field. Scanning through the spot after migration to a lower level gives a series of images containing zones of high alternating with zones of low conductivity (Fig. 60). Fractions can be detected by this means without any staining, even while the run is going. Combination of this technique with the use of isotopes, the test on fluorescence, or the application of staining methods is yielding new information on unknown components.

The development of the two-dimensional technique as a separative tool will depend on the quality of the supporting medium, and in this connection a new technique for preparing a cellulose powder medium for the two-dimensional technique has been proposed. Ethanol-treated cellulose pulp to which 5 % of starch is added is layered on a glass plate and dried. It adheres very strongly to the plate and can be used as a cardboard. Complete homogeneity, higher chromatographic speed, great sample capacity, better resolution and smaller diffusion of the fractions, and absence of adsorption are a few qualities of this substrate (P1).

Probably the most important advances in electrophoresis are to be expected from the "star" method, which has the advantage of giving pure and also new fractions, which could not be detected by other methods.

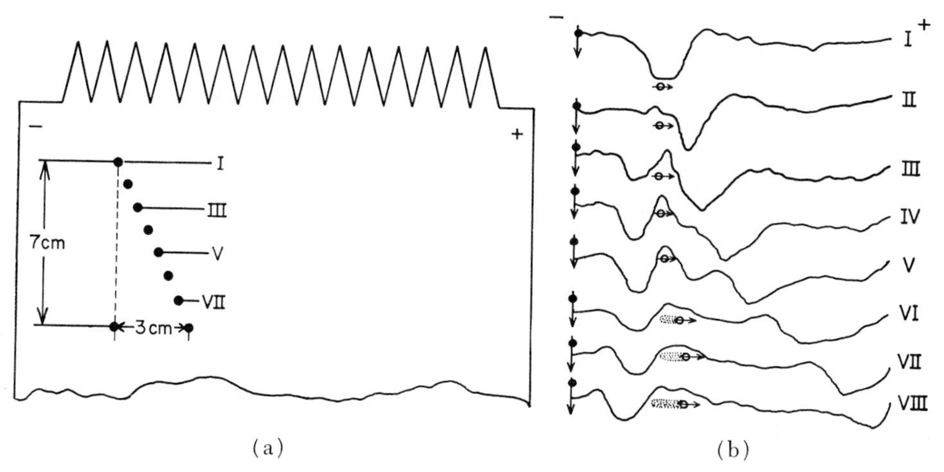

FIG. 60. Influence of the sample on the electrical field. Curtain, Whatman 3 MM, 22 (height) × 30 cm (width); Buffer, Veronal/sodium Veronalate, μ 0.033, pH 8.6; Sample, 0.04 ml of undialyzed serum prestained with bromophenol blue. Run, constant current 25 ma, voltage 580–360 volt. (a) The successive localization of the spot (●) on the curtain at the times of recording. (b) The field strength recorded at the level crossing the spot. Ordinate: the field strength. ●: 15 volts. **V**: 10 volts. Abscissa: the localization of the spot (O→) on the curtain.

As the technique yields results in a short time, it also minimizes denaturation and possibly other degradations. Part of the experience with supporting media, solvents, and dyes, which was gained through chromatography, will be useful for application in this field. Moreover, immunoelectrophoresis is being adapted to this two-dimensional method, and the combination of these important technical advances may prove of interest to the field of clinical medicine.

The immunoreaction of two-dimensional patterns is performed on agar gel after the transfer of the protein pattern from the paper into the gel. This is done in the following way. A sandwich is built up composed of three layers, the first being the paper on which the protein pattern was run, the second a layer of gelose, and the third a paper soaked with antiserum (P2b). Preliminary results allow the detection of most known fractions as being encircled by a precipitation line. After star electrophoresis a series of elliptoid precipitation lines appear. The trailing of albumin is clearly demonstrated by a long precipitation zone that can be followed up to the application point (Fig. 61).

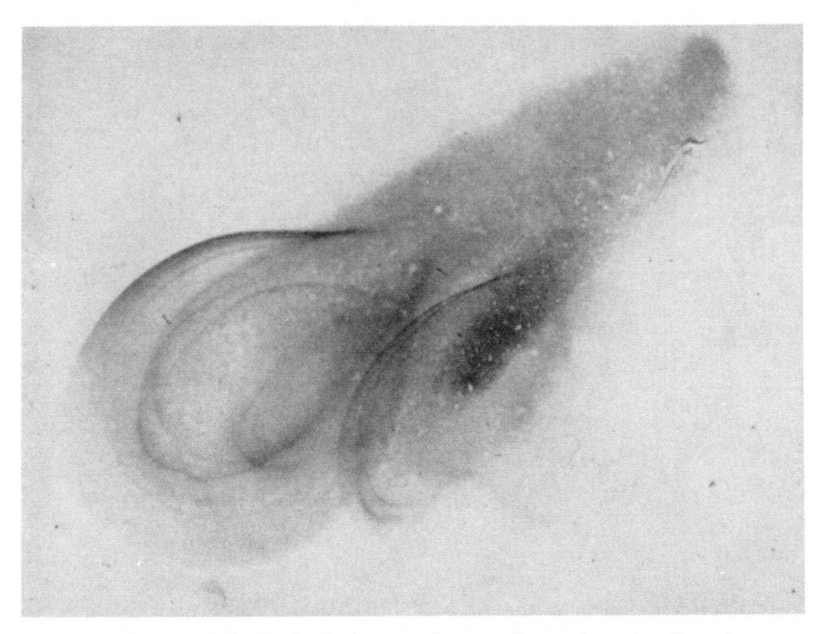

FIG. 61. Immunological development of star electrophoresis. Sample: normal human serum. Antiserum: horse antihuman serum obtained from Institut Pasteur Paris.

Some new techniques may be used to visualize fractions on curtains, and they reveal several unknown fractions. A first method is the spraying of a pH indicator[1] which not only indicates clearly acid values along the anode and alkaline values along the cathode, but also several constituents of the fractionated serum due to the "protein-error" of the indicator used (P2). This pattern shows clearly the albumin and globulins, and before the albumin a fast moving fraction (Fig. 62a), fluorescent in Woods light, and sensitive to ninhydrin. The appearance of a fluorescent fraction was also noted by Bodman (B1a). Fluorescence becomes more intense after correct heating and is due to the reaction of the free amino groups of the amino acids with free aldehyde groups of the paper (L2). This browning reaction occurs only with a very low percentage of the amino-acids present and does not interfere with subsequent elution. Using

[1] pH indicator: 0.8 g neutral red; 2.4 g α-naphtolphtalein; 1000 ml ethanol 96°.

specially sensitized films (Gevaert 'Scientia' 39C56 and 'Scientia' 22D50) with adequate filters and exposure time, some fluorescent fractions which are not visible with the naked eye, appear on the negative.

In our studies the ninhydrin-reaction[2] revealed a number of new fractions, one of which is fluorescent. There are two fractions with a mobility greater than albumin, and two with a mobility lower than γ-globulin (Fig. 62b). Some of these are present in many sera, some are rare. They are present in the ultrafiltrate of serum where also a central fraction can be seen, which has the mobility of the γ-globulin. They are also present in normal cerebrospinal fluid and were found in hemoglobin preparations. It is not clear if only amino acids are involved, because larger molecules also react with ninhydrin.

To facilitate the nomenclature of these new fractions, we introduced a reference system of four figures, the first two indicating the pH, the last two the angle of the migration path with the perpendicular. A plus or minus sign indicates the migration toward the anode or cathode. Thus a fraction moving in a buffer of pH 6.0 at an angle of 30° toward the anode is called + 60.30. This reference system will allow comparison of results between different authors. Evidently the field strength, ionic strength, and constitution of the buffer must be taken into account.

Recently Heremans demonstrated mucopolysaccharides in urine after zone electrophoresis by means of an Alcian blue-acid fuchsin staining technique (H1).[3] Applied in our laboratory to a curtain after a run of serum, two fractions, references + 86.30 and + 86.12, were demonstrated but do not seem to be regular components of serum, at least not in that high concentration (Fig. 62c).

The clinical significance of these new fractions and their physiological fluctuations and relation to disease is not yet clear. But these new techniques give a hopeful renewal to electrophoretic work. Ultimately some components of serum, difficult to determine with the usual methods, such as some organic anions, enzymes, vitamins, and hormones might be visualized, separated, and collected.

It is of interest to mention in this connection, that the electrophoretic

[2] Ninhydrin solution: 0.35 g ninhydrin; 4 ml 96% acetic acid; 100 ml butanol.

[3] Alcian blue dyes. *Dye A:* 1 g Alcian blue; 90% acetic acid in water. *ad* 100 ml. *Dye B:* 0.25 g acid fuchsin; 90% methanol in water. *ad.* 100 ml.

Staining technique: Dry at 110° C for 30 minutes; fix with 20% formalin in ethanol for 15 minutes; dry at room temperature; dye with dye A for 5 minutes; rinse by alternating washings with glacial acetic acid and running water (4 to 5 times); immerse in 10% acetic acid in methanol for 5 minutes; dye with dye B for about 30 minutes; rinse with 10% acetic acid in water.

separation of a reagent-grade chemical, namely glycerol, reveals the presence of several impurities even with the best qualities available. Separation of charged from uncharged fractions in reagents, drugs, and hormones, even if they are so-called chemically pure substances, gives excellent results.

Finally, attempts are being made to introduce a third differentiating force into two-dimensional techniques, namely, a pH gradient. Instead of using a buffer field with uniform pH, a series of vertical zones of different pH is built up (P1). For this purpose trickle feeding is adapted in such a way that each tooth of the upper edge of the paper receives buffer of a different pH. Thus the different fractions travel through sub-

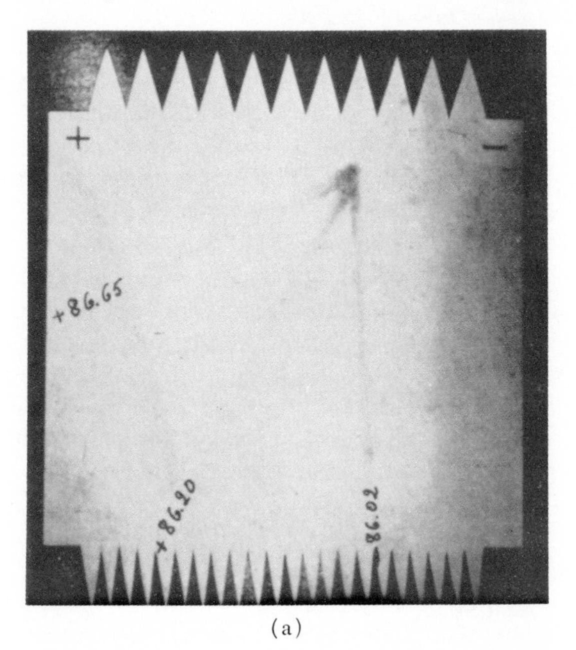

(a)

Fig. 62a, b, and c. Some staining techniques adapted to paper electrophoresis. Sample: human serum. Run: continuous electrophoresis. The fractions are labeled according to the proposed reference system (see text). The known fractions are: +86.20 Albumin; +86.18 α_1-globulin; +86.12 α_2-globulin; +86.07 β-globulin; —86.02 γ-globulin.

The new fractions are in: (a) After spraying with a pH indicator, fraction +86.65. (b) After infrared drying and spraying with ninhydrin. Fractions +86.45 and +86.30 lying before the albumin and —86.07 and —86.50 lying behind the γ-globulin. (c) After staining with Alcian blue-acid fuchsin according to Heremans (H1). B, blue; R, red. Fraction +86.30 and a fraction moving from +86.12 toward the albumin are stained in blue and may contain polysaccharides.

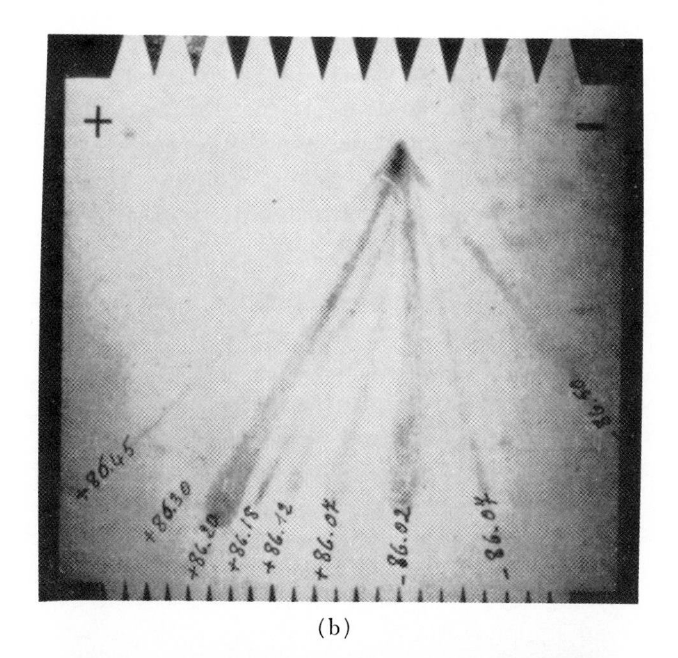

(b)

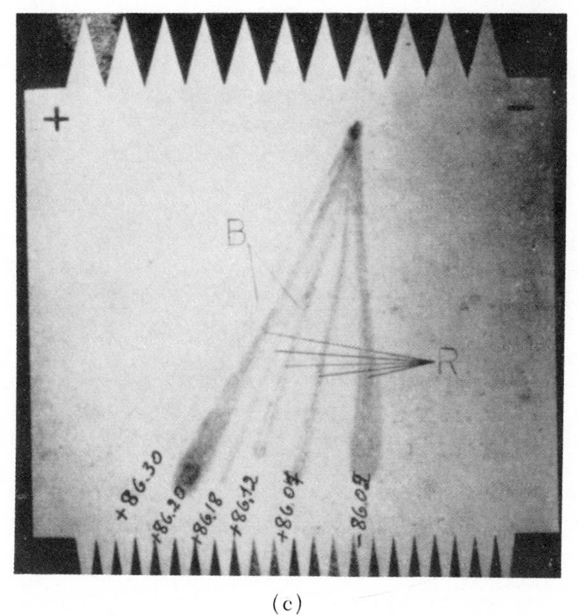

(c)

Fig. 62b and c. For legend see opposite page.

sequent zones of different pH. This can be considered as a series of electrophoreses at successive pH values but performed in one single experiment. The spraying of an efficient pH indicator (210 mg α-bromonaphthalene and 100 mg neutral red in 200 ml ethanol) over the field is of great help in visualizing the pH gradient (P2).

These different technical developments seem to give good promise for the future of electrophoretic investigation.

Acknowledgments

I wish to acknowledge the many helpful suggestions offered by Dr. C. P. Stewart, and also my indebtedness to my assistants, Dr. P. Vuylsteke and R. Noë, Techn. Eng., for the general layout of the text, to Sister Lioba, O.S.B., for reading the English manuscript, and to my secretary, Miss R. Rommelaere, for the careful typing. My sincere thanks go to all collaborators, who in the course of these last three years joined us in efforts to gain a better insight into electrophoretic techniques. Finally, I wish to thank the President, the Members of the Board, and the Secretary of the Commissie van Openbare Onderstand of the City of Brugge for the general equipment of the laboratory.

References

A1. Ackerman, P. G., Toro, G., and Kamitz, W. B., Zone Electrophoresis in the study of serum lipoproteins. *J. Lab. Clin. Med.* **44**, 517 (1954).

A2. Aronsson, T., and Grönwall, A., Improved separation of serum proteins in paper electrophoresis. A new Electrophoresis buffer. *2nd Intern. Congr. Clin. Chem., Stockholm* (1957).

B1. Bernsohn, J., Isolation of serum lipoproteins by zone electrophoresis. *J. Lab. Clin. Med.* **3**, 478 (1957).

B1a. Bodman, J., Round Table Conference on Protein Pattern in Normal and Pathological Cases, *In* "Protides of the Biological Fluids, Proceedings of the Sixth Colloquium, Bruges," p. 317. Elsevier, Amsterdam, 1958.

B2. Burtin, P., L'analyse immuno-, électrophorétique des sérums pathologiques. *4de Colloq. Sint Jans Hospitaal, Brugge* pp. 183-188. De Tempel, Brugge (1956).

B3. Burtin, P., Hartmann, L., Heremans, J., and Scheidegger, J. J., Westendorp, B. F., Wieme, R., Wunderly, C., Fauvert R., and Grabar, P., Etudes immunochimiques et immuno-électrophorétiques des macroglobulinémies. *Rev. franç. d'études clin. et biol.* **2**, 161 (1957).

C1. Crowle, A., A simplified agar electrophoretic method for use in antigen separation and serologic analysis. *J. Lab. Clin. Med.* **48**, 642 (1956).

F1. Flodin, P., and Kupke, D., Zone electrophoresis on cellulose columns. *Biochim. et Biophys. Acta* **21**, 368 (1956).

F2. Foss, O., I^{131} as an "amplifier" in paper electrophoresis of low concentrated protein solutions. *2nd Intern. Congr. Clin. Chem., Stockholm* (1957).

G1. Giri, K. V., Agar electrophoresis of serumproteins on cellophane and polyester films. *J. Lab. Clin. Med.* **48**, 775 (1956).

G2. Gordon, A. H., Keil, B., Schesta, K., Knessel, O., and Sonn, F., Electrophoresis of Proteins in Agar Jelly. *Collection Czechoslov. Chem. Communs.* **15**, 1 (1950).

G3. Grabar, P., and Williams, C. A., Méthode permettant l'étude conjugée des propriétés électrophorétiques et immunochimiques d'un mélange de protéine. Application au sérum sanguin. *Biochim. et Biophys. Acta* **10**, 193 (1953).

H1. Heremans, J., (Louvain, Belgium) Scientific Exhibit at the VIIth Colloquium, Bruges (1959) unpublished.

K1. Kohn, J., Membrane Filter Electrophoresis. "Protides of the Biological Fluids, Proceedings of the Fifth Colloquium, Bruges," pp. 120-125. Elsevier, Amsterdam, 1957.

K2. Kohn, J., Small-scale and "Micro" membrane filter electrophoresis and immunoelectrophoresis. "Protides of the Biological Fluids, Proceedings of the Sixth Colloquium, Bruges," pp. 74-78. Elsevier, Amsterdam, 1958.

K3. Kunkel, H. G., and Slater, R. J., Lipoprotein patterns of serum obtained by zone electrophoresis. *J. Clin. Invest.* **31**, 677 (1952).

L1. Lang, N., Quantitative Untersuchungen mit Überwanderungs-Elektrophorese über die Bindungskapazität von Plasmaproteinen. "Protides of the Biological Fluids, Proceedings of the Sixth Colloquium, Bruges," pp. 68-73. Elsevier, Amsterdam, 1958.

L2. Lederer, E., and Lederer, M., *In* "Chromatography: A Review of Principles and Applications," p. 202. Elsevier, Amsterdam, 1955.

P1. Peeters, H., and Vuylsteke, P., New trends in two dimensional electrophoresis. "Protides of the Biological Fluids, Proceedings of the Sixth Colloquium, Bruges," pp. 53-58. Elsevier, Amsterdam, 1958.

P2. Peeters, H., and Vulysteke, P., Coloured substances as an experimental tool in the study of electrophoretic techniques. "Protides of the Biological Fluids, Proceedings of the Sixth Colloquium, Bruges," pp. 115-118. Elsevier, Amsterdam, 1958.

P2a. Peeters, H., Vuylsteke, P., and Noë, R., Ionophoretic pattern in two-dimensional electrophoresis. *J. Chromatography* **2**, 308-319 (1959).

P2b. Peeters, H., and Vuylsteke, P., Immunoelectrophoresis of two-dimensional patterns. "Protides of the Biological Fluids, Proceedings of the Seventh Colloquium." Elsevier, Amsterdam, 1959. In press.

P3. Porath, J., "Zone Electrophoresis in Columns and Adsorption Chromatography on Ionic Cellulose Derivatives, as Methods for Peptide and Protein Fractionations." Almqvist and Wiksells, Uppsala, 1957.

R1. Rein, F. J., Neue Analytik durch exakte Stickstoff-Bestimmungen in Fraktionen des Blutserums. *4de Colloq. Sint Jans Hospitaal, Brugge* pp. 152-164. De Tempel, Brugge (1956).

S1. Scheidegger, J. J., Micro-Méthode d'immuno-électrophorèse. *Intern. Arch. Allergy Appl. Immunol.* **7**, 103-110 (1955).

S2. Smithies, O., Grouped variations in the occurrence of new protein components in normal human serum. *Nature* **175**, 307 (1955).

S3. Smithies, O., An improved procedure for starch-gel electrophoresis: further variations in the serum proteins of normal individuals. *Biochem. J.* **71**, 585-587 (1959).

T1. Tiselius, A., Electrophoresis: Past, Present and Future. "Protides of the Biological Fluids, Proceedings of the Fifth Colloquium, Bruges," pp. 1-10. Elsevier, Amsterdam, 1957.

T2. Tiselius, A., and Flodin, P., *Advances in Protein Chem.* **8**, 461 (1954).

U1. Uriel, J., Les réactions de caractérisation des constituants des liquides biologiques après électrophorèse de zone. "Protides of the Biological Fluids, Proceedings of the Fifth Colloquium, Bruges," pp. 17-23. Elsevier, Amsterdam, 1957.

U2. Uriel, J., and Grabar, P., Etude des lipoprotéines sériques par l'électrophorèse en gélose et l'analyse immunoélectrophorétique. *Bull. soc. chim. biol.* **38**, 1253 (1956).

U3. Uriel, J., and Scheidegger, J. J., Electrophorèse en gélose et coloration des constituants. *Bull. soc. chim. biol.* **37**, 1 (1955).

W1. Wieme, R. J., Description d'un colorimètre balayeur-intégrateur destiné à la lecture de taches irrégulières. "Protides of the Biological Fluids, Proceedings of the Sixth Colloquium, Bruges," pp. 64-67. Elsevier, Amsterdam, 1958.

W2. Wieme, R. J., Diagnostic application of enzymoelectrophoresis. "Protides of the Biological Fluids, Proceedings of the Sixth Colloquium, Bruges," pp. 236-240. Elsevier, Amsterdam, 1958.

BLOOD AMMONIA

Samuel P. Bessman*

Departments of Pediatrics and Biochemistry, School of Medicine, University of Maryland, Baltimore, Maryland

1. Historical Introduction

Ammonia became of significance to clinical investigation in the latter part of the nineteenth century when a group of researchers, studying the effect of the Eck fistula operation on dogs, found that high-protein diets caused a characteristic symptom complex to develop (H1, N5, N6). The forced feeding of meat to dogs in which a shunt had been established between the portal system and the peripheral venous system, bypassing the liver, invariably resulted in the death of the animals. The early

* Unpublished work referred to in the text has been supported by grants from the United States Public Health Service, the Playtex Park Foundation, and the National Science Foundation.

symptoms of "meat intoxication" (Fleischvergiftung) were dizziness, ataxic gait, anorexia, and blindness. Eventually stupor and death ensued. These results did not occur on a cereal diet, and the investigators postulated that the disease symptoms were due to the nitrogen content of the meat. Injection of ammonium salts intravenously confirmed their hypothesis that the symptoms could be due to ammonia intoxication. Attempts to measure the ammonia content of the blood by the techniques available at that time failed, and the hypothesis was discarded in favor of the assumption that carbamate ion, resulting from the combination of ammonia with carbon dioxide, was in fact the toxic material. Interest in the problem lagged when further chemical studies in other laboratories could not confirm the first reports on ammonia and carbamate. A number of different investigators proposed varying interpretations for the mental symptoms which occurred in animals operated on by the Eck technique (B3, S11).

The question of the role of ammonia in clinical symptoms was reopened by the classic experiments of Van Caulaert (V3, V4), who demonstrated a toxic effect of ammonium salts in patients with cirrhosis of the liver. He measured the blood ammonia in these patients and found that when the blood ammonia was elevated, symptoms of stupor appeared. This finding led to the exhaustive studies of Kirk (K2), who investigated the effect of oral administration of glycine and ammonium salts to patients with cirrhosis and hepatitis and to patients with no liver disease discernible. His experiments led him to the conclusion that the patient with hepatitis or with cirrhosis could metabolize ammonia perfectly well, provided that there was no shunt between the portal and peripheral venous systems. When protein or other nitrogenous material was fed to the individual, elevated levels of blood ammonia appeared only when such a shunt existed. In his studies, however, Kirk excluded all patients who had symptoms of coma, on the grounds that these patients already had elevated blood ammonia levels. As a result, most of the patients who might be expected to have inability to dispose of ammonia were not included in his observations. This work effectively diverted attention from ammonia as a significant factor in coma in spite of the reports by Burchi (B17) and by Monguio and Krause (M6) that the tolerance to ammonium salts was altered in disease of the liver.

Renewed interest in the relation of ammonia to hepatitic coma developed at the Boston City Hospital when Gabuzda (G1) and Phillips (P8) repeated and extended the experiments of Van Caulaert with patients with cirrhosis of the liver. They found, indeed, that any manipulation which produced an elevated blood ammonia would also cause symptoms of early hepatic coma or, as they decribed it, "impending hepatic coma." Their observations initiated a large series of investigations in many laboratories. It has become clear that ammonia has some relation to the development of mental symptoms, not only in hepatic disease, but in a number of other disease conditions.

2. Normal Metabolism of Ammonia

2.1. The Nature of Blood Ammonia

There is now general agreement that there is a small amount of material in normal blood which manifests itself on mild alkalinization as ammonia. There is also agreement that, under certain pathological conditions, the volatile base appearing in the blood is ammonia. At pH 7.4 ammonium is almost completely ionized according to the equation: $NH^+{}_4 \rightleftharpoons H^+ + NH_3$. The free ammonia, NH_3, penetrates cell membranes rapidly, whereas tissues are relatively impermeable to the ammonium ion (J1, J2). Since a hydrogen ion is involved in the reaction, the equilibrium is highly sensitive to pH. In the following pages the term "ammonia" is used throughout as a matter of convenience; it must, however, be interpreted as signifying the sum of NH_3 and $NH_4{}^+$ and therefore, for practical purposes, "ammonium ion."

Recent studies (V1) have shown that the un-ionized fraction of the blood ammonia might be altered markedly by the pH of the blood because the dissociation curve for ammonium ion has a fairly sharp flexure in this region. There is no doubt, however, that more than 99.9 % of the blood ammonia is in the form of ammonium ion, within the entire physiological range of pH. Evidence has been presented that the toxicity of blood ammonia might be enhanced by respiratory alkalosis because the ammonia fraction (diffusible fraction) would be increased, so that more ammonia would enter the cell. Our own studies with red blood cells (B10) reveal a small increase in intracellular ammonia in alkalinized plasma over the intracellular content in the same plasma which has been acidified. However, the data of the authors who propose this increase in permeability as a significant factor in ammoniagenic coma show a greater uptake of ammonia by tissues (arteriovenous difference) in acidotic animals than in the same animals when they were alkalotic. If the cellular content of ammonia were dependent upon the amount of un-ionized ammonia in the blood, there should be some correlation between the concentration of gaseous ammonia ($p$$NH_3$) and the uptake of ammonia by tissues. The data presented by the authors do not verify the prediction. Table 1 gives the mean calculated data on five dogs [Nos. 1, 2, 4, 5, 6 of Table 1 (L1)] with calculated arteriovenous differences.

TABLE 1
MEAN AMMONIA UPTAKE BY MUSCLE IN ECK FISTULA DOGS IN METABOLIC ACIDOSIS OR ALKALOSIS[a]

	pH		Total plasma NH_3 μg/ml		pNH_3 mm Hg × 10^4		Uptake of NH_3 μg/ml
	A	V	A	V	A	V	(A-V)
Control	7.40	7.34	2.45	2.02	2.28	1.64	0.43
0.5 *N* HCl infusion	6.98	6.94	3.76	2.80	1.35	0.91	0.96
4 % $NaHCO_3$ infusion	7.56	7.51	2.32	1.73	3.18	2.07	0.59

[a] Calculated from L1.

Dog No. 3 was omitted because of insufficient data. It is apparent that a marked uptake of ammonia occurs in tissues in metabolic acidosis. In alkalosis the uptake by tissues is only slightly greater than normal, and the arterial levels of ammonia are comparable with normal values. The elevation of blood ammonia which occurs on vigorous acidification of the animal might not be due to a drop in permeability with pH, but might be due to a phenomenon reported by Krebs and Henseleit (K4). Urea synthesis by liver slices is directly proportional to the pH and CO_2 content of the medium. Table 2, constructed from data reported in this paper, shows this dependence.

TABLE 2
DEPENDENCE OF UREA SYNTHESIS BY LIVER SLICES ON pH AND CO_2 CONTENT

CO_2 Content meq/liter	pH	Urea synthesis μmoles/g/hr
3.1	6.57	2.79
6.2	6.87	4.78
12.4	7.17	7.75
24.8	7.46	9.41
49.6	7.78	9.98

Thus the drop in ammonia content of plasma in alkalosis may be related to the pH effect on urea synthesis rather than to the alterations in permeability caused by changes in pNH_3. The treatment of hepatic coma by acidification with CO_2 would lower pH, inhibiting synthesis of urea. This would cause a rise in blood ammonia. The same type of change could be brought about by infusion of HCl, which would have a double effect on urea synthesis by lowering both pH and bicarbonate. It does not appear necessary to invoke a penetration hypothesis to explain the changes in blood ammonia resulting from alterations in pH. Warren (W5) has reported a direct relation between blood pH and toxicity of ammonium salts. Mice were given intravenous solutions of ammonium chloride, acetate, bicarbonate, carbonate, hydroxide, and citrate. The LD_{50} was estimated for each salt. The blood pH increased regularly from 7.05 for the chloride to 7.74 for the hydroxide in the order given, and the LD_{50} decreased from 6.6 meq/kg to 2.4 meq/kg. Ammonium citrate was most toxic, but its effect was diminished by calcium gluconate administration, suggesting that the calcium-binding property of the citrate was the cause of its disproportionate lethality. The fact that the lethal dose of ammonium salt is inversely related to the pH attained in the blood suggests that when a certain amount of ammonia has entered the cell, death ensues. The pH determines the rate at which the ammonia crosses the cell membrane. These data are consistent with the depletion theory of ammoniagenic coma described below.

2.2. Methods of Determining Ammonia

The techniques for the determination of quantities of ammonia such as are found in body fluids, that is, of the order of one microgram per milliliter, have been developed only recently. All methods for the determination of ammonia in animal fluids are based on two general processes. The first is the conversion of the ammonium ion to gaseous NH_3 and its segregation in a separate aliquot of solution. The second is the application of some technique for the quantitation of this isolated ammonia.

2.2.1. *Methods of Separation of Ammonia from Samples*

2.2.1.1. *Distillation.* The earliest methods for the separation of ammonia from solution were based on the alkalinization of the solution and the subsequent distillation of ammonia. The distillation can be carried out either by boiling, as were the original Kjeldahl distillations, or by more modern techniques, the first of which was introduced by Parnas and Heller (P5), for the determination of micro quantities of ammonia. This was a steam distillation technique in which steam was passed through the gently heated alkalinized solution, carrying off ammonia to an aliquot of standard acid. Although modifications of this technique are occasionally used for semimicro nitrogen determination, they are no longer used for direct estimation of ammonia in physiological material.

2.2.1.2. *Aeration.* Among the first of the aeration methods was that of Folin (F4, F5), who devised a technique whereby body fluid, such as blood, was alkalinized and the ammonia aerated into an acid solution.

The ammonia so separated was determined colorimetrically with Nessler's reagent. In this manner a very small quantity of ammonia could be concentrated into acid and the determination could be made with ease. The aeration period of 30 minutes chosen by Folin led, as we now know, to the inclusion of some ammonia resulting from the decomposition of certain labile compounds in blood, such as free glutamine. The exposure of these compounds to alkali was sufficient to result in their hydrolysis, with the liberation of significant amounts of ammonia. However, Folin's data on the blood levels of ammonia are similar to the values currently found in normal individuals. Folin studied the content of ammonia in the various vessels of the body and found that the highest concentration of ammonia was in the portal system. The use of these more gentle techniques permitted the investigators to discover that there were vanishingly small quantities of ammonia actually present in body fluids, while the heating of body fluid resulted in the degradation of ammonia-liberating compounds.

2.2.1.3. *Microdiffusion.* The most recent technique for the segregation of ammonia has been the microdiffusion process. This is nothing more than a room temperature distillation in which advantage is taken of the fact that ammonia, a volatile base, will diffuse from an alkaline solution to an acid one.

(*a*) *Conway.* Certain factors are critical to the success of this method, which was elaborated by Conway's group (C6). All of the current methods for the determination of the ammonia in body fluids are based on some variation of the microdiffusion technique. In general, two types of microdiffusion apparatus are used. The first type is that devised by Conway and consists of a small round dish divided into two compartments, an inner compartment and a concentrically arranged outer compartment, the vessel being covered by a glass plate. Acid is put into the inner compartment. The unknown is put into the outer compartment and alkalinized. After a suitable period the ammonia which has diffused into the inner solution is estimated by one of several methods.

(*b*) *Seligson.* The second microdiffusion technique involves the use of a small vial (Fig. 1) through the stopper of which a glass rod with an enlargement on the end projects. The unknown solution is placed in the vial and alkalinized, and the rod is wetted with dilute acid. The vial is then placed in a horizontal position and rotated on a wheel or other device in such a manner as to distribute the fluid uniformly and continuously over the sides of the bottle. The ammonia diffuses a distance of a few millimeters from the alkalinized outer fluid to the acid on the glass rod. This material is then washed off the rod into a suitable reagent for the estimation of ammonia. This technique was introduced by Seligson (S4).

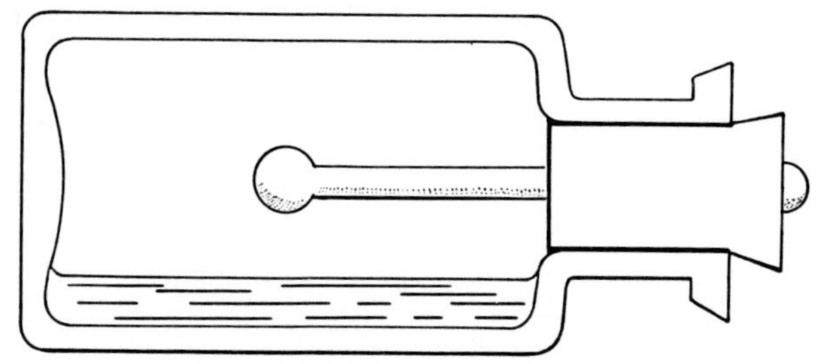

FIG. 1. Seligson microdiffusion vessel.

The Conway diffusion vessel is subject to a number of limitations, particularly if a short period of diffusion is required. The question of adequate mixing of the material in the outer chamber has been considered (S3). This shortcoming makes the results obtained with the Conway technique quite variable over short diffusion periods. In addition, unless some direct method of measuring ammonia in the solution in the central chamber of the Conway diffusion cup (e.g. titration) is used, the material in the center chamber must be transferred to a second container for colorimetric analysis. This involves quantitative pipetting of the material in the central chamber with its danger of exposure to air, which contains, in many clinical areas, considerable quantities of ammonia. The method is subject to this ubiquitous source of error and requires that all determinations done by the Conway technique be carried out in a scrupulously ammonia-free atmosphere. The Seligson diffusion technique, on the other hand, concentrates the ammonia diffused into a very tiny volume, a thin film covering the surface of the glass rod in the center of the diffusion vial. This rod can then be thrust very rapidly into a colorimeter tube with little exposure to air.

A further disadvantage of the Conway method is that the volume of solution in the center well is critical and must be pipetted with volumetric precision as well as dispatch. The Conway dishes must be prepared considerably in advance of their use, for the danger of contamination becomes excessive if they are prepared at the bedside. The Seligson technique requires no quantitative measurements except for the volume of the unknown specimen, since the ammonia diffused is concentrated in a very small volume, all of which is transferred by dipping it into the colorimeter tube. The volume of acid which coats the diffusion rod is not critical since it represents less than one per cent of the final volume of solution used for colorimetry.

Seligson has recently (S3) modified his original method because the total ammonia evolved from a mixture of blood and saturated sodium carbonate increases with time of diffusion. For this reason, a finding in all diffusion methods, the new method adds blood directly to a dry mixture of sodium carbonate and bicarbonate so that the pH of the mixture never rises above 11. The evolution of ammonia from blood under these circumstances is said to reach a plateau in about 20 minutes. The values obtained by this method, however, are somewhat higher than those obtained with the 10-minute diffusion from the sodium carbonate solution. Comparison of our modified form of the original method with the newer technique suggests no advantage for the newer method unless it is necessary to prolong the diffusion time.

2.2.2. *Methods of Estimating the Amount of Ammonia Diffused*

2.2.2.1. *Titration.* Since ammonia is a basic substance it can be titrated with standard acid or backtitrated from a standard volume of acid. The original Conway technique (C5) utilized a standard volume of dilute acid in the center well with a backtitration in the vessel with barium hydroxide solution. A microburet was required in order to measure the minute quantities of standard alkali used. This method was later changed by the use of a borate buffer maintained at a constant pH (A1), into which the ammonia was diffused. This borate buffer would be rendered alkaline by the ammonia and could be brought back to its original pH by the addition of standard acid. The endpoint was found by comparison with a control titration. This was an improvement over the previous method because it was quite difficult to maintain normality of a very dilute alkaline solution. The borate titration is probably the most commonly used of the titration methods now available (C5). It has, however, several theoretical and practical disadvantages. In the first place, it measures not only ammonia, but any diffusible volatile base. According to Richter's (R1) data, a number of volatile bases diffuse with approximately the same rapidity as ammonia; for example, *n*-butylamine and isoamylamine. Amines such as these have been found to be present in blood and would be included as "ammonia." These undoubtedly contribute to the values found as ammonia by the Conway method or by any method involving simple titration of alkalinity appearing in a diffusion setup. Any acidimetric technique should, therefore, always be reported as "volatile base" rather than as ammonia.

2.2.2.2. *Nessler's Reagent.* For this reason a number of specific color reactions have been applied for the detection of ammonia separated by microdiffusion. The most common one uses the Nessler reagent. This reagent has been employed with considerable success in the method of Seligson for the measurement of blood ammonia. Conditions have been established (B9) whereby the color developed by a microgram of ammonia, the amount normally present in one milliliter of blood, can be read in a satisfactory region of the colorimeter dial in a volume of 1.25 ml. The Nessler reagent will not determine aliphatic amines as ammonia. It will give a reaction of turbidity in the presence of amines, rather than the brown color which develops with ammonia. It is possible in this way to differentiate between volatile aliphatic amines and ammonia by using Nessler's reagent. This cannot be done with the titration technique.

2.2.2.3. *Other Methods.* The Berthelot reaction, in which a dye is formed by ammonia in the presence of phenol and hypochlorite, has also been used (V5). The dye color is proportional to the ammonia content of the unknown solution. This reagent can be made more sensitive by the addition of nitroprusside as a catalyst (L5). It has been applied to the Conway diffusion method (V5) and to a modification of the Seligson method (B16). The ninhydrin reaction has also been used both with the Conway method and the Seligson method for the detection of ammonia (N3). The reaction between hypobromite and ammonia has been applied to the microdetermination of ammonia in a unique manner by Stone (S9). The usual application of this reaction (S3) has been by adding a standard amount of hypobromite and determining the residual reagent. The Stone method utilizes the stoichiometric decolorization of phenosafranine by hypobromite. The unreacted hypobromite decolorizes a standard amount of dye, and the residual dye is proportional to the ammonia originally present. The sensitivity of this reaction is about as great as the phenol-hypochlorite or ninhydrin methods. A major disadvantage of this technique is its requirement for absolute accuracy in measurement of two critical reagents, the phenosafranine and the hypobromite. Since the hypobromite deteriorates fairly rapidly at room temperature, it is difficult to obtain very good checks at low values. For example, an error of 1 % in measurement of either the dye or the hypobromite results in a change of about 10 units on the Klett Colorimeter, equivalent to 0.1 μg. A summation of two such errors will cause double this variation, and at a level of 1.0 μg this is a 20 % error. When working with blood, the maximum allowable error should be no more than 5 % of the mean of duplicates. This can be attained easily with the other methods. The sensitivity of the various color reactions is shown in Table 3. Although the Nessler reagent is least sensitive of the specific tests for

TABLE 3
SENSITIVITY OF VARIOUS COLOR REACTIONS FOR THE ESTIMATION OF AMMONIA IN BLOOD

Reagent	Colorimeter	Filter[a] or wavelength[b]	Volume ml	Reading μg
Ninhydrin (Nathan-Rodkey)	Klett	56	10	70
Hypobromite-phenosafranine (Stone)	Klett	52	8.5	100
Phenol-hypochlorite-nitroprusside	Klett	60	2	400
Nessler's (Seligson-Bessman)	Klett	42	1.25	75
Micro-Nessler's (Bessman)	Beckman DU	410 mμ	0.2	0.590

[a] Filter for Klett Colorimeter.
[b] Wavelength for Beckman DU Spectrophotometer.

ammonia, it is adequate for measurement of blood content in normal and pathological states. It requires no heating or incubation and uses only one easily prepared and stable solution. Furthermore, it is subject to the least variable error of all of the color methods. The time required for the performance of a duplicate determination by the modified diffusion technique (B4) using Nessler's reagent is about 13 minutes. This is one-third to one-fifth the time necessary for the other colorimetric procedures.

A recent study of the Conway method applied to whole blood (R2) stresses the importance of the ammonia derived from the breakdown of blood constituents in the highly alkaline diffusion medium and recommends a correction for this. It is suggested that two diffusions be performed, one for 40 and one for 80 minutes. The difference between the two values would represent the actual preformed ammonia of the blood. This problem has been discussed by others (B6, S3), as well as by Conway (C5). The requirement for a 40-minute diffusion in order to obtain a complete yield of ammonia from the sample is a major handicap of the Conway apparatus. The Seligson method requires only 10 to 20 minutes, resulting in much less time for artifact. The use of the titration of basic (R2) leads to the inclusion of much volatile base which is not ammonia, and much of the apparent formation of "ammonia" during diffusion is probably volatile amines, for it is not seen so constantly using Nesslerization methods.

2.3. Sources of Ammonia in the Organism

2.3.1. *Gut*

The major source of blood ammonia was discovered by Folin, who showed that the blood leaving the intestinal tract in the portal system contained far greater amounts of ammonia than did the blood of the peripheral venous or arterial system (F6). Extensive studies with the gut contents in relation to the formation of amino acids from protein and to the liberation of ammonia during incubation led to the understanding that there were several sources from which ammonia developed from the nitrogenous material in the gut. In the first place the hydrolytic enzymes acting on protein liberate glutamine, the γ-amide of glutamic acid (D2). Glutamine is the form in which most of the glutamic acid is present in protein, and it represents the easily liberated nitrogen of protein (D1). This material hydrolyzes under the influence of bacterial glutaminases normally found in the gut to glutamic acid and ammonia. The second source of ammonia in the gut is from the hydrolysis of the urea of the gut secretion by the enzyme urease. Urea is normally present in all of the gastrointestinal secretions in amounts equal to its distribution in other body fluids, since urea easily penetrates all membranes. In the presence of large amounts of urease-containing bacteria in the gut, or conversely, in the presence of normal amounts of bacteria but increased quantities of urea, there is a marked evolution of ammonia from the contents of the gut. An example of this situation is found in uremia, in which tremendous amounts of ammonia are formed in the intestinal tract. These large amounts of ammonia probably exert a toxic action upon the gut mucosa, resulting in the ulceration and necrosis of the mucosa of the colon and small intestine so frequently seen as a complication of uremia.

It was originally thought that the secretion of hydrochloric acid by the stomach involved a simultaneous formation of ammonia to neutralize or balance the hydrochloric acid secretion (G3). However, it has been shown by the elegant work of Davies and co-workers that urease activity is not related to the formation of acid and that the urease activity of the wall of the stomach is a result of bacterial contamination (D4, K3). The bacterial origin of gastric urease was also shown by Glick and Von Korff (G4). It can be diminished or obliterated by the administration of antibiotics and yet the secretion of acid remains normal. This confusion of ammonia derived from urease activity with true secretion of ammonia also occurred in the case of saliva, in which there are occasionally considerable amounts of ammonia. It was shown by Hench and Aldrich (H4) that the sum of urea and ammonia nitrogen of saliva was equal to the urea nitrogen of the blood, indicating that the ammonia of saliva must have come from the breakdown of urea. The urease activity of the bacteria of the skin makes it very difficult to draw blood by skin prick for ammonia determinations by this route.

2.3.2. *Renal*

Another source of ammonia in the organism is the renal venous outflow. Although this may seem paradoxical, ammonia which is brought to the kidney is not excreted as such. Indeed the kidney puts a considerable amount of ammonia into the blood. The renal venous outflow was shown many years ago by Loeb, Atchley, and Benedict (L4) to contain more ammonia than the arterial inflow. This probably results from the back-diffusion of ammonia from the tubules of the kidney during the secretion of ammonia in the urine. The urinary ammonia derives not from the blood ammonia, but from glutamine (V6) and possibly from the oxidative deamination of other amino compounds brought to the kidney. The arteriovenous difference of glutamine across the kidney is sufficient to account, to a large extent, for the ammonia secreted by the kidney during that interval. Further experiments with acidotic and alkalotic animals showed that the glutaminase activity of the kidney was related to the acid-base balance of the animal (W12). Severely acidotic animals had a higher glutaminase activity of kidney tissue than did animals with normal blood pH, and animals with normal pH had higher glutaminase activity than did animals with alkalosis artificially induced. This implied an adaptive formation of the enzyme glutaminase by the kidney in response to alkalosis or acidosis (D3). Since the carbonic anhydrase inhibitor, acetazoleamide, inhibits the formation of ammonia by kidney, as evidenced by a diminished excretion of ammonia in the urine (W11), its toxic effect upon patients with liver disease was assumed to be due to this diminution of ammonia secretion in the urine. These patients developed symptoms of impending hepatic coma as their blood ammonia rose (W7). It was postulated by those studying the problem that this rise in blood ammonia was caused by the retention of ammonia formed in the kidney by normal glutaminase activity on the glutamine brought to this organ, owing to the prevention of the excretion of this ammonia by the carbonic anhydrase inhibitor (W8).

Figure 2 shows the author's study of the arteriovenous difference of ammonia across the kidney of a dog which was given a large amount of acetazoleamide, sufficient to diminish his urinary excretion of ammonia. As can be seen by this figure, the urinary excretion of ammonia drops, secretion of water increases, but the arteriovenous difference of ammonia across the kidney does not change. In fact, the arterial level of ammonia increases, suggesting that the source of the increased ammonia in the blood is not the kidney but is elsewhere. Some suggestion as to the mechanism of this ammonia formation comes from the work of Lipmann and his group (J3), who showed the synthesis of carbamyl phosphate from carbon dioxide, ammonia, and ATP (Fig. 3) to be dependent upon the carbon dioxide content of the medium. The enzyme carbonic anhydrase is required to liberate carbon dioxide from the bicarbonate ion, the form in which it is present at the pH of maximum activity of this enzyme. A carbonic anhydrase inhibitor should, therefore, prevent the

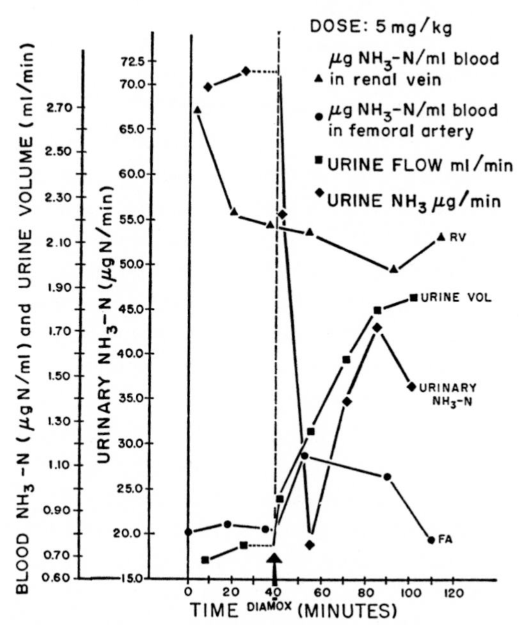

FIG. 2. Effect of intravenous acetazoleamide on renal ammonia metabolism (dog).

$$CO_2 + ATP + NH_3 \longrightarrow H_2\text{-}N\text{-}\overset{\overset{O}{\parallel}}{C}\text{-}O\text{-}\overset{\overset{O}{\parallel}}{\underset{\underset{OH}{|}}{P}}\text{-}OH + ADP$$

FIG. 3. Synthesis of carbamyl phosphate.

formation of carbamyl phosphate. This would lead to increased ammonia in the blood, due to the fact that the ammonia formed in oxidation of amino acids in muscle had not been utilized for the formation of carbamyl phosphate in the first step in the synthesis of urea. Further research on the by now well-documented specific toxicity of acetazoleamide for patients with liver disease promises to yield a great deal of information on metabolic interrelations.

2.3.3. *Muscle*

A third source of ammonia in the organism is muscle, and here, historically, arose the first interest of physiologists in the problem of ammonia metabolism. It was early noted in the laboratories of Parnas (P1) and Embden (E3) that on muscle contraction, large amounts of ammonia appeared in the venous blood. Experiments showed that the levels of ammonia in the arm veins could be increased from half a microgram of ammonia nitrogen per milliliter to 2 and 3 μg/ml during exercise equivalent to pressing a ball of clay between the fingers for a few minutes (P6). This outflow of ammonia was the subject of considerable discussion between the two groups of investigators because the source of the ammonia was obscure. Parnas discovered that the amount of ammonia which appeared in the muscle of animals when the muscle was ground with sand (he called it the "traumatic formation of ammonia"), was exactly equivalent to the purine amino groups which were available in the untraumatized muscle (P2). Gerhardt Schmidt then discovered that there was a specific muscle enzyme which deaminated adenylic acid (S2) to form inosinic acid, and it was shown that the amount of ammonia which was formed in the traumatized muscle was exactly equivalent to the inosinic acid which appeared and to the adenylic acid which disappeared (P3).

It was the contention of Parnas' group that once adenylic acid was deaminated, it would no longer be resynthesized in the muscle into adenylic acid (P4). On the other hand, Embden's group contended, from very meager evidence, that there was a resynthesis of adenylic acid from inosinic acid in frog muscle (E4). This problem is as yet unsolved. Lieberman's work with bacterial systems shows that there are at least two pathways for the amination of nucleotides. One is the formation of adenylosuccinate sparked by guanosine triphosphate (L3) and the second is apparently a direct amination of uridine triphosphate to cytidine triphosphate (L2). There is no indication that either of these systems is active in mammalian tissues. There are, thus, three major sources of ammonia in the organism: the resting gut with its normal hydrolytic activity and its normal content of organisms; the kidney, in relation to the secretion of ammonia in the urine; and the muscle, as a result of its contractile activity.

A word might be said at this point concerning the work reported from Vrba's laboratory (V7, V8). An increase of approximately 5% in the ratio of total protein nitrogen to protein amide nitrogen in rat brain following exercise is considered by these investigators to be a significant change and is interpreted as a liberation of energy from the "semimacroergic" amide bonds of protein. These bonds are considered to be a reservoir of available energy similar to the "macroergic bonds" of ATP. Glutamine synthesis and amide transfer to protein are proposed as the route of replenishment of this reservoir. By dependence upon values based on small differences in large numbers, this proposal can founder in the same manner as the ill-fated experiments of Bliss on the amide nitrogen of blood (B13). The work reported by Waelsch (C4) shows that protein amide groups can undergo turnover, and Vrba's suggestions could lead to a clearer understanding of the role of protein amide groups.

3. Pathological Alterations in Ammonia Metabolism

3.1. Hepatic Failure

The first human experiments which suggested a toxic effect of ammonia in disease were done by Van Caulaert. The revival and extension of this work by the group led by Davidson (G2) in the Boston City Hospital has stimulated widespread interest in ammonia as a factor in the production of mental symptoms in liver disease. The ability of various factors, such as urea feedings, high-protein diet, cation resins in the ammonia cycle, and amino acids, to induce symptoms of coma in patients with liver disease (G2, M3, M4, P7, S8) made it quite clear that ammonia was associated with the symptom complex called hepatic coma. The severe toxicity of ammonia in animals and the ability of intravenous or oral ammonium salts to provoke episodes of impending liver coma tended to substantiate the clinical impressions. Rapid confirmation of these observations was furnished by the experiments of other groups (B11, C2, E2, F1).

3.2. Cardiac Failure

The significance of ammonia in clinical disease was broadened markedly when it was discovered that ammonia poisoning might be the mechanism of the cerebral symptoms associated with chronic heart failure. It has been known for a long time that the mental symptoms associated with heart failure could not be correlated with the oxygen supply to the brain, which remains, in most cases, adequate. The work of A. N. Bessman (B4) demonstrated that the ammonia content of the blood was elevated in heart failure, probably due to the chronic passive congestion of the liver which prevented the liver from removing the normally formed ammonia from the portal system. When the blood ammonia fell, the mental symptoms of heart failure were relieved. This has been confirmed by Calkins and Delph (C1), who studied twenty-six cases of heart failure and found the blood ammonia to be elevated only in the two patients who had mental symptoms.

3.3. Shock

It has long been known that the adenylic deaminase of blood is increased in shock (T1). Seligson's experiments with animals in hemorrhagic shock demonstrated that the peripheral blood ammonia was markedly elevated (N4). Further work with this problem showed that the blood ammonia rises during hemorrhagic shock to tremendous levels and that these levels are compatible with the cerebral symptoms noted in shock (H2). Retransfusion of the bled animal does not cause this ammonia level to return completely to normal, and in fact it remains elevated to toxic levels until the death of the animal. A source of this ammonia has been shown to be the intestinal tract, for the highest rise of ammonia is found in the portal system.

3.4. Asparagine Intoxication

Another source of ammonia intoxication was found to be the ill-fated treatment of epilepsy with asparagine. Although the treatment was shown to be ineffective, and although asparagine administration caused practically no symptoms in most patients, there were some who developed severe nausea, vomiting, and mental symptoms during the course of the administration of large doses of oral asparagine. Experiments in this laboratory [(V2); Fig. 4, Curve *A*] have shown that the patients

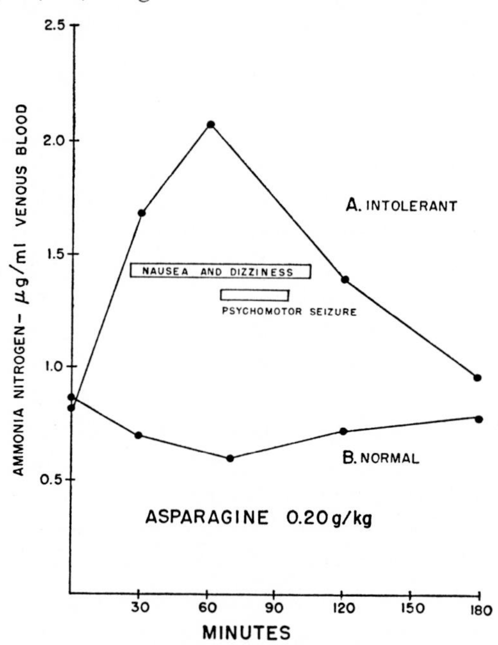

Fig. 4. Blood ammonia levels and oral asparagine toxicity.

who developed symptoms of intoxication with asparagine were the ones who developed a high blood ammonia level following oral administration of a single test dose of this medication. Normal individuals, as was shown by previous studies (B2), do not show a rise of blood ammonia following asparagine, but actually show a fall (Fig. 4, Curve *B*). This peculiar phenomenon, in which the administration of an amino acid containing a labile amide group plus an alpha-amino group, in other words, two "available" ammonia radicals, causes a fall in blood ammonia, is difficult to explain except in an indirect manner in its effect upon the urea cycle. The effect of asparagine in facilitating the utilization of ammonia could be interpreted on the basis of two experimental facts. The first is that aspartic acid, the amino acid resulting from the deamidation of asparagine, is essential for the synthesis of urea. The second fact is that asparagine is able to permeate into tissues more easily than is aspartic acid itself. There is only one difficulty with this explanation and that is that we know of no mechanism in tissues for the direct deamidation of asparagine.

3.5. Ureterocolic Anastomosis

Ammonia also increases in blood as a result of a pathologic alteration in physiology in cases of ureterocolic anastomosis in which the urinary outflow is surgically shunted into the colon because of some disease involving the bladder or urethra. This results in a return to the gut of the enormous quantities of urea normally formed in the organism. Usually this urea is excreted in the urine in a bacteria-free system which does not permit its reconversion to ammonia. The colon, however, contains large numbers of organisms which can break down urea to form ammonia. The consequence of such a shunt is the continued inundation of the organism with ammonia which has been synthesized into urea in the liver and excreted in the urine, but is again poured into the intestinal tract, there to be broken down and recycled to the portal system, into the liver again. The continued recycling of urea overcomes the normal urea-synthesizing system, and some of the ammonia coming from the colon could very well escape being converted again to urea by an overworked or diseased liver. This mechanism for the toxicity developed in ureterocolic anastomosis was postulated on theoretical grounds (B6), and clinical evidence for such a phenomenon has recently been adduced (M2).

3.6. Erythroblastosis Fetalis

The most recent addition to the clinical syndromes in which ammonia probably plays a dominant role is hemolytic disease of the newborn.

This disease usually results from autoimmunization of the mother because of an incompatibility between the fetal blood and the maternal blood, but can also result from severe sepsis in the early days of life. The jaundice which accompanies the hemolysis of blood frequently causes bile staining of the brain tissue, particularly the nuclei of the basal ganglia and Ammon's horn, producing the pathological condition known as kernicterus. The staining of the brain with bilirubin is invariably associated with mental deficiency and neurological symptoms. Attempts have been made to correlate the development of kernicterus with the level of bilirubin in the blood, but only a rough correlation exists. Patients with very high levels of blood bilirubin frequently do not get kernicterus, whereas some patients with relatively low levels do.*

An investigation of the ammonia metabolism of newborns with erythroblastosis fetalis has been conducted in this laboratory along the following lines (G5). All of the patients with severe or moderate erythroblastosis are subjected to a standard exchange transfusion with the administration and withdrawal of 500 ml of blood. Before, and during the course of, the transfusion ammonia determinations are made on the umbilical venous blood. In those patients who developed kernicterus or who died with cerebral complications, the umbilical venous ammonia exceeded 2 μg/ml. One patient developed a blood ammonia level of 2 μg/ml and survived. This patient had a sibling who died with kernicterus. All of the other 18 patients had normal or only slightly elevated blood ammonia levels, and these patients had no complications. A possible sequence of events leading to the development of kernicterus is presented in Fig. 5.

4. Theories of the Mechanism of Hepatic Coma

4.1. Thiol Intoxication

In 1949, C. J. Watson (W6) reported that methionine was peculiarly toxic in liver failure. Sherlock (S5) reported similar findings, and these observations led Challenger and Walshe (C3) to search for volatile thiols in the breath of patients with fetor hepaticus. The toxicity of mercaptans, presumably derived from methionine and cystine, was proposed as the cause of coma, and it was demonstrated that the breath of patients with fetor hepaticus contained methyl mercaptan. This observation, coupled with the finding of increased thiol levels in the blood of patients with hepatic disease, and a decreased ability to utilize cysteine (W4) gave some support to the proposal. The report that methionine

* See chapter by Barbara Billing, pages 275, 284.

exerted a toxic effect upon patients with liver disease stated that the effect was far greater than, and unrelated to, any ammonia intoxication which might derive from the amino group (S5). There are several defects in the thiol intoxication theory, however. In the first place, thiols are not seriously toxic per se. Furthermore, fetor hepaticus, considered

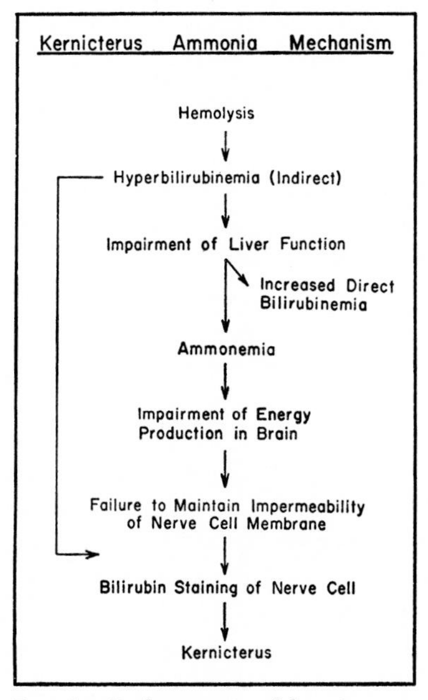

FIG. 5. Pathogenesis of kernicterus.

to be a major sign of thiol intoxication, is not a uniform concomitant of liver failure. Even the original report (W4) showed that most patients with liver coma did not have any significant increase of blood thiols. Recent observations by Webster (W7) on the ammonia content of portal collateral vessels in cirrhotics, following oral doses of various nitrogenous substances, show that the portal ammonia is markedly elevated by the oral administration of methionine (Fig. 6.). This figure also shows the dramatic effect of orally administered neomycin, which destroys most of the ammonia-forming organisms of the gut.

4.2. ATP Exhaustion Theory

A second hypothesis for the mechanism of hepatic coma is more specific about its site of action. It proposes a depletion of ATP available in brain, caused by the utilization of ATP in the detoxication of ammonia by brain by the synthesis of glutamine (S7). This depletion of ATP, proposed by Weil-Malherbe (W9) results in a deficiency of high-energy phosphate bonds for synthesis of other intermediates such as acetylcholine. Evidence which led to this hypothesis came from the observation of Quastel's group (M1) that glutamate plus ammonia caused a

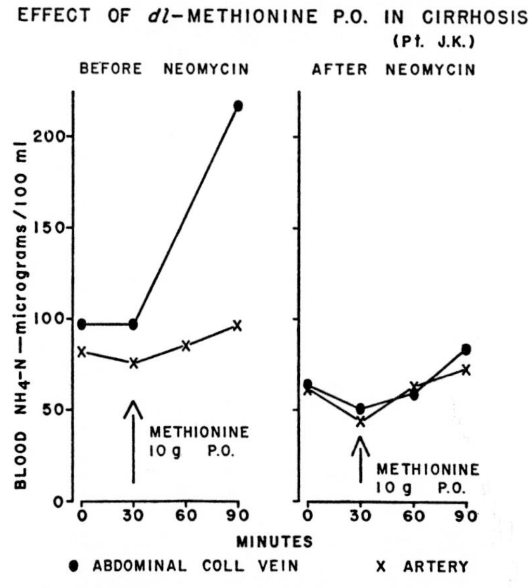

FIG. 6. Effect of methionine on gut formation of ammonia: (●–●–●) abdominal collateral vein; (x–x–x) artery.

drop in synthesis of acetylcholine by brain tissue reversed by addition of ATP.

More recently Braganca, Faulkner, and Quastel (B15) showed that this inhibition of acetylcholine synthesis in brain slices by ammonia is consistent only in the diminution of "bound" acetylcholine. They further showed that the addition of inhibitors of glutamine synthesis, such as methionine sulfoxide, ethionine sulfoxide, and methionine sulfoximine (the toxic product, causing convulsions, formed in flour chemically "aged" with nitrogen trichloride) would partially reverse the ammonium inhibition. These observations were confirmed and extended to a wide variety of ATP-requiring reactions by Weil-Malherbe. In support of this suggestion was the observation that ammonia is taken up by the brain in hepatic coma (B8). The observations are valid and have been confirmed, but the interpretation of the data and the hypothesis are questionable. A quantitative basis for the evaluation of this mechanism can be obtained from simultaneous measurement of oxygen and ammonia uptake by brain [Table 4; (B6)].

TABLE 4
CEREBRAL OXYGEN UPTAKE IN HEPATIC COMA

Condition	Arteriovenous difference of ammonia μg/ml	Oxygen uptake ml/100 g/min	Blood flow ml/100 g/min
Normal	0 (+0.05–0.05)	3.2 (3.0–3.4)	52 (47–57)
Coma	0.55 (0.03–2.40)	1.85 (0.8–2.4)	46.6 (34–55)

The normal brain takes up approximately 3 ml of O_2 per hundred grams per minute. This would correspond to about 268 micro gram-atoms of oxygen per hundred grams per minute. At a high energy phosphate-generating capacity of 4 phosphate bonds per atom of oxygen (P/O ratio of 4), about 1000 phosphate bonds would be generated per hundred grams per minute. At an efficiency of 60 % (P/O ratio of 2.4), this would be 600 bonds per minute. The oxygen uptake of the brain in hepatic coma is about half of normal (W10) and therefore would make, at the same efficiency, about 300 bonds per hundred grams per minute. Even at a lower efficiency of 30 % (P/O of 1.2), 150 bonds per minute would be made. Turning to the uptake of ammonia, we find that the greatest uptake of ammonia is about one microgram per milliliter across the brain, and this could correspond, at a blood flow of about 40 ml/100 g/min, to about three micromoles of ammonia. Since one bond of ATP is utilized per mole of glutamine synthesized (C3), it would require at most three micromoles of high-energy bonds to "use up" all of the ammonia taken up. The average brain in hepatic coma takes up less than half a microgram per milliliter. In the face of a generation of, at minimum, over 100 bonds/min a maximum loss of three would hardly be of significant proportions. This depletion might be of importance if there were some impairment of phosphate-energy generation at the same time. This would not be accounted for, however, by the hypothesis that hepatic coma is caused by depletion of ATP due to synthesis of glutamine.

4.3. Exogenous and Endogenous Coma

A third interpretation for the mental depression in hepatic failure is suggested by McDermott's group (M5). The impression gained by the clinical observations of these investigators is that the coma caused by administration of protein and the coma caused by bleeding into the gastrointestinal tract, both of which are called "exogenous coma," are more responsive to therapy than the coma which terminates most cases of severe liver disease. This second type of coma, called by the authors "endogenous coma," is implied to be caused by some agent other than ammonia. The therapeutic agent which was effective in the acute "exogenous" cases and ineffective in the "endogenous" cases was L(+)-glutamate. This separation of types of coma on the basis of a therapeutic test would have some validity if the agent actually were ineffective. But the data show all of the cases of spontaneous coma (endogenous) to have had a transient fall in blood ammonia and, in general, a parallel improvement. The difference between the two groups is not qualitative, but quantitative. Those who had the so-called exogenous coma had more prolonged effect of glutamate in lowering the blood ammonia than those who had "endogenous" coma.

It is clear, however, that the quantitative difference described by McDermott (M5) is proportional to the amount of destruction and repair present in a particular liver, rather than to a nebulous farrago of "complex biochemical changes." We are still dealing with ammonia intoxication in the cases reported. This is not to deny the very real and as yet minimally understood relation of the liver to brain metabolism (*vide infra*), but it is essential to the development of a lucid picture of disease as a biochemical phenomenon that we do not convert variations of the same process into a multiplicity of syndromes.

In the matter of bleeding into the gastrointestinal tract, work by Alice N. Bessman and G. S. Mirick (B5) shows that identical amounts of nitrogen in the form of protein hydrolysate or whole blood fed to the same patient produce different increases of ammonia in the blood. In all cases blood caused a marked rise in blood ammonia, far greater than that caused by the protein hydrolysate. This finding can explain the difference between the coma evidenced by gastrointestinal hemorrhage (Y1) without recourse to a different mechanism than ammonia poisoning. The cause of the difference in the formation of ammonia from a whole protein mixture like blood and a hydrolyzed protein mixture is obscure at present, but the investigators are of the opinion that it relates to the glutamate content of the materials given.

4.4. Krebs Cycle Depletion Theory

A lesion of the Krebs Cycle caused by ammonia has been proposed by Bessman and Bessman as the mechanism of ammoniagenic coma and some evidence has been advanced for it (B8). The site of attack of ammonia is considered to be the glutamic dehydrogenase reaction. By this reaction ammonia plus α-ketoglutaric acid form glutamic acid, of which the brain has a greater quantity than any other organ (W1). This formation of glutamate would be quantitatively minimal, as shown from the discussion of the ATP exhaustion theory above, but from another standpoint might be highly significant. There is only a small amount of α-ketoglutarate in brain, and it would not be difficult to deplete this intermediate if there were no way to replenish it, for there are only about one hundred micromoles per hundred grams of brain (E1). Figure 7 shows

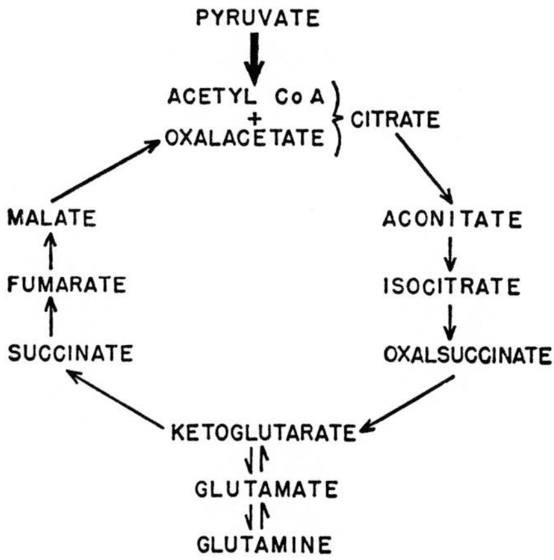

FIG. 7. Glutamate and glutamine synthesis in relation to the Krebs cycle.

that this intermediate is constantly regenerated, so that at first glance it appears impossible to deplete it. It should be noted, however, that the four-carbon chain is never degraded in the Krebs Cycle, and loss of any intermediate at any point would result in a net loss of four-carbon compounds from the cycle. Examination of the various pathways for replacement of Krebs Cycle intermediates reveals the singular fact that the only significant pathway for the replenishment of the Krebs Cycle *in brain* is the glutamic dehydrogenase reaction (B6) which is reversed by ammonia and is, in ammonia poisoning, the very pathway causing the depletion. Under such circumstances it is understandable that the ketoglutarate of the brain would suffer attrition by the constant loss to form glutamate. Confirming evidence of this depletion hypothesis comes from the experiments of Eiseman who demonstrated a 50 % fall in brain content of ketoglutarate in animals following intracarotid infusion of ammonium salts (E1). This has also been shown to occur in mouse brain following intraperitoneal injection of ammonium salts in the author's laboratory (B1).

An objection to the Krebs Cycle depletion theory of ammonia poisoning has been raised on the basis of the clinical observation (F2) that there is not a direct parallel between blood ammonia level and depth or incidence of coma. Although some of the discrepancy can be accounted for by the previously reported fact (B9) that the arterial levels of ammonia are more reliable indicators of the conditions affecting the brain, the implications of the Krebs Cycle depletion mechanism proposed must be examined. If a lesion develops as a result of depletion of a substrate it will manifest itself in relation to this substrate and only remotely in relation to the factors causing the depletion. The constant slow attrition caused by a small increase of blood ammonia over a long period could be more damaging than the loss produced by a sudden, greater, transitory rise of blood ammonia. Furthermore, even though blood ammonia may fall to normal, the repair processes to replenish brain α-ketoglutarate might require several days before a level of this intermediate compatible with normal mental function could be attained. The causal relation of many toxic substances to the symptoms evoked by them could be explained on such a basis of attrition and replenishment of essential metabolites.

4.5. Alkalosis and Coma

Recent observations have shown a direct relation between the pH of the blood and the depth of hepatic coma (V1). They are interpreted as evidence that the intracellular ammonia is increased in alkalosis because it is un-ionized ammonia and not ammonium ion which readily diffuses into cells. This finding may be a result of the central action of ammonium on the respiratory mechanism (F7), rather than the primary cause of the coma. In this sense it would seriously enhance the coma and set up a cycle which would be lethal. However this may be, there is no mechanism presented for the actual toxic effect of ammonia per se, and the data, except for the internal inconsistencies noted in a previous part of this discussion, would be compatible with the ketoglutarate depletion hypothesis.

4.6. Ammoniagenic Coma and Other Forms

It must not be assumed from the foregoing that ammonia is the only cause of hepatic coma, however. Recently evidence has been presented that hepatic failure can be due to a deficiency of certain trace amino acids, precursors of neurohumors normally synthesized in the liver and sent to the brain (B14). As the liver ceases to function, the supply of these trace amino acids becomes exhausted and the neurohumors of the brain diminish. The concept of hepatic disease as a deficiency syndrome has long been considered. It remains to determine exactly what is deficient.

5. Methods for the Alteration of the Blood Ammonia

5.1. Direct Methods

Direct methods tried for the diminution of blood ammonia include chemical therapy and vivodialysis.

5.1.1. *Glutamate*

The first chemical treatment was reported by Sapirstein (S1), who demonstrated in rabbits that sodium glutamate administered prior to or during ammonium chloride infusions delayed or prevented the convulsions regularly seen with untreated controls. Walshe (W2) suggested the use of sodium glutamate to remove ammonia in the ammoniagenic coma of liver failure because ammonia is normally combined with glutamate to form glutamine. Early reports of this therapy were promising (W3), but experience has shown it to be of only limited value (E2). A controlled study (B12) of the effect of glutamate on the removal of ammonia from the blood of cirrhotic patients showed that the uptake of ammonia by the muscle was enhanced by glutamate. This was interpreted as an adaptive mechanism developed in cirrhotics as a result of liver failure and moderately elevated blood ammonia. The muscle appears to increase total glutamine synthesis adaptively and thereby substitute for a small part of the decrease in urea synthesis by the liver. This implies that glutamate therapy would be most effective in cirrhosis with mild failure and least effective in ammonemia developing acutely in previously normal individuals who get liver failure.

5.1.2. *Arginine*

A second type of therapy was suggested by the experiments of Greenstein and his group (G6), who showed that highly toxic doses of amino acids and ammonia could be rendered innocuous by stimulation of the urea synthesis cycle in the liver by administration of arginine orally (B7) or parenterally (D5, H3, N1, N2). Compared to glutamate, arginine is much less effective in cirrhotics (B12), and has no effect on peripheral utilization of ammonia. This has been interpreted as indicating that cirrhotics who have had a long-standing elevation of blood ammonia have adapted their muscle glutamine synthesis. On the other hand, the use of arginine in acute hepatitis may well be indicated, for this type of patient has no adapted peripheral mechanism to utilize glutamate. It might be possible to stimulate the depleted liver enzymes somewhat with arginine, however, and an analysis of the cases reported suggests that arginine is most effective in acute liver failure.

5.1.3. *Arginine Glutamate*

Arginine is a strongly basic amino acid, and must be administered as the hydrochloric acid salt. Glutamic acid is a strong acid and must be given as the sodium or potassium salt. Recently, an ingenious combination of the two amino acids, arginine and glutamate, utilizing these properties to neutralize each other, has been prepared.* Preliminary clinical experience shows that it has about the effectiveness of either of the two medications alone, which is to be expected. It can be used whenever the salt forms of these amino acids would embarrass the patient with electrolyte imbalance.

5.1.4. *Vivodialysis*

Vivodialysis of patients with hepatic coma and ammonemia can be shown to remove a considerable amount of ammonia (K1), but the gradient is so small and the rate of the production of ammonia so great that at present vivodialysis is of no value in the treatment of hepatic coma. That it might even be deleterius is suggested by the evidence pointing to a deficiency syndrome in hepatic coma in addition to the ammonia intoxication (B14).

5.2. Indirect Methods

Indirect methods of lowering blood ammonia are, in the main, more effective than the chemical techniques. These include withdrawal of protein from the diet (M3, P7) and antibiotic therapy (F3, S6). A completely protein-free diet leaves little nitrogen in the gut for bacterial action. Antibiotic therapy, particularly by the oral route (S10, W5), so alters the gut flora as to diminish the number of ammonia-forming organisms. It must be kept in mind that urea is always present in the gut as a substrate for ammonia formation even on a nitrogen-free diet.

References

A1. Abelin, I., Über die Mikrobestimmung des Harnstoffes im Blute ohne Destillation und ohne Nesslerisation. *Biochem. Z.* **297**, 203 (1938).

B1. Bachur, N., and Bessman, S. P., unpublished (1957).

* Prepared by Gray Pharmaceutical Company, Worcester, Massachusetts.

B2. Baldwin, R., Van Buskirk, C., and Bessman, S. P., Observations in the author's laboratory (1957).

B3. Balo, J., Korpassy, B., The encephalitis of dogs with Eck fistula fed on meat. *A.M.A. Arch. Pathol.* **13**, 80 (1932).

B4. Bessman, A. N., and Evans, J. M., Blood ammonia in congestive heart failure. *Am. Heart J.* **50**, 715 (1955).

B5. Bessman, A. N., and Mirick, G. S., *J. Clin. Invest.* **37**, 990 (1958). Presented at The American College of Physicians, Richmond, Va., October 13, 1957.

B6. Bessman, S. P., Ammonia metabolism in animals. *In* "Nitrogen Metabolism" (W. McElroy and B. Glass, eds.), p. 433. Johns Hopkins, Baltimore, 1955.

B7. Bessman, S. P., The reduction of blood ammonia levels by certain amino acids. *J. Clin. Invest.* **35**, 69 (1956).

B8. Bessman, S. P., and Bessman, A. N., The cerebral and peripheral uptake of ammonia in liver disease with an hypothesis for the mechanism of hepatic coma. *J. Clin. Invest.* **32**, 622 (1955).

B9. Bessman, S. P., and Bradley, J. E., The uptake of ammonia by muscle, its implications in ammoniagenic coma. *New Engl. J. Med.* **253**, 1143 (1955).

B10. Bessman, S. P., and Stauffer, J., Factors affecting the appearance of ammonia in the gastric juice. *J. Clin. Invest.* **36**, 874 (1957).

B11. Bessman, S. P., Fazekas, J., and Bessman, A. N., Uptake of ammonia by brain in hepatic coma. *Proc. Soc. Exptl. Biol. Med.* **85**, 66 (1954).

B12. Bessman, S. P., Shear, S., and Fitzgerald, J., Effect of arginine and glutamate on removal of ammonia from blood in normal and cirrhotic patients. *New Engl. J. Med.* **256**, 941 (1957).

B13. Bliss, S., The amide nitrogen of blood. *Science* **67**, 515 (1928).

B14. Borges, F., Merlis, J., and Bessman, S. P., The effect of 5-hydroxytryptophane on the electroencephalogram in hepatic coma. *Proc. Soc. Exptl. Biol. Med.* **95**, 502 (1957).

B15. Braganca, B. J., Faulkner, P., and Quastel, J. H., Effects of inhibitors of glutamine synthesis on inhibition of acetylcholine synthesis in brain slices by ammonium ions. *Biochim. et Biophys. Acta* **10**, 83 (1953).

B16. Brown, R. H., Duda, G. D., Korkes, S., and Handler, P., A colorimetric method for the determination of ammonia; the ammonia content of rat tissues and human plasma. *Arch. Biochem. Biophys.* **66**, 301 (1957).

B17. Burchi, R., Prüfung der Leberfunktion durch Untersuchung der spontanen und provozierten Ammoniemie. *Kongr. inn. Med.* **47**, 80 (1927).

C1. Calkins, W. G., and Delph, M., Blood ammonia levels in congestive heart failure. *Ariz. Med.* **12**, 470 (1956).

C2. Carfagno, S. C., de Horatius, R. F., Tompson, C. M., and Schwartz, H. P., Hepatic coma. A clinical laboratory and pathological study. *New Engl. J. Med.* **249**, 303 (1953).

C3. Challenger, F., and Walshe, J. M., Foetor hepaticus. *Lancet* **91239** (1955).

C4. Clarke, D. D., Neidle, A., Sarkar, N. K., and Waelsch, H., Metabolic activity of protein amide groups. *Arch. Biochem. Biophys.* **71**, 277 (1957).

C5. Conway, E. J., "Microdiffusion and Volumetric Error." C. Lockwood, London, 1947.

C6. Conway, E. J., and Byrne, R., Absorption apparatus for microdetermination of certain volatile substances; microdetermination of ammonia. *Biochem. J.* **27**, 419 (1933).

D1. Damodaran, M., The isolation of asparagine from an enzymic digest of edestin. *Biochem. J.* **26**, 235 (1932).

D2. Damodaran, M., and Narayanan, E. K., Enzymic proteolysis III. Hydrolysis of asparagine, asparagine peptides and anhydroglycyl-asparagine. *Biochem. J.* **32**, 2105 (1938).

D3. Davies, B. M. A., and Yudkin, J., The origin of urinary ammonia as indicated by the effect of chronic acidosis and alkalosis on some renal enzymes in the rat. *Biochem. J.* **52**, 407 (1952).

D4. Davies, R. E., and Kornberg, H. L., Gastric urease and HCl secretion. *Biochem. J.* **47**, ii (1950).

D5. Du Ruisseau, J. P., Greenstein, J. P., Winitz, M., and Birnbaum, S., Free amino acid levels in the tissues of rats protected against ammonia toxicity. *Arch. Biochem. Biophys.* **68**, 161 (1957).

E1. Eiseman, B., *In* "The Chemical Environment of the Brain," 24th Ross Symposium, Baltimore, March 30, 1957.

E2. Eiseman, B., Bakewell, W., and Clark, G., Studies in ammonia metabolism I. Ammonia metabolism and glutamate therapy in hepatic coma. *Am. J. Med.* **20**, 890 (1956).

E3. Embden, G., and Wassermeyer, H., Über die Bedeutung der Adenylsäure für die Muskelfunktion; das Verhalten der Ammoniakbildung bei der Muskelarbeit unter verschiedenen biologischen Bedingungen. *Z. physiol. Chem.* **179**, 161 (1928).

E4. Embden, G., Carstensen, M., and Schumacher, H., Über die Bedeutung der Adenylsäure für die Muskelfunktion; Spaltung und Wiederaufbau der ammoniakbildenden Substanz bei der Muskeltätigkeit. *Z. physiol. Chem.* **179**, 186 (1928).

F1. Faloon, W. W., Auchincloss, J. H., Eich, R., and Gilbert, R., Ammonia metabolism in cirrhosis patients with portacaval shunt. *Clin. Research Proc.* **4**, 18 (1956).

F2. Fisher, C. J., and Faloon, W. W., Episodic stupor following portacaval shunt —Observations on etiology and therapy. *New Engl. J. Med.* **255**, 589 (1956).

F3. Fisher, C. J., and Faloon, W. W., Blood ammonia levels in hepatic cirrhosis; their control by the oral administration of neomycin. *New Engl. J. Med.* **256**, 1030 (1957).

F4. Folin, O., The determination of ammonia in blood. *J. Biol. Chem.* **39**, 259 (1919).

F5. Folin, O., and Denis, W., New methods for the determination of total NPN, urea, and ammonia in blood. *J. Biol. Chem.* **11**, 527 (1912).

F6. Folin, O., and Denis, W., The origin and significance of ammonia in the portal blood. *J. Biol. Chem.* **11**, 161 (1912).

F7. Forbes, G. B., and Erganian, J. A., Parenteral administration of ammonium chloride for alkalosis of congenital Pyloric stenosis. *A.M.A. J. Diseases Children* **72**, 649 (1946).

G1. Gabuzda, G. J., Jr., Phillips, G. B., and Davidson, C. S., Reversible toxic manifestations in patients with cirrhosis of the liver given cation exchange resins. *New Engl. J. Med.* **246**, 124 (1952).

G2. Gaustad, V., Transient hepatargy. *Acta Med. Scand.* **135**, 354 (1949).

G3. Glick, D., Studies in histochemistry; urease in human stomach with respect to acid secretion in ulcer and cancer. *J. Natl. Cancer Inst.* **10**, 321 (1949).

G4. Glick, D., Zak, E., and Von Korff, R., The role of urease in the gastric mucosa. *Am. J. Physiol.* **163**, 386 (1950).

G5. Gorten, M., Shear, S., Hodsdon, M., and Bessman, S. P., The complications of hyperbilirubinemia and their possible relation to the metabolism of ammonia. *Pediatrics* **21**, 1 (1958).

G6. Gullino, P., Winitz, M., Birnbaum, S., Cornfield, J., Otey, M. C., and Greenstein, J. P., The toxicity of essential amino acids with special reference to the protective effect of L-arginine. *Arch. Biochem. Biophys.* **58**, 255 (1955).

H1. Hahn, M., Massen, D., Nencki, M., and Pawlow, I., Die Eck'sche Fistel zwischen der unteren Hohlvene und der Pfortader, und ihre Folgen für den Organismus. *Arch. Exptl. Pathol. u. Pharmakol.* **32**, 161 (1893).

H2. Hankins, J., Bessman, S. P., Esmond, W., Mansberger, A., and Cowley, R. A., Origin and utilization of ammonia in shock. Submitted to Intern. Soc. Surg., Mexico City, October 1957.

H3. Harper, H. A., Najarian, J. S., and Silen, W., Effect of intravenously administered amino acids on blood ammonia. *Proc. Soc. Exptl. Biol. Med.* **92**, 558 (1956).

H4. Hench, P. S., and Aldrich, M., The concentration of urea in saliva. *J. Am. Med. Assoc.* **79**, 1409 (1922).

J1. Jacobs, M. H., The exchange of materials between the erythrocyte and its surroundings. *Harvey Lectures* **22**, 146 (1927).

J2. Jacobs, M. H., and Parpart, A. K., Osmotic permeability of the erythrocyte. X. On the permeability of the erythrocyte to ammonia and the ammonium ion. *J. Cellular Comp. Physiol.* **11**, 175 (1938).

J3. Jones, M. E., Spector, L., and Lipmann, F., Carbamyl phosphate, the carbamyl donor in citrulline synthesis. *J. Am. Chem. Soc.* **77**, 821 (1955).

K1. Kiley, J. E., Welch, H. F., Pender, J. C., and Welch, C. S., Removal of blood ammonia by hemodialysis. *Proc. Soc. Exptl. Biol. Med.* **91**, 489 (1956).

K2. Kirk, E., Amino acid and ammonia metabolism in liver disease. *Acta Med. Scand.* Suppl. 78 (1936).

K3. Kornberg, H. L., and Davies, R. E., Gastric urease. *Physiol. Revs.* **35**, 169 (1955).

K4. Krebs, H. A., and Henseleit, K., Untersuchungen über die Harnstoffbildung im Tierkörper. *Z. physiol. Chem.* **201**, 33 (1932).

L1. Lawrence, W., Jr., Jacquez, J. A., Dienst, S. G., Poppell, J. W., Randall, H. T., and Roberts, K. E., The effect of changes in blood pH on the plasma total ammonia level. *Surgery* **42**, 50 (1957).

L2. Lieberman, I., Enzymatic amination of uridine triphosphate to cytidine triphosphate. *Am. Chem. Soc.* **77**, 2661 (1955).

L3. Lieberman, I., Involvement of guanosine triphosphate in the synthesis of adenylosuccinate from inosine-5′-phosphate. *J. Am. Chem. Soc.* **78**, 251 (1956).

L4. Loeb, R. F., Atchley, D. W., and Benedict, E. M., Observations on the origin of urinary ammonia. *J. Biol. Chem.* **60**, 491 (1924).

L5. Lubochinsky, B., and Zalta, J. P., Microdosage colorimetrique de l'azote ammoniacal. *Bull. soc. chim. biol.* **36**, 1363 (1954).

M1. Mann, P. J. G., Tennenbaum, M., and Quastel, J. H., Acetylcholine metabolism in central nervous system; effects of potassium and other cations on acetylcholine liberations. *Biochem. J.* **33**, 822 (1939).

M2. McDermott, W. V., Jr., Diversion of urine to the intestines as a factor in ammoniagenic coma. *New Engl. J. Med.* **256**, 460 (1957).

M3. McDermott, W. V., Jr., and Adams, R. D., Episodic stupor associated with an Eck fistula in the human, with particular attention to ammonium metabolism. *J. Clin. Invest.* **33**, 1 (1954).

M4. McDermott, W. V., Jr., Adams, R. D., and Riddell, A. G., Ammonia metabolism in man. *Ann. Surg.* **140**, 539 (1954).

M5. McDermott, W. V., Jr., Wareham, J., and Riddell, A. G., Treatment of hepatic coma with L-glutamic acid. *New Engl. J. Med.* **253**, 1093 (1955).

M6. Monguio, M., and Krause, F., Über die Bedeutung des Ammoniagehaltes des Blutes für Beurteilung der Leberfunction. *Klin. Wochschr.* **13**, 1142 (1934).

N1. Najarian, J. S., and Harper, H. A., A clinical study of the effect of arginine on blood ammonia. *Am. J. Med.* **21**, 833 (1956).

N2. Najarian, J. S., and Harper, H. A., Comparative effect of arginine and monosodium glutamate on blood ammonia. *Proc. Soc. Exptl. Biol. Med.* **92**, 560 (1956).

N3. Nathan, D. G., and Rodkey, F. L., A colorimetric procedure for the determination of blood ammonia. *J. Lab. Clin. Med.* **49**, 779 (1957).

N4. Nelson, R. M., and Seligson, D., Studies of blood ammonia in normal and shock states. *Surgery* **34**, 1 (1953).

N5. Nencki, M., and Pawlow, I., Zur Frage über den Ort der Harnstoffbildung bei den Säugethieren. *Arch. exptl. Pathol. u. Pharmakol.* **38**, 215 (1896-97).

N6. Nencki, M., and Pawlow, I., Zur Frage über den Ort der Harnstoffbildung dei den Säugethieren. *Arch. exptl. Pathol. u. Pharmakol.* **37**, 26 (1895-96).

P1. Parnas, I. K., Über den Purinstoffwechsel des Muskels und über die Muttersubstanz des im Muskel entstehenden Ammoniaks. *Klin. Wochschr.* **7**, 2011 (1928).

P2. Parnas, I. K., Über den Purinstoffwechsel des Muskels und über die Muttersubstanz des im Muskel entstehenden Ammoniaks. *Klin. Wochschr.* **7**, 1423 (1928).

P3. Parnas, I. K., Über die Ammoniakbildung im Muskel und ihren Zusammenhang mit Funktion und Zustandsänderung; der Zusammenhang der Ammoniakbildung mit der Umwandlung des Adeninnucleotids zu Inosinsäure. *Biochem. Z.* **206**, 16 (1929).

P4. Parnas, I. K., Formation of ammonia during contraction of skeletal muscle. *Klin. Wochschr.* **11**, 335 (1932).

P5. Parnas, I. K., and Heller, J., Über den Ammoniakgehalt and über die Ammoniakbildung im Blute. *Biochem. Z.* **152**, 1 (1924).

P6. Parnas, I. K., Mozolowski, W., and Levinski, W., Über den Ammoniakgehalt und die Ammoniakbildung im Blute; der Zusammenhang des Blutammoniaks mit der Muskelarbeit. *Biochem. Z.* **188**, 15 (1927).

P7. Phillips, G. B., Schwartz, R., Gabuzda, G. J., Jr., and Davidson, C. S., The syndrome of impending hepatic coma in patients with cirrhosis of the liver given certain nitrogenous substances. *New Engl. J. Med.* **247**, 239 (1952).

R1. Richter, D., Adrenaline and amine oxidase. *Biochem. J.* **31**, 2020 (1937).

R2. Rosenoer, V. M., The measurement of ammonia in whole blood. *J. Clin. Pathol.* **12**, 128 (1959).

S1. Sapirstein, M. R., The effect of glutamic acid on the central action of ammonium ion. *Proc. Soc. Exptl. Biol. Med.* **52**, 334 (1943).

S2. Schmidt, G., Über fermentative Desaminierung im Muskel. *Z. physiol. Chem.* **179**, 243 (1928).

S3. Seligson, D., and Hirahara, K., The measurement of ammonia in whole blood, erythrocytes, and plasma. *J. Lab. Clin. Med.* **49**, 962-974 (1951).

S4. Seligson, D., and Seligson, H., A microdiffusion method for the determination of nitrogen liberated as ammonia. *J. Lab. Clin. Med.* **38**, 324 (1951).

S5. Sherlock, S., Summerskill, W. H. J., White, L. P., and Phear, E. A., Portal systemic encephalopathy; Neurological complications of liver disease. *Lancet* **ii**, 453 (1954).

S6. Silen, W., Harper, H. A., Maudsley, D. L., and Weirich, W. L., Effect of antibacterial agents on ammonia production with the intestine. *Proc. Soc. Exptl. Biol. Med.* **88**, 138 (1955).

S7. Speck, J. F., The enzymic synthesis of glutamine. *J. Biol. Chem.* **168**, 403 (1947).

S8. Stahl, J., Roger, S., and Witz, J., Essai sur l'interpretation de mechanisme de l'epreuve d'hyperammonemie provoquée chez les cirrhotiques. *Compt. rend. soc. biol.* **146**, 1787 (1952).

S9. Stone, W. E., A new colorimetric reagent for micro determination of ammonia. *Proc. Soc. Exptl. Biol. Med.* **93**, 589 (1956).

S11. Sweet, J. F., and Ringer, A. I., The influence of phlorrhizin on dogs with Eck fistula. *J. Biol. Chem.* **14**, 135 (1913).

T1. Trethwie, E. R., Ischaemia, anoxemia, and shock. *Australian J. Exptl. Biol. Med. Sci.* **26**, 291 (1946).

V1. Vanamee, P., Poppell, J. W., Glicksman, A. S., Randall, H. T., and Roberts, K. E., Respiratory alkalosis in hepatic coma. *A.M.A. Arch. Internal Med.* **97**, 762 (1956).

V2. Van Buskirk, C., Baldwin, R., and Bessman, S. P., Observations in the authors' laboratory, 1957.

V3. Van Caulaert, C., and Deviller, C., Ammonemie experimentale aprés ingestion de chlorure d'ammonium chez l'homme a l'état normal et pathologique. *Compt. rend. soc. biol.* **111**, 50 (1932).

V4. Van Caulaert, C., Deviller, C., and Halff, M., Troubles provoqués par l'ingestion de sels ammoniacaux chez l'homme atteint de cirrhose de Laennec. *Compt. rend. soc. biol.* **111**, 739 (1932).

V5. Van Slyke, D. D., and Hiller, A., Determination of ammonia in blood. *J. Biol. Chem.* **102**, 499 (1933).

V6. Van Slyke, D. D., Phillips, R. A., Hamilton, P. B., Archibald, R. M., Futcher, P. H., and Hiller, A., Glutamine as source material of urinary ammonia. *J. Biol. Chem.* **150**, 481 (1943).

V7. Vrba, R., A source of ammonia and changes in the rat brain during physical exertion. *Nature* **176** (1955).

V8. Vrba, R., On the participation of the glutamic acid-glutamine system in the metabolic processes in rat brain during physical exercise. *J. Neurochem.* **1**, 12 (1956).

W1. Waelsch, H., Glutamic acid and cerebral function. *Advances in Protein Chem.* **6**, 299 (1951).

W2. Walshe, J. M., The effect of glutamic acid on the coma of hepatic failure. *Lancet* **i**, 1075 (1953).

W3. Walshe, J. M., Glutamic acid in hepatic coma. *Lancet* **i**, 1235 (1955).

W4. Walshe, J. M., and Senior, B., Disturbances of cystine metabolism in liver disease. *J. Clin. Invest.* **34**, 302 (1955).

W5. Warren, K. S., The differential toxicity of ammonium salts. *J. Clin. Invest.* **37**, 497 (1958).

W6. Watson, C. J., Prognosis and treatment of hepatic insufficiency. *Ann. Internal Med.* **31**, 405 (1949).

W7. Webster, L. T., Davidson, C. S., and Gabuzda, G. J., Effect on portal blood ammonium of administered nitrogenous substances, etc. *J. Lab. Clin. Med.* **52**, 501 (1958).

W8. Webster, L. T., and Davidson, C. S., Production of impending hepatic coma by a carbonic anhydrase inhibitor. *Proc. Soc. Exptl. Biol. Med.* **91**, 27 (1956).

W9. Weil-Malherbe, H. *In* "Chemical Environment of the Brain," Ross Pediatric Conference, 24th Symposium on Psychological Implications etc., Baltimore. March 30, 1957.

W10. Wechsler, R. L., Crum, W., and Roth, J. L. A., The blood flow and oxygen consumption of the human brain in hepatic coma. *Clin. Research Proc.* **2**, 74 (1954).

W11. Welt, L. G., Thorup, O. A., Jr., and Burnett, C. H., A study of renal tubular phenomena under the influence of a carbonic anhydrase inhibitor. *Clin. Research Proc.* **2**, 63 (1954).

W12. White, H. L., and Rolf, D., Renal glutaminase and ammonia excretion. *Am. J. Physiol.* **169**, 174 (1952).

Y1. Young, T. C., Burnside, C. R., Knowles, H. C., and Schiff, L., Effects of intragastric administration of whole blood on the concentration of blood in patients with liver disease. *J. Lab. Clin. Med.* **50**, 11 (1957).

IDIOPATHIC HYPERCALCEMIA OF INFANCY

John O. Forfar and S. L. Tompsett

Departments of Pediatrics and Biochemistry, Edinburgh Northern Group of Hospitals and Departments of Child Life and Health, and Clinical Chemistry, University of Edinburgh, Edinburgh, Scotland

Page

1. Introduction

A decade or two ago textbooks of pediatrics devoted considerable space to the discussion of "marasmus," but over the years advancing knowledge has brought the recognition that many of the states of debility in infants, formerly embraced by that term, are in fact specific disease entities. Fibrocystic disease of the pancreas, various forms of amino aciduria, galactosemia, hyperelectrolytemia and hyperchloremic acidosis of infancy are but some of the conditions recently identified and separated. Their recognition has been largely the result of expanding biochemical knowledge and of the increasing application of biochemical methods of investigation. Thus, in the pediatric textbook of today, the term marasmus scarcely appears.

Idiopathic hypercalcemia of infancy was first described as such by Lightwood in 1952 (L3). In the same year Fanconi and Girardet (F2) and Schlesinger *et al.* (S1) described two children in whom hypercalcemia was first recognized at the age of approximately 20 months but who had exhibited symptoms at a much earlier age. In both of these children the hypercalcemia was associated with mental retardation, cranial abnormalities, and cardiac murmurs.

Is idiopathic hypercalcemia of infancy a new disease or is it one whose anonymity has for long been cloaked under terms such as marasmus? To this question no categorical answer can be given. In the earlier literature cases were reported which, from a clinical point of view, might well have been idiopathic hypercalcemia of infancy. Usually, however, the biochemical data essential for conclusive diagnosis have been absent.

Thatcher in Edinburgh in 1931 (T2) and again in 1936 (T3) reported two cases which in their age incidence, clinical picture, course, and pathological findings corresponded very closely with the cases subsequently described by Lightwood (L3). Malmberg in Sweden (M2) and Putschar in Germany (P2) described cases which could conceivably have been examples of this condition. Lightwood himself had in 1932 (L2) described a case similar to those later described by Fanconi and Girardet (F2) and by Schlesinger *et al.* (S1). In all of these cases recognition of the essentially hypercalcemic nature of the illness was based on the finding of nephrocalcinosis at post-mortem examination.

2. Clinical Types

Broadly speaking, idiopathic hypercalcemia has been divided into two types, the so-called "mild" type typified by the cases described by Lightwood in 1952 and the so-called "severe" type typified by the cases described by Fanconi and Girardet and by Schlesinger *et al.* in the same year. Originally the distinction between these two types was based on the self-limiting nature and uncomplicated recovery attributed to the mild type and the serious prognosis and associated congenital defects which featured the severe type. These associated congenital defects consisted of mental retardation, dwarfism, congenital heart disease, craniostenosis and osteosclerosis, squint, and facial paresis. It is now recognized that the distinction between these two types is not nearly so clear-cut as was once thought, and indeed there is doubt whether it exists at all. The conclusion that in the severe type congenital defects were present was usually an assumption based on clinical findings such as cardiac murmurs, small skull size, etc. Later post-mortem studies in severe cases have shown no evidence of congenital malformations of the heart, kidneys, or skeleton (S2). In fact, present evidence suggests that most of the symptoms and pathological changes could be the result of prolonged hypercalcemia and of the renal impairment which this may induce (S2). It is noteworthy that in most of these severe cases there has been a long interval between the onset of symptoms and the establishment of the diagnosis. Daeschner and Daeschner for instance (D1), reviewing severe cases, showed that the average interval between the onset of symptoms and recognition of the disease was 13 months. In this connection the case described by Stapleton *et al.* (S9) is of great interest. This infant had a cardiac murmur, hypertension, and a small skull—signs associated with the severe type of idiopathic hypercalcemia. Yet with adequate treatment the cardiac murmur disappeared, the blood pressure fell to normal, the skull increased normally in size, and no mental impairment occurred—surely evidence that it is delayed diagnosis and lack of adequate treatment which is responsible for the so-called severe cases. In some quarters, however, the impression still remains that in severe cases with mental impairment some degree of mental retardation may have been present before the onset of the symptoms of hypercalcemia (S2). As regards birth weight, too, there is, in severe as compared with mild cases, a statistically significant lower mean birth weight which is not related to prematurity. This suggests some degree of congenital dwarfing (S2).

In the present state of knowledge it appears unnecessary to attempt to define mild and severe cases. There is insufficient evidence that there is any fundamental pathological distinction between them. Essentially the difference between them is one of degree.

3. Clinical Features

3.1. Age Incidence

Children who have subsequently been diagnosed as suffering from idiopathic hypercalcemia have on rare occasions been reported to have had symptoms even from birth, but idiopathic hypercalcemia has not been recognized in the newborn. The usual age of onset of symptoms is during the period 3 months to 1 year, the peak being between 5 and 8 months (C3, F7, L6). The condition may be diagnosed at the time of onset of symptoms by the application of appropriate biochemical tests, but often there has been a very considerable time lag between the onset of symptoms and diagnosis. This has ranged up to a year and even longer, and the average for 43 cases reported in the literature was 7 months (C3, F7, L6, M1, M3, S2). With wider recognition of the condition the time lag has recently become much shorter than that.

3.2. Sex Incidence

If the various series of cases reported by a number of authors are pooled, no significant sex difference is observed.

3.3. Preceding Illnesses

In a number of cases a mild preceding illness has been observed. Thus Forfar *et al.* found a preceding respiratory tract infection in four out of six cases (F7). Creery noted respiratory infection in one, gastroenteritis in two, and otitis media in one, in four out of sixteen cases (C3). Lowe *et al.* noted a respiratory tract infection and a febrile response to vaccination as precursors in two out of their seven cases (L6).

3.4. The Clinical Picture

Characteristically, such infants are normal at birth and in the early weeks of life make satisfactory progress with a normal appetite and a normal weight gain. In the majority of cases vomiting is the initial symptom and is usually associated with anorexia and constipation. The vomiting may be unspectacular but may occasionally be severe and even projectile. In some cases constipation and in others refusal to feed are the presenting symptoms. Vomiting is usually most severe in the early part of the illness, tending to lessen or to cease after a week or two but frequently showing periodic exacerbations. Rarely, vomiting is absent altogether. Constipation is obstinate in most cases, the stools being small, hard, and infrequent. Anorexia frequently takes the form of a distaste for solid foods. In some cases mothers have gained the impression that their infants are passing an excessive amount of urine. In others thirst is mentioned. Not infrequently pallor and muscular weakness are complained of or are admitted on questioning. All infants fail to gain weight normally to a greater or lesser degree.

Physical examination reveals an infant who shows some evidence of wasting, who is usually pale and hypotonic, apparently mentally alert but physically somewhat apathetic. The wide-eyed, alert yet anxious facial expression, the pallor, and the hypotonic bodily habitus may be sufficiently striking to suggest the diagnosis.

Although obvious wasting is usually present, dehydration is slight or absent. Despite pallor, anemia may not be present or may be only moderate in degree. It is usually normocytic in type. Abdominal examination reveals fecal masses in the colon in about 50 % of cases. A tendency to intercurrent infection is often exhibited. A few pus cells have often been found in the urine, and on this account an erroneous diagnosis of renal tract infection has not infrequently been made. Such a conclusion has often obscured the real diagnosis. Electrocardiography has not shown any significant abnormality.

In severe cases the appearance has been described as elfin, and Schlesinger *et al.* have indicated specific features to be prominent epicanthic folds, an overhanging upper lip and underdevelopment of the bridge of the nose and the mandible (S2). As noted above, in severe cases described in the literature there has usually been a very considerable interval—of the order of 6 months to 2½ years—between the onset of symptoms and the diagnosis of hypercalcemia. In these cases dwarfism and mental retardation have been features, osteosclerosis has been much more marked, cardiac systolic murmurs have frequently been found, hypertension has occurred in some cases, squint has been common, and transient facial paresis has been noted. In such cases the skull circumference is usually less than normal and craniostenosis can occur.

3.5. Radiology

The essential biochemical investigations are dealt with below, but radiological examination may in some cases give strong confirmatory diagnostic evidence. In mild cases, however, there may be no radiological abnormality. There may be a general increase in bone density, particularly at the lower ends of the radius, the ulna, the tibia, and the fibula where there may be bands of increased density at the metaphyseal margin. There may also be a broad band of increased density round the periphery of the carpal and tarsal bones, giving them a target appearance. Osteosclerosis at the base of the skull and around the orbit may also occur. Nephrocalcinosis, if present, may be revealed by a straight radiograph of the abdomen. Pyelography may be unsatisfactory. Barium meal examination of the alimentary tract is usually negative.

3.6. Course and Prognosis

In the majority of cases of idiopathic hypercalcemia of infancy the course is a self-limiting one. The disease on the one hand may clear up fully after a few weeks; on the other hand it may drag on for months with intermittent remissions and exacerbations. In some cases the infant suffers from a comparatively trivial transient illness. In others, and fortunately a minority, he may be critically ill and suffer permanent sequelae. In a group of 45 cases reported in the literature (B5, C3, D3, F7, L6, M1, M3, S2) there were six deaths, but of course the real death rate will be only a fraction of this because many milder cases will not have been reported or will not have been recognized.

There is good evidence that hypercalcemia continued over a long period can cause severe and often permanent damage, especially cerebral, skeletal, and renal. In the later stages of severe cases an irreversible azotemia, possibly with hypertension and with secondary renal dwarfism, may dominate the clinical picture. Thus prognosis is materially influenced by early diagnosis and the early institution of adequate treatment.

The majority of patients recover and in these the further prognosis is good. Bonham Carter *et al.* (B5) noted normal growth and development during a 2-year follow-up. Forfar *et al.* (F7) noted similar full physical and biochemical recovery with no subsequent impairment of intelligence. Recovery may, however, occur in the physical sense but with mental impairment (C3).

4. Morbid Anatomy and Histology of Fatal Cases

The evidence that is available regarding the morbid anatomy of idiopathic hypercalcemia of infancy is naturally derived from severer cases which have proved fatal (D3, L6, R1, S2). In these, pathological changes are found predominantly in the skeleton, kidneys, and cardiovascular system, but other systems may be involved to a lesser degree.

4.1. Kidneys

These may be normal or considerably reduced in size. The surface is smooth. On section the cortex may be seen to be narrowed. The cut surface may show minute calcium deposits in the medulla, especially the outer medulla, and in the cortex. Microscopically the lesions are found most frequently to involve the tubules. Vacuolation of the cells of the proximal tubules with occlusion of the lumen may be present. The distal and collecting tubules may contain eosinophilic hyaline casts. Amorphous medullary deposits of calcium salts, if present, are closely related to the loops of Henle and are associated with focal degeneration of both interstitial tissue and tubules. Aggregates of calcium salts may occur within the wall of the tubule and have been reported within the tubule itself. Rhaney (R1), however, has suggested that the latter appearance may be false and due to flattened mesenchymal cells surrounding calcium deposits, giving the appearance that the calcium salt is within the lumen of the tubules. Calcium may be present also in concretions rather than amorphous deposits, occurring in this form within the interstitial tissue but producing tubular compression and distortion. In the cortex, glomerular changes may be present, varying from patchy hyaline degeneration to ischemic necrosis and fibrous replacement. Endothelial proliferation and capsular adhesions may be present. Deposits of calcium salts are less frequently seen in the cortex than in the medulla. They are found predominantly in the interstitial tissue but may involve adjacent tubular epithelium.

The renal lesions which have been described in idiopathic hypercalcemia are not specific. Similar changes have been observed in hyperchloremic renal acidosis of infancy (B7, D7), in various other diseases of infancy and in hypervitaminosis D. It is interesting to note that a recent re-examination of the histological sections of Thatcher's cases of a quarter of a century ago showed that the changes there were essentially the same as those occurring in idiopathic hypercalcemia of infancy (R1).

4.2. Cardiovascular System

The heart may be enlarged with left ventricular hypertrophy, particularly in cases which have exhibited hypertension during life. Metastatic calcification of the valves and of the myocardium may be present, explaining no doubt the cardiac murmurs which may have been detected clinically in such cases. Occasionally, medullary proliferation of large arteries and, in the smaller arteries, fragmentation of the intimal elastic lamina has been reported, in one case with impregnation with calcium salts. Local calcification of the media of the middle cerebral artery has been described.

4.3. Skeletal System

The bones may appear normal but may show increased sclerotic change coinciding with the areas of increased radiological density, i.e., predominantly in the base of the skull and ends of the long bones. Microscopically, the bone architecture is largely preserved, suggesting excessive deposition of calcium rather than any primary bone affection.

4.4. Other Organs

The parathyroid glands have, naturally, been carefully studied but have always been found to be normal. The liver has been noted to be the subject of fatty changes and centrilobular focal necrosis. In some cases the lungs have shown evidence of terminal bronchopneumonia or purulent bronchiolitis. Other organs have usually appeared normal.

5. Biochemistry

5.1. Blood

The blood picture obtained depends, naturally, on the stage at which diagnosis is first made. The most striking feature is an elevation in the level of the serum calcium. Other typical changes are an increase in blood urea and plasma cholesterol. Although the sera are not lipemic, a perusal of the literature would suggest that a detailed examination of the lipoid fractions might assist in the elucidation of this little-understood condition.

5.1.1. *Serum Calcium*

The increase in serum calcium above the normal range would appear to have been the initial stimulus to the study of this condition. Values up to 18.7 mg/100 ml have been reported (S2), although the usual range of maximum levels is from 12.5 to 15 mg/100 ml. There is, however, good evidence that in the disease entity which we call idiopathic hypercalcemia the serum calcium can be lower than 12.5 mg/100 ml. The disease has been reported in the presence of levels of 11.2 mg/100 ml and 12.0 mg/100 ml (F7, L6). Recovery from idiopathic hypercalcemia is accompanied by a return to within the normal range. It is probably true to say that an elevated serum calcium may be regarded as a diagnostic aid, and it is, of course, implied in the title of the condition.

It is quite possible that until this particular condition with its specific biochemical abnormalities was recognized, the occurrence of "hypercalcemia" in infants could have been dismissed as due to "laboratory errors."

There is little definite information with regard to the magnitude of the ionized fraction of the serum calcium.

5.1.2. *Blood Urea*

A moderate increase in the concentration of urea is nearly always present. In cases without permanent renal damage levels of up to 100 mg/100 ml have been reported. Where significant renal damage has occurred levels considerably in excess of 100 mg/100 ml are usual.

5.1.3. *Serum Inorganic Phosphorus*

Widely fluctuating levels have been reported, which may be below, within, or above the normal range.

5.1.4. *Serum Phosphatase*

Values for serum alkaline phosphatase tend to be low. Acid phosphatase does not appear to have been studied.

5.1.5. *Plasma Bicarbonate*

Values are usually within the normal range, but low results have been obtained.

5.1.6. *Serum Sodium, Potassium, and Chloride*

Values are within the normal range.

5.1.7. *Plasma Cholesterol*

Hypercholesterolemia has been reported by Lowe *et al.* (L6), Creery and Neill (C3), Dawson *et al.* (D3), MacDonald and Stapleton (M1), and Forfar *et al.* (F7). It would appear to be a usual feature of the condition. During the recovery phase, it has been noted that the serum calcium may return to normal levels before the cholesterol (F7). The increase in cholesterol appears to be mainly in the free cholesterol fraction, so that there is an abnormal free cholesterol to cholesterol ester ratio (F7). Difficulties in the study of cholesterol in this condition include lack of data with regard to the normal ranges of this substance in the plasma of infants at different age periods and the different methods of estimation employed.

Using the technique of electrophoresis on paper, Forfar *et al.* (F7) found that cholesterol was abnormally concentrated in the β-lipoprotein fraction. Using reverse phase chromatography, an attempt was made to demonstrate sterols other than cholesterol, but this was unsuccessful. If they are actually present, such substances must have an R_f similar to that of cholesterol or else be present in quantities too small to be demonstrated by the technique employed.

5.1.8. *Vitamin A*

Fyfe (F10) has investigated plasma vitamin A levels. He observed that the fasting levels in infants with idiopathic hypercalcemia were higher than in the normal and that after a large oral dose of vitamin A, plasma levels returned to the fasting range much more slowly than in the normal.

5.1.9. *Citrate*

Harrison (H5) has reported that serum citrate is increased above the normal range in hypervitaminosis D. Winberg and Zetterstrom (W1) found it to be low in an 11-month-old infant suffering from vitamin D intoxication. Forfar *et al.* (F7) have reported that during the active phase of idiopathic hypercalcemia, serum citrate levels are low.

5.1.10. *Plasma Proteins*

The total plasma protein level has been found to be normal or raised (C3, F7, L6, P1). Payne found an increase in the α_2- and γ-globulins (P1). Creery and Neill found an increase in the α_2- at the expense of the β-globulin, with an occasional rise in the γ-globulin (C3). Forfar *et al.* found a rise in the α_2- and β-globulin in two out of six cases (F7).

5.2. Urine

With regard to urine, the principal observations that have been made are concerned with pH and calcium and phosphorus content.

5.2.1. *pH*

The ability to produce a urine with an acid reaction is a characteristic of this condition.

5.2.2. *Calcium*

Hypercalcuria is undoubtedly an important feature. Table 1 gives a list of figures reported by various authors.

TABLE 1
URINARY CALCIUM EXCRETION IN IDIOPATHIC HYPERCALCEMIA

Urinary calcium (mg/day)	Reference
23–122 (mean 69)	Lowe *et al.* (L6)
40–67 (mean 53)	Creery and Neill (C3)
34–128 (mean 82)	Bonham Carter *et al.* (B5)
30–127 (mean 83)	Forfar *et al.* (F7)
45–55	Morgan *et al.* (M3)

In healthy infants of the same age group, on about the same intake of calcium, the urinary calcium output has been reported to be 8–55 (mean 27 mg) per day (D2, F5, H4, H8, J1a, S10, T1). Knapp (K2) has reported a range of 3 to 86 per day.

5.2.3. *Phosphorus*

On a milk diet the average output of phosphorus in the urine at this age is of the order of 430 mg/day (F7). The administration of calciferol normally results in hyperphosphaturia (F1, H9).

In idiopathic hypercalcemia hypophosphaturia has been found (F7, M3), but hyperphosphaturia has also been reported (S6).

5.2.4. *Amino Acids*

Amino acid excretion appears to be within the normal range and no characteristic pattern has been described.

5.2.5. *Citrate*

Forfar *et al.* (F7) have examined urinary citrate excretion in one case of idiopathic hypercalcemia. During the active phase of the condition, citrate excretion was 9.5 and 10.6 mg/day rising to 96 mg/day during the recovery phase. The latter represents a low normal value. It would appear that in this particular case, a hypocitruria existed during the active phase of the condition.

The urinary excretion of citrate in hypervitaminosis D has been reported to be increased above the normal range (H5).

5.3. Metabolic Studies

Balance studies involve the measurement of both the intake of a substance and the excretion (urine and feces). The interpretation of some published studies is therefore difficult since the intake has been calculated merely from tables. A reliable balance study should include a chemical analysis of a comparable aliquot of the intake. The vomiting which is a feature of idiopathic hypercalcemia of infancy may also make accurate balance studies very difficult technically.

5.3.1. *Calcium Balance Studies*

In the normal infant a wide range of calcium retention has been reported by Hoobler (H8), Telfer (T1), Daniels and Stearns (D2), Flood (F5), Stearns (S10), Jeans *et al.* (J1a), and Harrison (H4). From these authors we have obtained the results of 24 calcium balances in healthy infants (7–12 months), fed on cow's milk or on evaporated milk, with, so far as can be ascertained, a normal intake of vitamin D. On an average intake of 1.17 g calcium daily, the mean retention of calcium was 32 % (range 0–79 %) or 0.38 g. This agrees very closely with the daily retention of calcium in a similar age group reported by Sherman and Hawley (S3) from their study of German published reports and with the calculated requirements of the body at this age (L1).

Forfar *et al.* (F7) have reported the calcium balance in 3 active cases of idiopathic hypercalcemia. The mean calcium intake was 1.02 g/day and the mean retention was 49% or 0.45 g/day. Morgan *et al.* (M3) observed retentions of 53 % and 54 % in two cases, the actual daily retentions being 0.46 g/day and (on a low calcium intake) 0.31 g/day. These results suggest that there may be some increased calcium retention in idiopathic hypercalcemia. As there is at the same time hypercalcuria without evidence of loss of calcium from bones, the most likely explanation is that there is increased alimentary absorption of calcium.

5.3.2. *Phosphorus Balance Studies*

In 24 published phosphorus balance studies in healthy infants, aged 7–12 months, having, so far as can be ascertained, a normal intake of vitamin D (B4, D2, H4, H8, J1a, S10, T1), the mean intake was 0.89 g/day and the mean retention was 23 % (0.20 g/day).

Forfar *et al.* (F7) have studied phosphorus metabolism in 2 active cases of idiopathic hypercalcemia. The average retention of phosphorus on a mean intake of 0.9 g/day was 43 % (0.39 g). Morgan *et al.* (M3) have reported retentions of 46 % and 37 % in two infants, the actual daily retentions being 0.30 g and 0.18 g, respectively.

There thus appears to be an increased retention of phosphorus in idiopathic hypercalcemia.

A considerable amount of experimental work has been carried out with the aim not only of obtaining information regarding the nature of this condition but also with regard to possible means of treatment. Particular attention has been paid to agents that might reduce the serum calcium to within the normal range since it was believed that this might be accompanied by an alleviation of the symptoms.

5.3.3. *Low-Calcium Diet*

A low-calcium diet lowers the serum calcium level in idiopathic hypercalcemia of infancy (B5, F4, F7, M1, R4, S8). It results in a marked reduction in calcium retention and can convert a strongly positive calcium balance to a less positive or even negative one (F7, M1).

5.3.4. *Unrestricted Diet with Additional Calciferol*

Houet (H9) gave a single dose of calciferol, viz. 600,000 units, to a healthy infant aged 6 months, whose retention of calcium before treatment was 45 %. Over 4 consecutive 3-day periods retentions of 67, 47, 70 and 45 % were recorded. Forfar *et al.* (F7) found that the addition of 50,000–100,000 units of calciferol daily to the diet over a period of a week increased the retention of calcium in 3 cases of idiopathic hypercalcemia, to 73 % in an active case, to 40 % in an early recovery case, and to 42 % in an inactive case. It was concluded that calciferol did not appear to increase the retention of calcium to a degree greater than in the normal infant. In the active case the serum calcium level fell with calciferol administration; in the early recovery case the level rose.

5.3.5. *Cortisone and Prednisolone*

Anderson *et al.* (A2) have shown that cortisone reduces the serum calcium level in the hypercalcemia which may accompany sarcoidosis. They suggested that cortisone might act as a direct antagonist to vitamin D.

Forfar *et al.* (F7) examined the effect of cortisone (25 mg/day) in one case of idiopathic hypercalcemia. Cortisone was administered on four occasions. On each occasion the serum calcium fell, rising again after the cortisone was withdrawn. Similar results have been obtained by other workers (C3, F3, S9).

In an active case cortisone also antagonized the effect of calciferol in that when the latter was administered in a dosage of 100,000 units daily by mouth along with 25 mg of cortisone, the expected increase in calcium retention did not occur. Indeed the calcium retention was lower than that found on a similar diet without any added cortisone or calciferol. The retentions were 48 % without treatment, 73 % on calciferol alone, and 25 % on calciferol plus cortisone (F7).

Morgan *et al.* (M3) examined the effect of cortisone (25 mg daily) in two cases. It was noted that the raised serum calcium levels returned to normal. In one case the serum calcium fell from 19.0 to 12.6 mg/100 ml and in the other from 14.7 to 10.6 mg/100 ml. Balance experiments indicated that such treatment resulted in the fecal output of calcium and phosphorus being doubled and that the retentions of these elements were reduced considerably. In one case, when the administration of cortisone was stopped, the serum calcium rose from 11.5 to 14.5 mg/100 ml. A dosage of 25 mg of cortisone was found to be much more effective than a dosage of 12.5 mg.

It is suggested that the action of cortisone is to increase the fecal output of calcium and phosphorus, presumably as a result of reduction in the absorption of these elements by an antagonizing effect of the hormone on vitamin D or vitamin D-like substances.

Forfar *et al.* (F7) examined the effect of oral prednisolone (5 mg/day) upon the calcium balance in the acute phase of the condition. A positive calcium balance of 69 % (average daily retention, 0.45 g) was converted into a positive balance of 46 % (average daily retention, 0.37 g). A marked fall in serum calcium and cholesterol was noted, rises being encountered subsequently after discontinuance of the treatment. A positive phosphorus balance of 59 % (average daily retention, 0.18 g) was reduced to one of 40 % (average daily retention, 0.16 g) by such treatment. Prednisolone would thus appear to have a similar action to cortisone.

5.3.6. *Ethylenediaminetetraacetic Acid*

Foreman and Trujillo (F6) have shown that disodium ethylenediaminetetraacetate forms chelate compounds with calcium which are soluble and nonionized and therefore biologically inactive. These are virtually unabsorbed from the alimentary tract. It has been shown that 95 % of intravenously administered calcium EDTA is excreted in the urine in 24 hours.

Morgan *et al.* (M3) examined the effect of the administration of EDTA in idiopathic hypercalcemia. Given orally, a dosage of up to 3 g daily resulted in a fall in the serum calcium level. Given subcutaneously in a dosage of 1 g daily, there was a fall in calcium retention in a balance study, a retention of 51 % being converted to a retention of 28 % on similar dietary intakes of calcium. There was an increase in the urinary excretion of calcium.

6. Diagnosis

From the reports on severe cases it is evident that in these there has been on the average a delay of over a year between the onset of symptoms and the establishment of diagnosis. In milder cases the delay in diagnosis has been very much less than this but has often been considerable. Thus a significant time lag between the onset of symptoms and diagnosis is not uncommon in idiopathic hypercalcemia. Yet early diagnosis is probably important. The longer hypercalcemia persists in an infant, the more serious do its deleterious effects become. A condition which in its early stages can be comparatively easily held in check pending spontaneous recovery may, if allowed to run on, reach a point at which the kidney in particular has suffered a degree of permanent damage which, no matter what happens to the hypercalcemia, will ultimately kill the child.

Awareness of the existence of idiopathic hypercalcemia of infancy is probably the most important factor in early diagnosis. Thereafter, in the majority of cases, a knowledge of the clinical and biochemical features of the disease makes diagnosis fairly easy.

The age period of the disease is clearly defined, and even in the earliest stages suggestive symptoms are usually present. Radiology may help, although in the early stages bone changes may be absent. Establishment of diagnosis depends on biochemical examination. The serum calcium may be so high as to leave no doubt about the essentially hypercalcemic nature of the disease. It has been pointed out, however, that there may be some dissociation between the severity of the disease and the serum calcium level. Infants may be severely affected and yet have a serum calcium in the region of 12 mg/100 ml. Others may appear moderately affected and yet show a serum calcium of 15 mg/100 ml. There will be little difficulty in diagnosis in the latter case; there will be considerable difficulty in the former. In view of the fluctuations of the serum calcium level in this disease, calcium estimation should certainly be repeated on several occasions in a suspected case with an apparently normal serum calcium level. Stapleton and Evans (S8) have taken the diagnostic level of serum calcium as above 12.5 mg/100 ml but there is no doubt that lower levels than that may occur (F7, L6). Thus levels above this figure are strongly suggestive of this disease, levels below it do not exclude it.

It has recently been appreciated that the estimation of serum cholesterol is also diagnostically important in view of the fact that the serum cholesterol level is usually raised in this condition. We have found the average normal serum cholesterol level between the ages of 2 months and 15 months to be 186 mg/100 ml with a standard deviation of 39. In ten cases of idiopathic hypercalcemia the average of the highest serum cholesterol levels reached was 270 mg/100 ml with a range of 216–370. The rise in cholesterol appears to be due predominantly to a rise in free cholesterol (F7). We have found that calcium and cholesterol concentrations can be used together, their product giving a diagnostic index. In ten estimations on infants aged 6–15 months who were convalescent from or in the active stage of other diseases, the range of this product was 1007–2726 with a mean of 1717, whereas a figure of over 2800 is not uncommon in idiopathic hypercalcemia even when the calcium or cholesterol levels are not significantly beyond the normal range. Thus this product is probably a more sensitive diagnostic index than either serum calcium or cholesterol concentration taken alone. Because of this association between calcium and cholesterol and the possibility that the serum calcium level may really be within the normal range, it has been suggested that "idiopathic hypercholesterocalcemia" rather than idiopathic hypercalcemia would be a more appropriate title for this condition (F7).

The 24-hr output of calcium may also be of some diagnostic value if it is above the upper limit of normal of 86 mg/day or persistently above the normal average of 27 mg/day.

As the blood urea nitrogen is usually raised in idiopathic hypercalcemia, its estimation is important in diagnosis.

7. Etiology

To the pediatrician and to the biochemist, idiopathic hypercalcemia presents an immediate clinical problem, but the disease has a wider interest—to the nutritionist, to the epidemiologist, to those working in the field of preventive medicine, to the radiologist and the pathologist. Much of the interest stems from the uncertainty surrounding the etiology of the disease and from the light which it may throw on the metabolism of vitamin D. Does the apparent restriction of this disease to the northern half of Europe indicate that it occurs only there or is it being missed in other countries? What, if any, is the relationship between vitamin D ingestion and this disease? What role do excess dosage, specific hypersensitivity, or individual intolerance play? Is it possible that irradiated ergosterol can break down in the food or in the alimentary tract to give rise to toxic substances? Can an endogenous disturbance of cholesterol metabolism give rise to toxic substances having a vitamin D-like effect? Does infection play a part in causation? To what extent do the constituents of the diet, other than vitamin D, influence or cause the disease? These and other questions have been asked, and to them few conclusive answers are available.

The various etiological theories which have been advanced in respect of idiopathic hypercalcemia are given in Table 2.

7.1. Relationship between Idiopathic Hypercalcemia and Vitamin D

Since vitamin D has been implicated in infantile idiopathic hypercalcemia, some reference to our knowledge concerning the chemistry and physiology of this group of substances appears justified.

The term vitamin D is applied to that group of steroids showing antirachitic activity. Differences in chemical structure are present, although slight. A slight change in chemical structure may result in tachysterol, a substance with no antirachitic activity but with a parathyroidlike action, or toxisterol, a substance with toxic calcifying properties. In view of the effect of slight differences in chemical structure on physiological properties within the "cortisone" group of steroids, one might expect discrete differences in physiological action within the vitamin D group, but so far none have been described.

Whereas we have considerable, although incomplete knowledge concerning the degradation and elimination of the "cortisone" group of steroids, our knowledge concerning the vitamin D group is unfortunately

TABLE 2
IDIOPATHIC HYPERCALCEMIA — SUGGESTIONS ADVANCED AS TO CAUSE

Theory	Reference
Specific hypersensitivity to vitamin D	Lightwood (L4); Lowe *et al.* (L6); Bonham Carter *et al.* (B5); Morgan *et al.* (M3); Rhaney and Mitchell (R1)
Excessive intake of vitamin D	Morgan *et al.* (M3)
Toxic effect of vitamin D_2 as opposed to vitamin D_3	Evans (E1)
Disorder of cholesterol metabolism resulting in production of vitamin D-like substance	Forfar *et al.* (F7)
The result of infection	Forfar *et al.* (F7); Creery and Neill (C5)
Excessive calcium intake in diet	Creery and Neill (C3)
Deficiency of essential (unsaturated) fatty acids in diet	Sinclair (S5)
Toxic effect of alkali medication	Creery (C2)
Toxic effect of sulphadimidine sulfamethazine	Harris (H3)

lacking. This is probably due mainly to the absence of suitable techniques.

In many respects idiopathic hypercalcemia of infancy resembles the clinical picture resulting from excessive vitamin D dosage. Wherein do hypervitaminosis D and idiopathic hypercalcemia of infancy resemble each other and wherein do they differ? Fatal vitamin D poisoning in infancy has been described by Ross and Williams (R3) and by Ross (R2). Their patients had had 20,000–40,000 units of vitamin D daily for several months, and one on whom an autopsy was performed had widespread visceral calcification. Nonfatal vitamin D poisoning in infancy has been described by Hess and Lewis (H7), Ross and Williams (R3), Jelke (J3), Debré and Brissaud (D4), and Fanconi and de Chastonay (F1). Features which may occur both in vitamin D poisoning in infancy and in idiopathic hypercalcemia are: vomiting, anorexia, constipation, weight loss, hypotonia, pallor, hypercalcemia, hypercalcuria, azotemia, lowered serum alkaline phosphatase, radiological osteosclerosis, increased absorption of calcium from alimentary tract, similar histological changes.

Fyfe (F10), too, in showing that the fasting plasma vitamin A levels and the levels 4 hours after a standard dose of vitamin A are significantly higher in idiopathic hypercalcemia than in the normal, has adduced this as additional evidence that the metabolism of the other fat-soluble vitamin, vitamin D, is abnormal in idiopathic hypercalcemia of infancy.

There is thus a good deal of common ground between the two conditions but there are also important differences.

Perhaps the most immediately evident difference is in vitamin D dosage. It is clear that the dosages of vitamin D administered in the majority of cases of idiopathic hypercalcemia are in no way comparable with those which have been reported to cause frank vitamin D poisoning. Creery and Neill (C3) noted that of 15 affected infants two were having 700 international units daily; six, 1100–1600; seven, 1800–2400; and one, 3200 units daily. Forfar *et al.* (F7) noted an average daily intake of from 400 to 1350 units in their six cases, and for the month prior to the onset of symptoms the daily dosage ranged from 400 to 2300 IU/day. In two cases reported by Morgan *et al.* (M3) the daily vitamin D intake was of the order of 2000 units daily. At the same time most of these estimations of vitamin D intake are based on the minimum vitamin D content of the preparations estimated. Regulations for manufacture demand a certain minimum content and to ensure this and to allow for a fall in potency with processing and with the passage of time, manufacturers commonly oversupplement their products. Thus to allow for deterioration, the amount of calciferol added to National Dried Milk in Britain may be 80 % above the minimum content. In proprietary Dried Milk the amount of vitamin D present, if the milk is used shortly after manufacture, may be 75 % above the minimum content (B6). After storage for 6–18 months the vitamin D content may be only a little above the minimum (B6).

Such considerations, however, have only limited relevance as regards comparison of the dosage recognized to cause frank vitamin D poisoning, on the one hand, and the intake of vitamin D in idiopathic hypercalcemia, on the other, as there is no comparison between the dosages in the two groups of cases. They do, however, raise the question of the optimal intake of vitamin D. Stearns (S11) has contrasted two groups of infants, one receiving approximately 400 units of vitamin D daily and the other approximately 2000–4500 units daily. She found that the retention of calcium per kilogram of body weight daily was the same in the two groups, but the intake of calcium in the higher dosage group was appreciably lower, i.e., there was a drop in appetite. This distinct drop in appetite tended to show itself at about the fifth month of age. Jeans and Stearns (J2) have also shown in a few individual infants that with an intake of 2000–4500 units of vitamin D per day from birth, slowing in growth tends to occur at about 5 months of age. This arrest in growth can be reversed and a normal growth rate reestablished by reducing the vitamin D intake to 400 units/day. Further, Stearns (S11) has shown that maximum calcium retention (approx. 50 mg/kg/day) is obtained in the healthy infant fed on cow's milk with a vitamin D intake of approximately 400 units/day. Increasing the intake above this level up to 4500 units/day does not increase retention. Thus Stearns has concluded that the optimum daily intake of vitamin D for an infant is 300–400 IU/day and that 2000 units or possibly less than this may result in delayed symptoms of chronic toxicity. The British Paediatric Association has recommended 700 IU daily.

It is thus clear that the intake of vitamin D in idiopathic hypercalcemia may be no higher than the amount recommended by the British Paediatric Association but that in many cases the intake is supraoptimal. Yet in Britain intakes of over 1000 units of vitamin D daily are the rule rather than the exception, and of the great number of infants who receive them only the minutest fraction develop idiopathic hypercalcemia. Creery and Neill (C4) have shown that in Britain 4 % of infants receive over 2000 IU daily; 51 %, 1000–2000; 19 %, 600–1000; 15 %, less than 600 units; and 11 %, no vitamin D supplement or fortification of any sort.

Thus as regards the actual dosage of vitamin D the most that can be said is that the average intake of vitamin D in Britain may be somewhat above the optimum. In many individual cases of idiopathic hypercalcemia no direct relationship between dosage and disease can be established.

Lacking evidence of gross overdosage the concept of an undue hypersensitivity to vitamin D has been put forward by Lightwood (L3, L5). This has received support from Lowe *et al.* (L6), Bonham Carter *et al.* (B5) and Morgan *et al.* (M3). In view of the age incidence of the disease, such a hypersensitivity would need to be a strictly age-defined one; yet it is not impossible to imagine that this could be so. It is recognized, for instance, that renal function as regards calcium excretion is somewhat different before the age of 2 years than it is thereafter (K2), and as regards blood chemistry the blood cholesterol level is somewhat lower at this period than it is in later life (B1). In the normal healthy child at the age at which idiopathic hypercalcemia usually disappears spontaneously, it is possible that certain renal and biochemical functions reach maturity of development and that the attainment of this stage in a child suffering from idiopathic hypercalcemia may be a factor in determining the spontaneous termination of the disease.

There are, however, certain features of idiopathic hypercalcemia which are out of keeping with the concept of hypersensitivity. In the presence of hypersensitivity to vitamin D the administration of this vitamin would be likely to aggravate the symptoms. This does not always occur. Creery and Neill (C3) gave one of their patients 2000 units of vitamin D daily for 3½ weeks and obtained a steady clinical improvement despite a transient rise in the serum calcium level. Forfar *et al.* (F7, F8) gave two infants 50,000–100,000 units daily for a week (i.e., in a week, about 5 times the total amount of vitamin D which, according to the vitamin D hypersensitivity hypothesis, was supposed to have caused the illness). Neither patient showed any accentuation of the symptoms of anorexia, vomiting, hypotonia, or constipation from which characteristically they were suffering. One showed some loss in weight, but in this disease intermittent weight loss is common. The other showed no loss of weight. On the other hand, Bonham Carter *et al.* (B5) administered 10,000 units of vitamin D daily by mouth for 10 days to an infant in whom the disease was in moderate remission. This was followed by an exacerbation of the disease. The administration of large doses of calciferol to children who have recovered from idiopathic hypercalcemia has had no adverse effect (F7, H10).

Knowledge of phosphorus and citrate metabolism in vitamin D poisoning and in idiopathic hypercalcemia of infancy is very incomplete and somewhat contradictory, but in certain respects the changes in these electrolytes in these two disorders may differ.

Hypercholesterolemia is a feature of idiopathic hypercalcemia of infancy (D3, F7, S2). Although in the adult and older child hypercholesterolemia is a feature of vitamin D poisoning, it does not appear to be a usual feature of such poisoning in infancy. Fanconi and de Chastonay (F1) drew attention to this distinction. Of four children who had suffered gross vitamin D overdosage, they found no rise in the serum cholesterol level in three infants aged 9½ months, 6 months, and 11 months, but found a significant rise in a 4½-year-old child. Jelke (J3) also found no hypercholesterolemia in vitamin D poisoning in an infant. Ross and Williams (R3) refer to hypercholesterolemia in a preliminary report on vitamin D overdosage in infancy without giving any figures, but in a later full report hypercholesterolemia is not even mentioned (R2). Thus, the high serum cholesterol level occurring in idiopathic hypercalcemia is probably at variance with what would be expected in a simple hypersensitivity to vitamin D.

Another argument against simple vitamin D overdosage or hypersensitivity in idiopathic hypercalcemia is the response to withdrawal of all vitamin D. In even the grossest cases of vitamin D poisoning in infancy the longest duration of symptoms after withdrawal is of the order of 2 months (F1). In idiopathic hypercalcemia the duration of symptoms after withdrawal of all vitamin D may be considerably longer than that (F8).

Lastly, the uneven geographical occurrence of the disease does not suggest hypersensitivity which would presumably be distributed in an even way, but rather suggests that some other regional factor, e.g., infection, is operative.

If indeed idiopathic hypercalcemia is related to vitamin D, the question has been raised as to whether any specific form of the vitamin is responsible. Vitamin D_2 or calciferol has been most suspect. A few years after the discovery of the antirachitic properties of irradiated ergosterol it became evident that imperfect irradiation, particularly as seen in the product Vigantol, resulted in a preparation containing large amounts of toxic substances, particularly the overirradiation product toxisterol (B2). It has been suggested that the breakdown of calciferol in food or in the alimentary tract into toxic products, e.g., toxisterol and lumisterol, may be an etiological factor in idiopathic hypercalcemia of infancy (E1). No direct evidence of such a change has been produced. Some cases of idiopathic hypercalcemia have had a very small intake of vitamin D_2, probably less than 200 units per day. In such, the vitamin has been given predominantly in the form of vitamin D_3 (F7). It seems clear, however, that the great majority, if not all, cases of idiopathic hypercalcemia have received vitamin D_2 in greater or lesser amount. Even those infants in Britain fed on cow's milk and receiving as their only vitamin D supplement the Government issued cod-liver oil are receiving nearly half of their vitamin D intake as vitamin D_2, as the cod-liver oil is fortified to this extent.

7.2. Cholesterol Metabolism and Idiopathic Hypercalcemia of Infancy

It has been shown that there is a statistically significant correlation between the serum calcium and cholesterol levels in idiopathic hypercalcemia of infancy (F7). At the same time it is recognized that a cholesterol derivative, 7-dehydrocholesterol, is a precursor of vitamin D_3. In view of the clear association between hypercalcemia and hypercholesterolemia in idiopathic hypercalcemia, and in view of the possible relationship between cholesterol and vitamin D, Forfar *et al.* (F7) have suggested that idiopathic hypercalcemia may be due to a disorder of cholesterol metabolism, in which an excess of vitamin D-like substance is produced. They further postulate that this substance, although hypercalcemic, is probably essentially toxic. Such a hypothesis would explain those cases in which hypercalcemia is not a marked feature and in which the severity of the symptoms is out of keeping with the serum calcium level. In support of such a hypothesis, too, is the fact that the majority of cases develop symptoms between the months of April and October when solar irradiation of the excess provitamin would be expected to be maximal. The same authors suggest that such a disorder of cholesterol metabolism might be the result of infection.

Whether the synthesis or storage of provitamin D_3 can occur in the human is uncertain, but it can certainly occur in animals. Glover *et al.* (G3) have shown that this provitamin is present in the tissues even when the animals are fed on a practically sterol-free diet. The conclusion is reached that the animal is not dependent on dietary provitamin D but has the power to synthesize it from cholesterol. The small intestine is especially rich in provitamin. The same authors noted that, in the guinea pig, infection of the liver with *Pasteurella pseudotuberculosis* is associated with an increased concentration of provitamin in the intestinal mucosa, an interesting observation in view of the possible role of infection in idiopathic hypercalcemia in infancy.

7.3. The Role of Infection in Idiopathic Hypercalcemia

In suggesting that infection might be a factor in idiopathic hypercalcemia, Forfar *et al.* noted that in four out of their six cases there was apparently related, preceding, respiratory infection. They pointed out that in Finland another infection, interstitial plasma cell pneumonia, had been recognized as producing prolonged hypercalcemia in certain cases (H1, H2). This disease occurs in other northern European countries and in Central Europe as well, affecting premature and young infants particularly. There is considerable support for the view that the infecting agent is *Pneumocystis carinii* (G1). This infection has now been found to occur in Britain, and the age group of the patients involved has been similar to that of idiopathic hypercalcemia (B3). As far as is known the serum calcium has not been elevated in the British cases. Creery and Neill have noted a preceding infective illness in 10 (respiratory in 6 and enteral in 4) out of 33 cases of idiopathic hypercalcemia. As further evidence of the possible relationship between infection and hypercalcemia they quote the case of an infant who exhibited a serum calcium level of over 12 mg% after an attack of diarrhea and during an attack of pertussis, the levels before and after these peaks being in the region of 10–11 mg% (C5). An infective etiology would be in keeping with the varying incidence in different parts of the country.

7.4. Dietary Factors in Idiopathic Hypercalcemia

It has been stated that idiopathic hypercalcemia has not been known to develop in a wholly breast-fed infant (B6). Morgan *et al.* (M3), however, have described an infant who was still breast fed at the time of diagnosis at 8 months of age and whose intake of cow's milk consisted only of that present in puddings and other cereal foods. The association between artificial feeding and idiopathic hypercalcemia has been related to the fact that the calcium content of cow's milk is four times that of human milk. On the other hand the phosphorus content of cow's milk is also higher, and this has been suggested as a cause of hypocalcemia in infants (G2).

Another possible dietary factor concerns the essential fatty acid content of human and artificial milk. It has been postulated by Sinclair that many modern dietaries are deficient in the essential polyethenoid fatty acids (EFA) and that in consequence there is a rise in unesterified (and more active) vitamin D and in unesterified cholesterol. He has suggested that a part of the etiology of infantile idiopathic hypercalcemia may be attributed to EFA deficiency (S5). He has pointed to the lower content of certain unsaturated fatty acids in cow's milk as compared with human milk as a factor in the development of idiopathic hypercalcemia in artificially fed infants. He considers that dried milk has an even lower content of essential fatty acids than liquid cow's milk and that the longer it is stored the lower does the essential fatty acid content become. On the basis of some observations on rats, he suggests that a dietary deficiency of the essential fatty acids increases susceptibility to the possible toxic effects of vitamin D. The age of the rats, the duration of the essential fatty acid deficient diet, or the dosage of vitamin D is not mentioned, and there would appear to be no other experimental data to support these views.

7.5. Drugs as Etiological Factors

Creery (C2) has suggested that the syndrome of idiopathic hypercalcemia may be of the same nature as the "milk and alkali" syndrome of adults where an excessive intake of milk and alkali causes renal damage. These cases show muscular weakness and hypotonia, thirst, polyuria, ocular calcinosis, hypercalcemia without hypercalcuria, normal or high serum inorganic phosphorus, and a raised serum protein level. Some cases of idiopathic hypercalcemia have received alkali medication, particularly magnesium hydroxide, before the onset of the disease but certainly not all have done so. In this suggested etiology it was assumed that the renal damage caused by milk and alkali resulted in impaired calcium excretion and thus hypercalcemia. It is now known, however, that hypercalcuria is a feature of idiopathic hypercalcemia.

Harris (H3) has suggested that sulfadimidine may aggravate idiopathic hypercalcemia. This report is based on three cases only and does not substantiate that it was the sulfadimidine which contributed to the severity of the cases.

7.6. Other Etiological Theories

Hypercalcemia naturally suggests increased parathyroid activity. It is generally agreed however that hyperparathyroidism is not a feature of idiopathic hypercalcemia of infancy (L6).

Although the kidney obviously plays an important role in idiopathic hypercalcemia, there is also wide agreement that the disease is not primarily a renal one (L6, S6).

8. Geographical Distribution

In its geographical distribution idiopathic hypercalcemia of infancy has been largely limited to northern Europe, occurring particularly in Great Britain, Switzerland, and Scandinavia.

In Great Britain in a 2½ year period from 1953 to 1955, 204 cases are known to have been recognized (B6). The total number of cases would be in excess of this. In London between 1952 and 1955, one pediatric team had knowledge of 55 cases, but this team was especially interested in the disease (S8). In Belfast in the north of Ireland, 16 cases were seen during a 2-year period from 1952 to 1954 (C3).

Even within the confines of one country, however, the distribution is not an even one. On the east coast of Scotland, in the city of Dundee, over 4½% of all medical hospital admissions between the ages of 6 months and 1 year suffer from this condition and among causes of hospitalization it ranks fourth in frequency after respiratory infection, feeding difficulties, and otitis media (M3). In Sheffield in the North of England, however, only two cases were seen over a 3-year period in a 236-bed children's hospital serving a large, densely populated area (C1).

8.1. Idiopathic Hypercalcemia in the United States and in Canada

From the literature it would appear that idiopathic hypercalcemia of infancy is very uncommon in the United States or in Canada. From the United States two cases were reported by Sissman and Klein (S6) and recently there has been a report of a severe case (D1). One or two possible explanations for the rarity of the condition on the American continent could be advanced in the light of some of the theories on etiology.

As canned milk in the United States contains 400 units of vitamin D per reconstituted quart, whereas most of the dried milk in Britain contains over 1000 units per reconstituted quart, the intake of vitamin D from artificially prepared milk in Britain is obviously considerably higher. Yet in the large population of the United States there must surely be many infants who are receiving vitamin D supplements in an amount sufficient to offset the difference between the two countries.

Another possible explanation is that in artificial feeding of infants in Britain dried milk powder is predominantly used. In the United States canned evaporated milk is much more popular. Sinclair (S5) states that the essential fatty acid content of dried cow's milk is lower than that of liquid cow's milk and has put this forward as a possible etiological factor. Yet in Britain idiopathic hypercalcemia has occurred in infants who have never received dried milk powder.

It has also been suggested that the condition is being missed or misdiagnosed in America. It seems unlikely that this would be the case were the condition to occur with anything like the frequency with which it occurs in Britain.

Lastly, is it possible that an infection which has been present in Europe for some years is now spreading westwards? The recognition of severe cases both on the continent of Europe and in Britain heralded the more widespread recognition of idiopathic hypercalcemia. Does the recent recognition of a severe case in America similarly presage the recognition of more cases there? Is it possible that this is the early manifestation of an infection which will spread more widely throughout the American continent? Can there be any relationship with the fact that interstitial plasma cell pneumonia, a disease which it is recognized can be complicated by hypercalcemia (H1, H2) has recently been described in America for the first time (L7)? These are interesting speculations.

9. Prevention

So far as present knowledge goes, there would appear to be only one good method of preventing idiopathic hypercalcemia and that is to feed an infant wholly from the breast. As the age period of the disease extends beyond the period when continued breast feeding is normally possible, this is advice which it will seldom be possible fully to accept. Two possible protective qualities of breast milk have been considered. These are its lower calcium content (C3) and its higher essential fatty acid content (S5) as compared with cow's milk.

Restriction of vitamin D intake so that it does not exceed 700 units per day is now being canvassed as a preventive measure, although it has not, in fact, been shown that this will reduce the incidence of the disease in a community. That such a restriction of dosage will not prevent the development of the disease in individual cases has been shown repeatedly.

Natural, as opposed to synthetic, vitamin D has its advocates (F9), i.e., vitamin D_3 as opposed to vitamin D_2. Again there is suggestive evidence that if ingested vitamin D plays any part in the etiology of idiopathic hypercalcemia of infancy, it is vitamin D_2 particularly which comes under suspicion.

10. Treatment

Three main lines of treatment are generally accepted in idiopathic hypercalcemia of infancy: the withholding of all vitamin D supplement and vitamin D-fortified food, the reduction of the intake of calcium, and the use of cortisone.

10.1. Withholding Vitamin D

Whatever the true nature of idiopathic hypercalcemia, most are agreed that there is evidence of increased vitamin D-like activity. We know too that the administration of vitamin D in this condition may sometimes further raise the serum calcium level B5, C3, F7). The removal of all added vitamin D from the diet is therefore logical and in practice appears beneficial. So many proprietary milks and prepared foods are fortified with vitamin D that it is necessary to be vigilant in selecting a dietary.

10.2. Low-Calcium Diet

The withholding of vitamin D in itself, although it may modify the disease, will seldom result in cure. Something more is usually necessary. The effect of a vitamin D-like substance on hypercalcemia can be modified in another way—by the reduction of intake of calcium. The amount of calcium absorbed from the gut is obviously dependent on the amount ingested. If the intake is low enough, even 100% absorption (as opposed to the normal $<50\%$) will not result in hypercalcemia no matter how great the dosage of vitamin D-like substance. This principle has been recognized in the treatment of idiopathic hypercalcemia of infancy.

Ferguson and McGowan (F4) first advocated the use of a low-calcium milk. They passed cow's milk through a column of ion exchange resin in such a way that most of the calcium was removed in exchange for potassium, sodium, and magnesium in about normal proportions. The resin was conditioned by running through it a solution containing potassium, sodium, and magnesium at about 40 times the concentration found in cow's milk, followed by the same solution diluted 40 times. A column of resin 10 inches high and 1 inch in diameter can deal with a pint of milk at a rate of 150 drops per minute. Milk treated in this way has a calcium content of 1–5 meq/liter as opposed to the normal 60 meq/liter. The same authors also compounded a synthetic low-calcium milk using as a basis casein which had been rendered almost calcium-free and adding sugar, fat, and electrolytes. The resulting milk was on analysis akin to human milk except for the calcium content, which was 2 meq/liter as opposed to the 15 meq/liter in human milk (H6). Using these milks these authors found that there was improvement in the general well-being of the child in a few days, gain in weight, and a lowering of the serum calcium level. The preparation of low-calcium milk has also been described by Dent (D5).

Low-calcium milk is now available commercially in Britain, so too is low-calcium cereal, and this has made the prolonged use of a low-calcium diet much easier to accomplish.

Clinical and experimental support for the use of low-calcium feeding has come from Creery and Neill (C3), Russell and Young (R4), MacDonald and Stapleton (M1), Bonham Carter *et al.* (B5), Forfar *et al.* (F7), and Morgan *et al.* (M3). A low-calcium diet generally lowers the serum calcium level significantly within a few days to a few weeks. This is frequently accompanied by a gain in weight and other symptomatic improvements, but it is the role of the low-calcium diet in preventing the serious long-term effects of hypercalcemia on the kidney which is most important. Unless the blood urea nitrogen also returns to normal, it is probable that irreparable renal damage has occurred. Hence the importance of early diagnosis and the reduction of the serum calcium level to normal as soon as possible. A normal diet should not be reinstituted until the serum calcium and blood urea concentrations are normal, until a rise in serum alkaline phosphatase has occurred, and until weight is being gained steadily (S9). Any reversal of these improvements on reintroducing a normal diet would be an indication for return to a low-calcium diet. Stapleton *et al.* (S9) have shown the value of a low-calcium diet in a patient who exhibited the features of the so-called severe type of idiopathic hypercalcemia in that he showed a cardiac murmur, a raised blood pressure, and some degree of arrest of cranial development. A low-calcium diet was commenced 3 months after the onset of symptoms and continued for 4 months. During this time his cardiac murmur disappeared, his blood pressure returned to normal, and his skull circumference increased nearly 1½ inches in diameter. He ultimately made a good recovery with no evidence of mental impairment.

10.3. Cortisone and Prednisolone

It has been recognized since 1951 that cortisone can lower the serum calcium level in the hypercalcemia which accompanies sarcoidosis (D6, L5, S4). In explanation of this Anderson *et al.* (A2) have put forward the view that cortisone may be antagonistic to vitamin D. Creery and Neill therefore used cortisone in one case, noting clinical improvement, a gain in weight, and a fall in serum calcium (C3). MacDonald and Stapleton (M1) confirmed that cortisone can cause a fall in serum calcium in idiopathic hypercalcemia and showed further that serum phosphorus and blood urea also fell. They also noted that cortisone had a more rapid and extensive effect but also a more transitory one on serum calcium concentrations than did a low-calcium diet. Further evidence of the value of cortisone has been given by Morgan *et al.* (M3). Forfar *et al.* showed that cortisone and prednisolone caused a fall in the serum calcium level and a reduction in calcium retention. They found, however, that even while causing these biochemical effects its administration did not always result in symptomatic improvement and that during its administration there could even be an exacerbation of symptoms (F7).

Thus, whereas the value of withholding vitamin D and of a low-calcium diet are in little dispute as methods of treatment in idiopathic hypercalcemia, the value of cortisone is more debatable. Its main use would appear to be in the early stages of treatment before the full effect of a low-calcium diet has had time to manifest itself, and it will also be indicated where it is not possible to reduce the serum calcium level to normal by dietary means alone. In milder cases in which the serum calcium level is little raised, it will probably not be required.

10.4. EDTA

EDTA, disodium ethylenediaminetetraacetic acid, a chelating agent which forms soluble complexes with multivalent metallic ions which are nonionized and readily excreted (S7), has also been used in treatment. Morgan *et al.* (M3) added EDTA powder to milk to produce a milk in which the calcium was rendered virtually inabsorbable because of its chelation. Unfortunately EDTA binds only one-tenth of its weight of calcium, and approximately 10 g would be necessary to produce a calcium-free diet of cow's milk. One-fifth of this dosage daily causes toxic effects in the form of loose stools. EDTA can also be given subcutaneously to combine with the calcium of the body fluids and thus reduce the plasma-calcium level. One gram in 100–200 ml of $M/5$ solution has been given over four hours. No really ill effects developed. The plasma calcium level fell to normal but rose again the following day. The urinary excretion of calcium increased considerably but not to the extent expected. Because of its toxic effects in the gut and other possible toxic effects, Morgan *et al.* do not recommend treatment with EDTA.

Addendum

Since this chapter was completed, several further papers on idiopathic hypercalcemia have appeared.

Further cases in Europe have been reported by Rinvik (R1a), Kendall (K1), Anderson *et al.* (A1), and Joseph and Parrott (J4). From the United States of America five more cases have been reported (Bongiovanni *et al.*, B4a; Farber and Craig, F3a; Snyder, S6a). A case has also been reported from Israel (K3).

From a study of the response to ACTH of the urinary excretion of 17-ketosteroids, Anderson *et al.* (A1) suggested that impaired adrenal function may be a factor in idiopathic hypercalcemia of infancy.

James *et al.* (J1) investigated the possible role of deficiency of essential fatty acids in idiopathic hypercalcemia. They found no evidence that an essential fatty acid deficiency is a cause of, or is present in, infants with idiopathic hypercalcemia. Much further investigation is still required to determine the role of essential fatty acids in hypercalcemia. Existing knowledge on the general aspects of the subject has been surveyed at the 4th International Conference on Biochemical Problems of Lipids (I1).

Joseph and Parrott (J4) analyzed the characteristic facies seen in severe degrees of the disease.

Kowarski (K3) has shown that the serum calcium level in idiopathic hypercalcemia can be reduced by the administration of sodium sulfate. Kendall (K1) has pointed out that prolonged low-calcium feeding may involve a low sodium intake and lead ultimately to severe sodium depletion.

A change has recently been made in the extent to which British National Dried Milk is fortified with vitamin D. The vitamin D content has been reduced from, not less than 280 units per ounce, to an average content of 100 units per ounce. Manufacturers of proprietary dried milks have been advised to make comparable changes. The vitamin D content of government-provided cod liver oil has been halved (H6a).

Concerning the etiology of idiopathic hypercalcemia, an interesting report has come from Fellers and Schwartz (F3b). By bio-assay methods they estimated the vitamin D-like activity of the serum of two severe cases of idiopathic hypercalcemia and found increased activity. They further found that this increased activity persisted unchanged for periods up to 17 months after all exogenous vitamin D had been removed from the diet. The degree of activity was greater than that found in 2 children receiving 100,000 units of vitamin D daily. These findings appear to support the suggestion that an endogenous vitamin D-like substance is present in idiopathic hypercalcemia of infancy.

References

A1. Anderson, J., Brewis, E. G., and Taylor, W., Adrenocortical function in hypercalcaemia of infancy. *Arch. Disease Childhood* **32**, 114-119 (1957).

A2. Anderson, J., Dent, C. E., Harper, C., and Philpot, G. R., Effect of cortisone on calcium metabolism in sarcoidosis with hypercalcaemia. *Lancet* ii, 720-724 (1954).

B1. Behrendt, H., "Diagnostic Tests for Infants and Children," p. 142. Interscience, New York, 1949.

B2. Bills, C. E., *In* "The Vitamins" (W. H. Sebrell, and R. S. Harris, eds.), Vol. I, p. 199. Academic Press, New York, 1954.

B3. Bird, T., and Thomson, J., Pneumocystis carinii pneumonia. *Lancet* i, 59-64 (1957).

B4. Blauberg, cited by Hoobler, B. R., The role of mineral salts in the metabolism of infants. *Am. J. Diseases Children* **2**, 107-140 (1911).

B4a. Bongiovanni, A. M., Eberlein, W. R., and Jones, I. T., Idiopathic hypercalcaemia of infancy with failure to thrive. *New Engl. J. Med.* **257**, 951-958 (1957).

B5. Bonham Carter, R. E., Dent, C. E., Fowler, D. I., and Harper, C. M., Calcium metabolism in idiopathic hypercalcaemia of infancy with failure to thrive. *Arch. Disease Childhood* **30**, 399-404 (1955).

B6. *Brit. Med. J.* **ii**, 149 (1956). Hypercalcaemia in infants and vitamin D.

B7. Butler, A. M., Wilson, J. L., and Farber, S., Dehydration and acidosis with calcification at renal tubules. *J. Pediat.* **8**, 489-499 (1936).

C1. Colver, T., and Illingworth, R. S., Hypercalcaemia in infancy. *Lancet* **ii**, 97 (1956).

C2. Creery, R. D. G., Idiopathic hypercalcaemia of infants. *Lancet* ii, 17-19 (1953).

C3. Creery, R. D. G., and Neill, D. W., Idiopathic hypercalcaemia in infants with failure to thrive. *Lancet* **ii**, 110-114 (1954).

C4. Creery, R. D. G., and Neill, D. W., Intake of vitamin D in infancy. *Lancet* **ii**, 372-374 (1955).

C5. Creery, R. D. G., and Neill, D. W., Infection in the aetiology of infantile hypercalcaemia. *Lancet* ii, 1357-1358 (1956).

D1. Daeschner, G. L., and Daeschner, C. W., Severe idiopathic hypercalcaemia of infancy. *Pediatrics* **19**, 362-371 (1957).

D2. Daniels, A. L., and Stearns, G., The nitrogen and mineral balances in infants receiving cow's and goat's milk. *Am. J. Diseases Children* **30**, 359-366 (1925).

D3. Dawson, I. M. P., Craig, W. S., and Perera, F. J. C., Idiopathic hypercalcaemia in an infant. *Arch. Disease Childhood* **29**, 475-482 (1954).

D4. Debré, R., and Brissaud, H. E., Action toxique de la vitamin D_2 administrée à doses trop fortes chez l'enfant. *Ann. méd.* **50**, 417-489 (1949).

D5. Dent, C. E., How to make decalcified milk. *Helv. Paediat. Acta* **10**, 165-166 (1955).

D6. Dent, C. E., Flynn, F. V., and Nabarro, J. D. N., Hypercalcaemia and impairment of renal function in generalised sarcoidosis. *Brit. Med. J.* **ii**, 808-810 (1953).

D7. Doxiadis, S. A., Idiopathic renal acidosis in infancy. *Arch. Disease Childhood* **27**, 409-427 (1952).

E1. Evans, D. M., Hypercalcaemia of infancy. *Lancet* **ii**, 998 (1956).

F1. Fanconi, G., and de Chastonay, E., Die D-Hypervitaminose im Säuglingsalter. *Helv. Paediat. Acta* **5**, 5-36 (1950).

F2. Fanconi, G., and Girardet, P., Chronische Hypercalcämie, kombiniert mit Osteosklerose, Hyperazotämie, Minderwuchs und kongenitalen Missbildungen. *Helv. Paediat. Acta* **7**, 314-334 (1952).

F3. Fanconi, G. and Spahr, A., Beiträge zur Frage der idiopathischen Hypercalcämie. *Helv. Paediat. Acta* **10**, 156-164 (1955).

F3a. Farber, S., and Craig, J. M., Clinical pathological conference. *J. Pediat.* **51**, 461-473 (1958).

F3b. Fellers, F. X., and Schwartz, R., "Vitamin D activity" in idiopathic hypercalcaemia. *Am. J. Diseases Children* **96**, 476-477 (1958).

F4. Ferguson, A. W., and McGowan, G. K., Idiopathic hypercalcaemia of infants; Low calcium treatment. *Lancet* **i**, 1272-1274 (1954).

F5. Flood, R. G., Calcium balance with hydrochloric acid milk. *Am. J. Diseases Children* **32**, 550-553 (1926).

F6. Foreman, H., and Trujillo, T. T., The metabolism of C^{14} labeled ethylenediamine tetra-acetic acid in human beings. *J. Lab. Clin. Med.* **43**, 566-571 (1954).

F7. Forfar, J. O., Balf, C. L., Maxwell, G. M., and Tompsett, S. L., Idiopathic hypercalcaemia of infancy: Clinical and metabolic studies with special reference to the aetiological role of vitamin D. *Lancet* **i**, 981-988 (1956).

F8. Forfar, J. O., and Tompsett, S. L., Idiopathic hypercalcaemia of infancy. *Lancet* **ii**, 306-307 (1956).

F9. Frazer, A. C., Cod liver oil and idiopathic hypercalcaemia. *Lancet* **ii**, 1215 (1956).

F10. Fyfe, W. M., Vitamin A levels in idiopathic hypercalcaemia. *Lancet* **i**, 610-612 (1956).

G1. Gajdusek, D., Pneumocystis carinii—Etiologic agent of interstitial plasma cell pneumonia of premature and young infants. *Pediatrics* **19**, 543-565 (1957).

G2. Gardner, L. I., MacLachlan, E. A., Pick, W., Terry, M. L., and Butler, A. M., Etiologic factors in tetany of newly born infants. *Pediatrics* **5**, 228-240 (1950).

G3. Glover, M. J., Glover, J., and Morton, R. A., Provitamin D_3 in tissues and conversion of cholesterol to 7-dehydrocholesterol in vivo. *Biochem. J.* **51**, 1-9 (1952).

H1. Hallman, N., Calcium metabolism in interstitial plasma cell pneumonia in infants. *Helv. Paediat. Acta* **10**, 119-122 (1955).

H2. Hallman, N., Tahka, H., and Ahvenainen, E. K., High plasma calcium and influencing factors in interstitial plasma cell pneumonia in infants. *Ann. Paediat. Fenn.* **1**, 34-39 (1954).

H3. Harris, L. C., Biochemical changes and osteosclerosis after sulphamezathine therapy in idiopathic hypercalcaemia of infancy. *Arch. Disease Childhood* **29**, 232-237 (1954).

H4. Harrison, H. E., Retention of nitrogen, calcium and phosphorus of infants fed on sweetened condensed milk. *J. Pediat.* **8**, 415-419 (1936).

H5. Harrison, H. E., Mechanism of action of vitamin D. *Pediatrics* **14**, 285-295 (1954).

H6. Hawk, P. B., Oser, B. L., and Summerson, W. H., "Practical Physiological Chemistry," 12th ed. Churchill, London, 1947.

H6a. Her Majesty's Stationery Office, Report of the Joint Sub-Committee on Welfare Foods (1957).

H7. Hess, A. F., and Lewis, J. M., Clinical experience with irradiated ergosterol. *J. Am. Med. Assoc.* **91**, 783-788 (1928).

H8. Hoobler, B. R., The role of mineral salts in the metabolism of infants. *Am. J. Diseases Children* **2**, 107-140 (1911).

H9. Houet, R., Récherches sur le metabolisme du calcium et du phosphore dans l'enfance: Effets de l'administration per os de 15 mg. vitamin D sur les bilans du phosphore et de calcium chez le nourrisson. *Ann. Paediat.* **166**, 177-192 (1946).

H10. Hubble, D., Hypercalcaemia in infancy. *Lancet* ii, 143-144 (1956).

I1. *4th Intern. Conf. on Biochem. Problems of Lipids, Oxford, 1957*, pp. 1-286.

J1. James, A. T., Webb, J., Stapleton, T., and Macdonald, W. B., Essential fatty acids and idiopathic hypercalcaemia of infancy. *Lancet* i, 502-504 (1958).

J1a. Jeans, P. C., Stearns, G., McKinley, J. B., Goff, E. A., and Stinger, D., Factors possibly influencing retention of calcium, phosphorus, and nitrogen by infants given whole milk feedings; Curdling agent. *J. Pediat.* **8**, 403-414 (1936).

J2. Jeans, P. C., and Stearns, G., Effect of vitamin D on linear growth in infancy: Effect of intakes above 1,800 U.S.P. units daily. *J. Pediat.* **13**, 730-740 (1938).

J3. Jelke, H., Über D-Vitamin Vergiftung. *Acta Med. Scand.* Suppl. 170, 345-363 (1946).

J4. Joseph, M. C., and Parrott, D., Severe idiopathic hypercalcaemia with special reference to the facies. *Arch. Disease Childhood* **33**, 385-395 (1958).

K1. Kendall, A. C., Infantile hypercalcaemia with keratopathy and sodium depletion. *Brit. Med. J.* **ii**, 682-683 (1957).

K2. Knapp, E. L., Factors influencing urinary excretion of calcium in normal persons. *J. Clin. Invest.* **26**, 182-202 (1947).

K3. Kowarski, A., Idiopathic hypercalcaemia. Treatment with sodium sulphate. *Pediatrics* **22**, 533-537 (1958).

L1. Leitch, I., Determination of calcium requirements of man. *Nutrition Abstr. & Revs.* **6**, 553-578 (1937).

L2. Lightwood, R., A case of dwarfism and calcinosis. *Arch. Disease Childhood* **7**, 193-208 (1932).

L3. Lightwood, R., Idiopathic hypercalcaemia in infants with failure to thrive. *Arch. Disease Childhood* **27**, 302 (1952).

L4. Lightwood, R., Signification des troubles du metabolisme dans la genèse du marasme. *Arch. franç. pédiat.* **10**, 190-193 (1953).

L5. Lightwood, R., Artificial feeding of infants. *Lancet* ii, 1230 (1954).

L6. Lowe, K. G., Henderson, J. L., Park, W. W., and McGreal, D. A., Idiopathic hypercalcaemic syndromes of infancy. *Lancet* ii, 101-110 (1954).

L7. Lovelock, F. J., and Stone, D. J., Cortisone therapy in Boeck's sarcoidosis. *J. Am. Med. Assoc.* **147**, 930-932 (1951).

L8. Lunseth, J. H., Kirmse, T. W., Prezyna, A. P., and Gerth, R. E., Interstitial plasma cell pneumonia. *J. Pediat.* **46**, 137-145 (1955).

M1. Macdonald, W. B., and Stapleton, T., Idiopathic hypercalcaemia of infancy. *Acta Paediat.* **44**, 559-578 (1955).

M2. Malmberg, N., Some histological organic changes after cod liver oil medication. *Acta Paediat.* **8**, 364-374 (1928).

M3. Morgan, H. G., Mitchell, R. G., Stowers, J. M., and Thomson, J., Metabolic studies on two infants with idiopathic hypercalcaemia. *Lancet* ii, 925-931 (1956).

P1. Payne, W. W., The blood chemistry in idiopathic hypercalcaemia. *Arch. Disease Childhood* **27**, 302 (1952).

P2. Putschar, W., Über Vigantolschädigung der Niere bei einem Kinde. *Z. Kinderheilk.* **48**, 269-281 (1929).

R1. Rhaney, K., and Mitchell, R. G., Idiopathic hypercalcaemia of infants. *Lancet* i, 1028-1033 (1956).

R1a. Rinvik, R., Idiopathic Hypercalcaemia. .Ann. Paediat. Fenn **3**, 464-468 (1957).

R2. Ross, S. G., Vitamin D intoxication in infancy: Report of 4 cases. *J. Pediat.* **41**, 815-822 (1952).

R3. Ross, S. G., and Williams, W. E., Vitamin D intoxication in infancy. *Am. J. Diseases Children* **58**, 1142 (1939).

R4. Russell, A., and Young, W. F., Severe idiopathic infantile hypercalcaemia: Long term response of two cases to low calcium diet. *Proc. Roy. Soc. Med.* **47**, 1036-1040 (1954).

S1. Schlesinger, B. E., Butler, N. R., and Black, J. A., Chronische Hypercalcaemie, kombiniert mit Osteosklerose, Hyperazotämie, Minderwuchs und kongenitalen Missbildungen. *Helv. Paediat. Acta* **7**, 335-349 (1952).

S2. Schlesinger, B. E., Butler, N. R., and Black, J. A., Severe type of infantile hypercalcaemia. *Brit. Med. J.* i, 127-134 (1956).

S3. Sherman, H. C., and Hawley, E., Calcium and phosphorus metabolism in childhood. *J. Biol. Chem.* **53**, 375-399 (1922).

S4. Shulman, L. E., Schoenrich, E. H., and Harvet, A. M., Effects of adrenocorticotropic hormone (A.C.T.H.) and cortisone in sarcoidosis. *Bull. Johns Hopkins Hosp.* **91**, 371-415 (1952).

S5. Sinclair, H. M., Fats and disease. *Lancet* ii, 101-102 (1956).

S6. Sissman, N. J., and Klein, R., Idiopathic hypercalcaemia of infancy. *Clin. Research Proc.* **4**, 36-37 (1956).

S6a. Snyder, C. H., Idiopathic hypercalcaemia of infancy. *Am. J. Diseases Children* **96**, 376-380 (1958).

S7. Spencer, H., Vankinscott, V., Lewin, I., and Lasslo, D., Removal of calcium in man by ethylene diamine tetra-acetic acid: Metabolic study. *J. Clin. Invest.* **31**, 1023-1027 (1952).

S8. Stapleton, T., and Evans, I. W. J., Idiopathic hypercalcaemia of infancy. *Helv. Pediat. Acta* **10**, 149-155 (1955).

S9. Stapleton, T., Macdonald, W. B., and Lightwood, R., Management of idiopathic hypercalcaemia in infancy. *Lancet* **i**, 932-934 (1956).

S10. Stearns, G., Soya bean flour in infant feeding: Study of relation of comparative intakes of nitrogen, calcium and phosphorus on excretion and retention of these elements. *Am. J. Diseases Children* **46**, 7-16 (1933).

S11. Stearns, G., *In* "Infant Metabolism," pp. 64-71, 80-89. Macmillan, New York, 1956.

T1. Telfer, S. V., Studies on calcium and phosphorus metabolism. 1. The excretion of calcium and phosphorus. *Quart. J. Med.* **16**, 45-62 (1922).

T2. Thatcher, L., Hypervitaminosis D with report of fatal case in child. *Edinburgh Med. J.* **38**, 457-467 (1931).

T3. Thatcher, L., Hypervitaminosis D. *Lancet* **i**, 20-22 (1936).

W1. Winberg, J., and Zetterström, R., Cortisone treatment in vitamin D intoxication. *Acta. Paediat.* **45**, 96-101 (1956).

AMINO ACIDURIA

E. J. Bigwood, R. Crokaert, E. Schram, P. Soupart, and H. Vis[1]

Department of Biochemistry, Faculty of Medicine, Brussels University, Brussels, Belgium

1. Introduction

1.1. Terminology and Scope of the Chapter

The term amino aciduria is not consistently used with the same meaning. According to certain clinical biochemists, it refers to pathological conditions only, just as the word uremia refers to a disorder and not to the blood urea level in general. Others use the term in a broader sense, referring to amino acid excretion in general, whether it is normal or not, in the same way as the word glycemia is used. We adopt the latter view, since there are normal conditions in which the amino acid output may be different from the usual output, although the condition is not patho-

[1] R. Crokaert and E. Schram have contributed to Sections 1 and 2, P. Soupart to Section 3, and H. Vis to Section 4.

logical. An instance of this is found in the hyperhistidinuria which occurs in the course of normal pregnancy.

The scope of this chapter refers essentially to the quantitative excretion of free amino acids. Whereas in the past, the available methods could give only an indication about the phenomenon in terms of a total estimate of amino nitrogen output, today a detailed estimate can be made of each amino acid excreted. Now that this can be achieved, it becomes possible to notice that the calculated value for the total nitrogen or amino nitrogen corresponding to all the free amino acids excreted does not agree with, and falls in fact below, the value actually measured by any of the classic procedures described for amino nitrogen determination. A critical analysis of the methods themselves and of the exact nature of the complex mixture of substances involved in the measurements leads to the conclusion that such agreement is not necessarily to be expected. Moreover, what is referred to by the expression "free urinary amino acids" requires close consideration if a clear and exact quantitative definition of it is to be given. It by no means concerns a simple and logical concept; it is a conventional and provisional one, depending largely on the nature of the methods actually available for the purpose. This will emerge from the discussion in Section 1.3.1.

As our knowledge of amino aciduria has developed chronologically, a striking observation has arisen, namely, that much information was gathered in the study of amino aciduria in disease long before an exact picture of the normal situation regarding amino acid excretion became available. This has often led to misleading interpretation of the pathological findings. It is only recently that a rather exact and complete account of amino acid excretion has been given for normal subjects; our paragraphs dealing with normal amino aciduria will therefore form one of the major parts of our subject.

1.2. Urinary Nitrogen Partition and Amino Nitrogen Excretion

When the protein supply in the diet varies between 70 and 125 g daily, the human adult nitrogen balance adjusts itself somewhere between 10 and 20 g of total nitrogen of which 9–18 g is excreted in the urine, chiefly in the form of urea (at least 80 %). Under these conditions, less than 3 % of human urinary nitrogen is normally in the form of amino acid nitrogen. In fact, the excretion range usually varies between 80 and 300 mg of amino acid nitrogen (M9). This includes total nitrogen of free as well as combined amino acids. If α-amino nitrogen alone is estimated and if the figure refers to free amino acids exclusively, the range is of course lower (see Section 3.1).

1.3. Amino Acid Excretion

1.3.1. *Substances Included in the Expression Amino Acid Excretion*

Free Amino Acids. In the past, free amino acid output in 24 hours was given either in terms of total amino acid nitrogen when colorimetric methods for estimating amino nitrogen were used, or in terms of α-amino acid nitrogen when the gasometric method of Van Slyke *et al.* was used (ninhydrin reaction and CO_2 determination). These authors indicate that the daily excretion of free amino acid nitrogen usually amounts to less than 150 mg (V6), whereas Hawk *et al.* give a range of 100 to 150 mg for the same method (H10), and according to Eckhardt and Davidson it is reported to be usually close to 118 mg (E1). In fact, total amino nitrogen or α-amino nitrogen determinations on unhydrolyzed urine samples refer not only to the nitrogen of free amino acids but also to that of the free amino groups of peptides or other substances, in which case the data obtained are necessarily too high for free amino acids alone; this is what is actually observed now that free amino acids can be individually estimated. According to Stein (S32), the average total daily excretion of free amino acids in six normal adult males amounted to 1.1 g, a figure which includes some 100 mg of glutamine; it also includes taurine. The corresponding calculated figure for α-amino nitrogen is 120 mg.

According to data obtained in our laboratory by Soupart (S24) (see Section 3.1), the results are lower still among 15 normal adults of both sexes; they vary between 0.35 and 1.18 g/24 hr, with an average level of the order of 0.8 g; in terms of α-amino nitrogen, the corresponding excretion levels are 41–133 mg/24 hr, with an average of the order of 90 mg.

We stress the point that the group we refer to here includes free amino acids, substituted free amino acids, or amino acid derivatives, in which the amino group stays free. Examples of such substances are taurine, glutamine, or asparagine (the last two being amino acids combined with ammonia) or tyrosine-*O*-sulfate (T2), an amino acid combined with sulfuric acid by a linkage which does not involve the amino group. Chromatographic separation of these compounds occurs when unhydrolyzed urine is used, and they are detected by the means employed for amino acids proper. The group also includes alcohol amines which are other types of amino acid derivatives in which the amino group remains free; for instance ethanolamine, which corresponds to decarboxylated serine.

Combined Amino Acids. This expression refers to a group of substances which has never been accurately defined. Since it concerns the extra amounts of free amino acids which appear in the urine after acid hydrolysis, it is logical to confine the group exclusively to *hydrolyzable compounds in which the amino acid moiety is combined by an amino group to another substance, whether the latter is another amino acid, in which case the compound is a peptide, or some other substance.* The group is therefore a rather heterogeneous one comprising chiefly substances of the second category, such as, for instance, hippuric acid or phenylacetylglutamine. Whether, in the case of urine, it also comprises peptides, except for minute amounts, is still a debated subject. In dealing more intensively with amino acid excretion in normal subjects (see Section 3.1) we shall see that a very great variety of substances is involved in the expression amino aciduria for unhydrolyzed as well as hydrolyzed urine specimens.

The amounts of combined amino acids to be found in normal urine may be very differently recorded, depending on what the group is considered to include. We shall see for instance, that Stein (S32) evaluates the normal amount of combined amino acids as about twice that of the free amino acids, but others report it to be as high as four times that amount (H12), depending on whether the monoamides of the dicarboxylic amino acids are or are not considered as combined amino acids; once again, we consider them as free amino acids. Other data on combined amino acids are referred to below (see Section 3.1.). Concluding these comments, we stress the importance of definitions, particularly the one indicated above with regard to the group of combined amino acids. On the other hand, *the group of free urinary amino acids consists, in our opinion, of the ninhydrin-reacting amino acids and closely related substances individually identifiable on chromatograms of unhydrolyzed urine specimens.* We propose this as a definition of amino aciduria and as the best for use with the most satisfactory methods of estimation available at present. For all practical purposes, this rather widened definition of the group is further justified from a physiological and pathological point of view, even though it may seem a little illogical from the purely chemical standpoint. The group does not include ammonia and urea although these are ninhydrin-reacting substances appearing on urine chromatograms. Moreover, the group may not necessarily include all the free amino acids which are eventually present in the sample. Should it contain, for example, β-amino-*n*-butyric acid, this will not appear in the chromatographic determinations because it gives no color reaction in solution with ninhydrin[2] (C20). Such a case, however, appears most

[2] When the color reaction is developed on paper, the situation is a different

probably to be a very exceptional one, and we do not even know whether the output of β-amino-*n*-butyric acid in urine does actually ever occur. It is important, however, to stress the point that such a case is possible.

1.3.2. *Methods of Expressing Data on Amino Acid Excretion*

In the case of adult subjects, the most suitable expression seems unquestionably to be one indicating the amounts (in mg) per 24-hr collection, both for free and for combined amino acids. When relative amounts are of interest, it is of course preferable to compute the amounts on a molar basis (S24). For the sake of possible comparison with other data than one's own, it is always highly desirable to indicate at the same time, total nitrogen output, 24-hr urine volume, body weight, and, if possible, body height.

When observations concern infants and children and are discussed in relation to the situation in adults, data are preferably expressed per kilogram of body weight or as the amino acid nitrogen in percentage of total nitrogen excretion. In principle, the most suitable procedure consists in comparing data with those usually observed in normal infants or children of corresponding age. Before this can be done, further knowledge concerning normal amino aciduria in growing children is required.

Some clinical chemists have found it suitable to adopt another unit basis of comparison, namely, a conventionally adopted amount of creatinine excretion. We feel that it is not advisable to encourage this procedure as one to be advocated as a standard for purposes of computation, particularly if it eventually becomes confirmed that creatinine excretion may not be as constant as it has been claimed to be, according to a recent publication (V7). In case the creatinine basis of reference is adopted, however, by certain workers, it is certainly highly desirable to indicate at the same time the urine volume output in 24 hours, the nitrogen output, as well as the creatinine output, and finally also the body weight.

In recent years, amino acid excretion has been more and more investigated in terms of blood clearance, and this method has contributed usefully to our knowledge of the mechanism involved in pathological conditions (C13, E5, P2). Ratios of a given amino acid output to its blood plasma level undoubtedly yield instructive information. There are three points however that deserve close consideration if misleading interpretations are to be avoided.

(*a*) In many cases our information regarding amino acid blood levels

one. A very slight purple spot is obtained in comparison with the strong reaction and dark purple spot obtained with α- or γ-amino-*n*-butyric acid at the same concentration (still unpublished observations).

and their eventual constancy seems to be still insufficient for the purpose of sound clearance measurements.

(*b*) The validity of blood clearance determinations in artificial conditions, such as those produced by overloading the blood stream by amino acid infusions, is debatable.

(*c*) When blood clearance applies to a waste product such as urea and when it may be taken that during a short period of observation on renal excretion the blood level of the considered substance stays sufficiently constant, the blood clearance concept is a fairly simple and clear one since the kidney is virtually the sole agent by which the substance is removed from the blood. In such a case, the procedure consisting in expressing the amount excreted per unit time in terms of the blood volume containing that amount has a clear meaning from the physiological point of view. But when, on the other hand, the substance considered is not a waste product but a metabolite, such as an amino acid, and when several highly metabolizing tissues compete with the kidney in withdrawing the substance from the blood, then surely it becomes more difficult to interpret safely renal clearance data.

Quite independently of these considerations concerning blood clearance, it is obvious of course that investigations dealing with renal reabsorption or tubular secretion are most important in the study of amino acid excretion in the urine.

2. Methods

2.1. General Trend of Their Evolution

At first, our knowledge of amino acid excretion was based exclusively on aggregate estimates of total amino nitrogen, or more specifically of α-amino nitrogen, both in unhydrolyzed and in hydrolyzed urine samples. Methods of individual chemical estimation of certain amino acids were also available although their accuracy in such a complex medium as urine was a matter of debate. This same observation holds also for more recent procedures, such as those based on polarography. The need was pressing in physiology and in pathology for more specific information on each of the amino acids excreted, giving an accurate and complete picture of amino aciduria. It gave rise to procedures involving enzymatic, microbiological, and chromatographic methods. Our comments will refer particularly to the third of these, partly because this chapter deals essentially with chemical or physical chemical procedures and also because recent contributions to the subject have been based predominantly on these methods. We shall therefore refer much more briefly to the others.

With the introduction of chromatographic methods, our knowledge of the subject entered into a second and important phase of development. Paper chromatography developed by Martin and Synge, has been applied to urine analysis in Dent's outstanding pioneer work (D7, D8). Moreover, it is also less involved, less expensive and less time consuming than column chromatography. It leads to qualitative identification of a group of amino acids which are most abundantly excreted and gives therefore some semiquantitative information on the relative importance of their output. It must be borne in mind, however, that ninhydrin reactions developed on paper are difficult to control and are therefore not suitable for accurate quantitative work. This explains why, notwithstanding the most valuable information obtained from paper chromatography data, a third phase of development of our knowledge soon became necessary, leading to a more accurate and more complete picture of amino aciduria based on more trustworthy quantitative information. Moore and Stein's contributions are conspicuously outstanding in importance for the development of this third phase. It is true to say that column chromatography still remains a rather too elaborate procedure to be considered as a standard method for routine purposes in a clinical laboratory. The most recent developments of their completely automatic devices (see below) will undoubtedly lead in a short time to widely applicable clinical investigations. Column chromatography will then become standard equipment for clinical investigation in combination with paper chromatography used both for preliminary screening and as a complementary technique for proper interpretation of certain column chromatography findings.

2.2. Quantitative Methods Not Involving Fractionation

2.2.1. *Total Amino Nitrogen, Free and Combined*

We shall not review the literature on this part of our subject; this has already been done in the classic textbooks and also in previous reviews (B37, H4, H10, N13, P6). The classic gasometric methods of Van Slyke *et al.* (V5, V6), as well as Soerensen's formaldehyde titration and Albenese and Irby's copper methods (A3) are the standard ones in use at the present time. Of these, the method of Van Slyke *et al.* (V5) is specific for free α-amino nitrogen groups. Moubasher (M21) and Moubasher and Sina (M22) proposed a similar procedure, in which ninhydrin is replaced by peri-naphthindan-2, 3, 4-trione hydrate, but questioned its application in the case of urine. Van Slyke and Kirk's nitrous acid method (V4) and Northrop's application of Soerensen's formaldehyde titration to urine analysis (N2) lead to higher results than the preceding ones; they include a greater number of amino nitrogen groups than the ones located only in the α-position; moreover, they include substances such as taurine. Colorimetric procedures for total free amino nitrogen determinations, such as Folin's naphtoquinone sulfonate procedure (F8) or Antener's modification (A7), require the removal of ammonia prior to the determination, with Permutit or similar adsorbents. Sobel *et al.* (S18) remove ammonia by ion exchange. They have also suggested modifications in Albanese and Irby's copper method. For peptides, Balikov and Castello (B3) have proposed a colorimetric method based on the biuret reaction. We have decided that methods for estimating metabolic derivatives of amino acids which no longer contain amino groups, such as phenylpyruvic acid for instance (K4, K10), do not fall strictly within the scope of this chapter.

2.2.2. *Individual Amino Acid Determinations*

Attempts to determine histidine in urine colorimetrically have been made by Kapeller-Adler (K2), Cheval and Hans (C10), and many others, but unsatisfactorily because of lack of specificity of the bromine reaction used (S22).

Pentz *et al.* (P5) estimate taurine with fluorodinitrobenzene in urine passed through Dowex 50 H^+ columns, but there are doubts as to whether this procedure is really specific for taurine (B38). Dent *et al.* have compared results obtained for the estimation of sulfur-containing amino acids in urine of cystinuric patients, by polarographic and microbiological methods (D18, D19). Hier (H12) and Schreier and Plückthun (S10, S11) have published data on amino acid excretion as determined microbiologically. Enzymatic methods have been used with success in the case of histidine in urine with specific decarboxylase preparations (S23).

2.3. Quantitative or Semiquantitative Methods Involving Fractionation

2.3.1. *Fractionation by Chromatography*

2.3.1.1. *Into Subgroups.* Because of the chemical complexity of urine composition, certain researchers have found it useful for both preparative and analytical purposes to separate the constituents by groups first. When this is done with ion exchange resins, allowing for partial desalting at the same time, the procedure is of interest when the further steps are obtained by paper chromatography. This involves successive treatments on anion and on cation exchangers (B37). In general, a first fractionation is carried out on urine with columns of sulfonated resins in the H^+ form. The filtrate contains fairly strong acids (taurine, cysteic acid, etc., as well as peptides). The filtrate is then passed through anion exchangers. Progressive elution of the first column then allows for the separation of the constituents retained on that column. Boulanger and Biserte's preliminary fractionation into groups of constituents was specially used for the study of urinary peptides and has served as a first step toward further separation of the latter by paper chromatography and paper electrophoresis. It must be stressed though, that in order to obtain such peptides from urine, a very large volume (20 liters or more) must be used, showing in what minute amounts they are to be found in urine.

2.3.1.2. *Into Individual Constituents*

Paper chromatography. As emphasis is put on the simplicity of this method, we must stress at the same time how cautious one must be about the proper interpretation of some of the results—we might even say, about their hazardous if not misleading character—in certain instances. The main drawback lies in the fact that they are essentially qualitative. They may eventually become semiquantitative when the samples analyzed are first desalted. Even then, merely developing the ninhydrin reaction on paper is not suitable for safe quantitative work, although satisfactory quantitative data may be obtained under optimal conditions. In the case of routine urine analysis this is not the case, certainly not if there is no desalting prior to analysis. Paper chromatography is, however, an essential complementary procedure to column chromatography for the identification of certain constituents obtained by the latter procedure. This has been the case, for instance, for taurine, 1- or 3-methylhistidine, β-aminoisobutyric acid, etc.

Paper chromatography of urine samples rests fundamentally on Dent's pioneer work (D7, D8). We shall refer essentially to reviews and monographs previously published on this part of our subject, which include full details of technique (B11, B23, B33, C19, L3, S5, T5). In a complex medium such as urine, the classic two-dimensional procedures are indispensable. The one-dimensional and also the circular procedures have only been used exceptionally (H2). It is hardly necessary to recall that large containers, holding several large-size sheets of paper at a time, are necessary. The temperature must be kept constant at $\pm 1°C$, notwithstanding adequate ventilation of the room allowing for proper outlet of vapor of the organic solvents used.

In order not to confuse the two-dimensional pictures with too many overlapping spots, only very small samples of urine can be used on paper and then only the most abundant amino acids will show up. This disadvantage cannot be overcome.

The sample must vary between 25 and 100 μl of urine depending on the 24-hr volume of urine. In assessing the amount required, it is safer, according to Parry (P4), to rely on total nitrogen output rather than on the density of the sample; the total nitrogen content of the sample should be of the order of 0.25 mg; Berger (B11) recommends a total nitrogen content of between 0.25 and 0.5 mg in order that the individual quantities of amino acids may be of the order of 10 to 30 μg. Dent (D7) uses samples of 25 μl.[3] These various recommendations depend on the solvents used. Dent uses urine as such for the sake of convenience and practicability in routine analyses, but other workers recommend preliminary separation of substances which interfere with adequate resolving power and with the satisfactory identification and separation of spots. Desalting by electrodialysis, using the procedure of Consden *et al.* (C16) may lead to destruction of certain amino acids, such as arginine (S33). Different desalting processes have been recommended. Zweig and Hood (Z2) desalt on the chromatographic paper itself, and by their method 5–25 μl of urine can be treated in 1 to 5 minutes, avoiding the usual dilution and reconcentration processes; Dowex-2 has also been used for this purpose, with 4 *N* acetic acid for elution (A12), or Amberlite IR120, with ammonia for elution (B41). In case of albuminuria, the protein is separated by heat coagulation (S16), or dialysis (B23), or alcohol precipitation (B16). Extraction methods are not quantitative but may be used to separate important amounts of foreign interfering substances. Various acids in acetone solutions have been tried, such as hydrochloric (B32), trichloroacetic, benzene-sulfonic, or *dl*-camphorsulfonic acids (R2).

The use of H_2O_2 for oxidation of methionine to sulfone and of cystine to cysteic acid seems to have been widely adopted.

The phenol-water, and collidine-lutidine process of separation is very widely used both for the ascending or descending technique, thanks to the corresponding reference charts of Dent (D8). Parry (P4) prefers the acetic acid-butanol mixture and has also published corresponding reference charts for 56 amino compounds. Busson and Guth (B41) use the same; Braun *et al.* (B39) use propanol, whereas Berger (B11) uses butanol-formic acid and has determined the R_f of 38 substances for this mixture.

For color reactions, ninhydrin or isatin are most widely used in acetone solution. Other reagents have however been proposed for specific purposes: a diazo reaction for histidine (B19, F13); Sakaguchi's reagent for

[3] According to H. Loeb, when the size of the sample is based on its α-amino nitrogen content rather than on its total nitrogen content, the chromatograms look very different (unpublished observations).

arginine; Ehrlich's reagent for hydroxyproline; and Gerngross' reagent for tyrosine (B23).

The spots are frequently stabilized by a copper nitrate spray (B11) in acetone solution (P4) rather than alcohol in which amino acid derivatives are more soluble.

For semiquantitative purposes, colorimetric estimates have been made directly on the spots or after extraction. In both cases, the spots must obviously be well separated from each other and show no evidence of diffuse spreading, conditions which are rarely met. Evaluation based on color intensity of the spot or its size are misleading. Even when compared to spots obtained in parallel with given substances, such as taurine at known concentration (B23) or with color scales, errors of 20 to 30 % are accepted as frequent. Boissonas and Lo Bianco (B35) claim that with colorimetric measurements after extraction the errors are reduced to about 10 %, but then the method becomes an elaborate one, the essential benefit of paper chromatography is lost, and one might as well employ the more elaborate and safer quantitative estimates obtained by column chromatography. In order to control color yields with ninhydrin properly, the reaction must take place under very definitely controlled conditions. When it develops on paper, the color yield cannot be safely controlled for several reasons. Differences in color yield, hence of color factors of various ninhydrin-reacting substances as compared with each other, with reference to those obtained under standard and optimal conditions, may vary extremely widely. For histidine they may drop 100-fold, for phenylalanine 25-fold, for arginine 20-fold compared with alanine or valine (S5). Although one may be well aware of this and try to take account of it, the fact remains that a paper chromatogram may give a very misleading picture of reality; we have often checked this when comparing results obtained by paper chromatography and by column chromatography on parallel samples of the same urine. The color factors are also very easily influenced by slight deviations from optimal temperature in the course of development of the ninhydrin reaction, and this is particularly the case for substances such as taurine, ethanolamine, histamine, etc. (P4). Biserte *et al.* (B30) have studied systematically the factors influencing the ninhydrin color reaction. They have also studied the possibility of combining electrophoresis and chromatography on paper, using volatile buffers.

Ion exchange chromatography. Apart from one exception (W15) of a partition chromatography on starch, the estimation of amino acids in urine by column chromatography has always been done on ion exchange columns. Like paper chromatography, it has the advantage of avoiding the effects of interfering substances and of detecting unexpected ones. This was the case for instance with 3-methylhistidine (T1). Many urinary constituents other than the ninhydrin-reacting ones are also separated by this procedure, but with ninhydrin, the ones that come into account are those containing a free amino group, and this includes peptides.

Column chromatography yields quantitative results even for constituents present in only small amounts (with the exception of tryptophan, methionine, and glutamine, in which cases the recovery is not complete; see below, (*b*) Dowex 50-X4 columns). It lends itself also to extraction on a preparative scale for further analysis of a particular constituent; it does not require previous desalting of urine. The major inconvenience of ion exchange column chromatography for routine analyses is that the technique is elaborate and time consuming, and requires experienced personnel and expensive equipment. Recent developments, however, have allowed for considerable reduction in time consumption, and the latest completely automatic devices open the way to serial analyses.

Moore and Stein's contributions are conspicuously outstanding in this field of research. They have described four successive procedures which must be briefly recalled with specific reference to their application to urine analysis.

(*a*) *Dowex 50-X8 columns, 1951 (M14)*. This procedure is suitable in many cases, namely, for protein analyses. It has been used in urine analysis (see below, Section 3), but because of the very numerous ninhydrin-reacting substances present in urine, their complete resolution is not obtained and certain peaks are mixtures. The procedures subsequently developed appear to be more suitable than this original one, since they have a higher resolving power. Even when additional eluting buffers are used for urine analysis with the original 1951 procedure (S32), serine, asparagine, and glutamine, α-alanine and α-aminoadipic acid, β-alanine and β-aminoisobutyric acid, 1-methylhistidine and lysine are not adequately separated; they can, however, be separated ultimately and estimated quantitatively by subsequent chromatography on new columns of adequate length with appropriate conditions of elution. Paper chromatography may serve also most usefully for analysis of the fractions forming a peak of doubtful interpretation.

Moore and Stein's original standard 1951 technique involves the use of 100×0.9 and 15×0.9-cm columns. Huisman (H27) uses the original technique, whereas van der Schaaf and Huisman (V2) have used 30×0.9-cm columns. Evered (E5) uses the technique with Dowex 50-X12 for urine analysis. His chromatograms are similar to those obtained with Dowex 50-X8. They show poor results in the separation of several amino acids. We have, in collaboration with Moore (S7) introduced certain modifications, including lower ninhydrin consumption, an automatic siphon device for automatic collection of fractions instead of the drop counting procedure, and other details of procedure (S6, S7).

(*b*) *Dowex 50-X4 columns, 1954* (*M15*). The chromatograms are more widely spread, hence the resolving power is increased. The number of 1-ml fractions is larger. Instead of the two columns of the 1951 procedure which required about 10 days to produce one complete analysis, the 1954 procedure requires the operation of one single column 150 $\times$ 0.9 cm with 100 % yield ($\pm$ 3 %) and requires about one week. A pH gradient is used; it replaces the use in succession of different buffer mixtures of different pH and ionic strength. It involves a new adjustment of the ninhydrin reagent which no longer requires the neutralization of the fractions prior to their colorimetric analysis (M16). *This 1954 procedure is in our opinion very definitely more suitable for urine analysis than the preceding one published in 1951.*

A certain tendency to overlapping of peaks is not always entirely overcome, however; this is especially the case for glutamine, asparagine, sarcosine, 1-methylhistidine, carnosine, and anserine (Fig. 2). It is true to say that Moore and Stein do describe methods for estimating individually each of these incompletely separated constituents, but then serial analyses of complex solutions, such as urine, are no longer practicable. This must, however, be borne in mind when interpreting urine chromatograms. It must also be stressed that in their 1954 method, Moore and Stein's proposed conditions of procedure are a compromise among various conditions, each of which is optimal for a given area of the chromatogram. When temperature and pH adjustments are introduced to improve the separations in a given area, they will produce overlapping and deterioration of the results in other areas of the chromatogram.

With regard to the three amino acids which are not quantitatively recovered within $\pm$5 %, the situation with this method is as follows. Recovery for tryptophan varies between 40 and 60 %; for methionine, it is of the order of 90 %; finally, part of the glutamine is hydrolyzed to glutamic acid and NH_3 (M15), but since fresh urine contains only glutamine, we express the amount in terms of total glutamine by adding glutamic acid and glutamine.

A drawback of Dowex 50-X4 preparations lies in the fact that with such a low percentage of cross linking as 4 %, there are relatively important variations in the volume of the resin in the course of operation, hence the column must be entirely repoured each time it has been used. Moreover, when new batches of Dowex 50-X4 are used, they must be standardized against a known mixture of amino acids, and they may need eventual adjustment by incorporation of a certain percentage of Dowex 50-X5.

Huisman and van der Schaaf (H30) have applied Moore and Stein's 1954 procedure, but with another resin, namely, a cation exchanger Zeocarb 225 (Zerolit 225). This modification does not seem to provide better results than the original procedure.

(*c*) *Amberlite IR 120-X8 columns, 1958* (*M17*). This procedure is a still more rapid one than the two preceding ones. The sulfonated resin (cation exchanger) which is used is in a micropowder form (Amberlite IR 120-X8) produced by the Rohm & Haas Co., Philadelphia. The reproducibility of its resolving power is remarkably constant from one lot to another. It is the small and homogeneous size of the resin particles which allows for a much more rapid separation and elution of the amino acids. Columns of 150×0.9 cm of 56 ± 9-μ particles are used for the neutral and acid amino acids, whereas columns of 50×0.9 cm of 40 ± 7-μ particles are used for the basic amino acids. The rate of flow reaches 15 ml/hr for the 150-cm column (complete operation in 60 hr), and 16–20 ml/hr for the 50-cm column (complete operation in 48 hr) under pressure of 20–40 cm of mercury. Notwithstanding this reduction in time, the number of 1-ml fractions collected stays the same, although 2-ml fractions may also be collected with satisfactory results. The changes of buffer solutions and of temperature are also reduced, limiting the inconveniences deriving from them. As far as resolving power is concerned certain difficulties are not overcome. Glutamine and asparagine are not separated from each other; the galactosamine peak interferes with tyrosine and phenylalanine. This recent method has not been used yet for urine analyses.

(*d*) *Completely automatic method, 1958* (S25). Fraction collectors and serial spectrophotometric measurements are no longer necessary. All the operations are automatic. Amberlite IR 120 is used. The method is applicable to urine (see Moore *et al., Federation Proc.* **17**, 1107, 1958). Full descriptions of the equipment and technique are given in the paper of Spackman, Stein, and Moore (S25). The rate of flow is doubled in comparison to the previous method (see above, (*c*) Amberlite IR 120-X8 columns), a complete operation taking 24 hours on the 150-cm column while one on the 50-cm column is completed overnight. Time and labor consumption are reduced very considerably. The method can also be used on Dowex 50-X4 columns, provided the original flow rate is maintained.

Thanks to this latest very remarkable improvement in technique, column chromatography is definitely likely to become a standard clinical laboratory method, notwithstanding the heavy initial investment it involves, because of the economy in labor consumption it allows for at the same time.

In summary we underline the fact that as far as urine analysis is concerned, and this is also true for blood, the most satisfactory results published so far are those obtained with Moore and Stein's 1954 procedure (see (*b*) Dowex 50-X4 columns); whereas the two latest improvements in technique open the way to much wider investigation in this field of biochemical and medical research. The technique is also applicable for the determination of a single amino acid, adequate shortcuts and simplifications in the procedure being introduced accordingly. As examples, we shall mention that appropriate rapid operations have been described for the separation on ion exchange columns and subsequent determination of β-alanine (C20), β-aminoisobutyric acid (E5) or γ-aminobutyric acid (W1). Similar adjustments may also be made for other amino acids, such as the basic amino acids, for instance, with Moore and Stein's 1954 procedure (M15).

Prior to the chromatographic separation of amino acids on Dowex 50 columns, Carsten (C5) first desalts the urine sample on Amberlite IR 100 or Duolite C 3 and removes most of the nitrogenous bases on Amberlite IRA 400. This preliminary treatment allows for amino acid separations at ordinary temperatures using $2\,M$ and $4\,M$ HCl on H^+ columns for elution, instead of buffer mixtures; a single column of 25 g of Dowex 50 is sufficient for all amino acids and 350–375 one-milliliter fractions are collected. The resolving power of this method does not seem to be as satisfactory as Moore and Stein's procedures, and it is not less time nor labor consuming.

We shall not endeavor to describe in this chapter the various types of fraction collectors that have been used in column chromatography. For urine analysis 1-ml fractions are necessary ($\pm$ 1 % error). In Moore and Stein's 1958 procedure (see above (*c*) Amberlite IR 120-X8 columns), 2-ml fractions can also be collected.

2.3.2. *Fractionation by Ionophoresis*

Little has been done with this procedure in the case of urine. Paper electrophoresis under high potential (10,000 volts) has yielded excellent separations of amino acids (W9). It has been applied to urine analyses by Nöller *et al.* (N1). Biserte *et al.* have used it for group fragmentation (B30) under potentials of 300 to 400 volts in volatile buffer mixtures at pH 2.4, 3.9, 6.5, and 8.9 and find it suitable as a first step prior to further, more complete separations by paper chromatography.

2.4. Conclusions

Up to the present time, Moore and Stein's 1954 procedure has given the most satisfactory results, both qualitative and quantitative, for the study of amino aciduria. Paper chromatography is useful as a screening process for routine urine analysis prior to the use of column chromatography for more accurate and complete analysis. It is also most useful for the analyses of fractions containing unknown constituents obtained by column chromatography or to check whether such fractions are single substances or mixtures.

3. Amino Aciduria in Normal Subjects

3.1. In Adults

In children above 2 years of age, adolescents, and adults, the situation is much the same. An over-all indication of the situation in normals has already been given in Section 1.3.

The most accurate and complete picture of daily excretion of free amino acids in normal adults irrespective of dietary habits is to be found in six publications (E4, E5, S22, S23, S24, S32) in the last of which all the data are computed. Stein and Evered have used Moore and Stein's 1951 procedure, whereas Soupart's data were obtained with Moore and Stein's 1954 method. Table 3 and Fig. 1 give an account of the situation for both sexes as described by Soupart (S24). In Table 3 the relative amounts of free amino acids excreted are given in 10^{-2} mM/24 hr. Glutamic acid is absent from freshly voided urine (S32). The peak found with stored urines may come in part from hydrolysis of glutamine. Glutamine, asparagine, and sarcosine, when the latter is present, are found in the same peak. Any attempt therefore to arrive at exact values for glutamine in urine from the chromatographic results is of doubtful value. Some cyclization of glutamic acid into pyrrolidone-carboxylic acid which does not react anymore with ninhydrin is also a possible error in the determination of this amino acid.

Soupart's data for histidine excretion are lower than those of Stein and of Evered because they do not include 3-methylhistidine. The data of Stein and Evered include 3-methylhistidine because their figures were obtained with Moore and Stein's 1951 method and not their 1954 method which was used by Soupart (S24).

Normal urine contains only traces of γ-aminobutyric acid and of tryptophan (only a few milligrams in 24 hours). We have already referred to the fact that tryptophan is an amino acid poorly recovered by Moore and Stein's 1954 technique.

Comparison with values from the literature obtained by microbiological methods has been discussed by Stein (S32). The main objection to microbiological methods lies in the fact that amino acid derivatives or combined amino acids may be as readily available to the microorganisms used in these methods as are the parent amino acids from which they

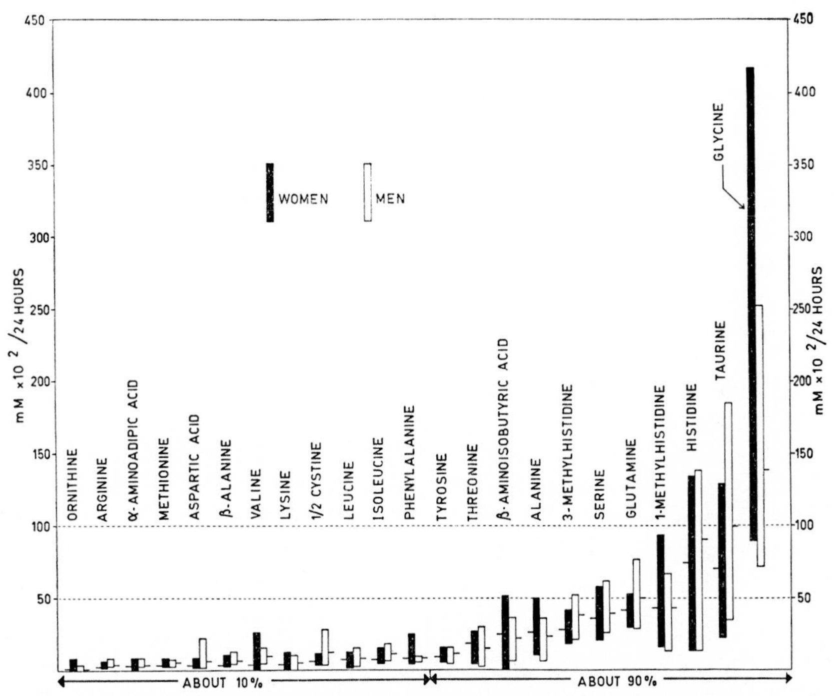

FIG. 1. Free urinary amino acids of healthy adults. This figure illustrates the relative amounts in which free amino acids are excreted by the kidney, giving a characteristic picture of amino aciduria in normal adults. The data are expressed in 10^{-2} mM free amino acids in 24-hour urine specimens collected from *nine* females and *six* males; they are drawn from Table 3 (on the ordinate scale the figure 200 corresponds to 2 mM). For the amino acids with an output of less than 0.2 mM (twenty 10^{-2} mM), the amounts are too small to be estimated with accuracy; they correspond to traces. The bulk excretion of the first twelve amino acids amounts to less than 10 % of total excretion of free amino acids. The marks on the side of each range indicate arithmetic mean values; they show *no* significant differences according to sex. This is so even in the case of glycine. The figures for the output of this amino acid varied between 71 and 252 10^{-2} mM in both sexes except for one discrepant figure of 416 in the case of one of the females. This exceptionally high output fell out of the range of the arithmetic mean ± twice the standard deviation. Since, however, the subject was clinically an apparently healthy person, we have not discarded this marginal figure from the normal range.

derive, and these microorganisms may therefore not distinguish the free amino acids from their derivatives for the purpose of their growth. This difficulty does not seem, however, to apply to the case of histidine. The influence of low-protein diets on amino aciduria has been studied (H12) with the use of microbiological methods.

Proline, hydroxyproline, glucosamine, and lanthionine have regularly been found absent from normal urine. Hydroxylysine and ethanolamine have not been looked for systematically. Moore and Stein's 1954 procedure enables one to recognize some 40 ninhydrin-reacting constituents in normal urine. Twenty-nine of them correspond to free amino acids, including taurine, asparagine, and glutamine. The remaining unidentified peaks, about 12, correspond to unknown amino acids, or to their metabolic derivatives, or to peptides. Ninhydrin-reacting peptides containing up to 8 or 10 amino acid residues should show up with Moore and Stein's 1954 procedure as well-defined peaks, provided their color factor allows for it. The dozen unidentified peaks are well defined but also very small in size, either because their output is minute, or because their color factor is low. Several of them are acid stable. Acid hydrolysis of normal urine produces a marked increase in some, and a slight increase in nearly all, of the free and identified amino acids. The bulk excretion of combined amino acids is about twice that of the free amino acids (S32). They consist predominantly of substances such as hippuric acid or phenylacetylglutamine and include only traces of peptides (B37, W10).

3.1.1. *Amino Aciduria in Normal Men*

The data are grouped in Table 1 as well as in Fig. 1 drawn from Soupart's publication (S24; diets not controlled).

3.1.2. *Amino Aciduria in Normal Women*

The data are given in Tables 2 and 3 as well as Fig. 1 drawn from the same publication (S24; diets not controlled). No gross sex-linked differences are noticeable, although in females, there seems to be a sex-induced regulation controlling the excretion of some of the amino acids during the menstrual and reproductive cycle (S22, S23). Soupart's data for normal women correspond therefore to 24-hr urine collections obtained between the 14th and 20th days of the menstrual cycle because for histidine at least there is a maximum urinary output at that time.

It appears from Soupart's data that normal men excrete 0.35–1.18 g of free amino acids in 24 hours, whereas in the case of women the range is 0.48–1.11 g, the average outputs being 0.80 and 0.73 g, respectively. According to Stein, the mean output in adult males was of the order of

1.1 g. Calculated in terms of corresponding α-amino nitrogen, Soupart's data amount to 41–133 mg in males and 59–132 mg in females (average outputs, 91 and 87 mg respectively). These figures check fairly well with the value ordinarily stated in the literature, 1 % of the total nitrogen output (8–18 g of urinary total nitrogen daily).

Among the 23 substances listed in Table 3, and usually found in normal adult urine, 22 are free amino acids, the twenty-third being an amino acid derivative, taurine, the abundance of which ranks second with histidine, after glycine, in decreasing order of abundance. The average levels of free amino acid output may be listed as follows in per cent of the total amount, on a molar basis: glycine, 29 %; taurine, 14 %; histidine, 13.5 %; 1-methylhistidine, 7.5 %; serine, 6.4 %; 3-methylhistidine, 5.5 %; α-alanine, 4.4 %; β-aminoisobutyric acid, 4.3 %; threonine, 3 %. These 9 amino acids amount to 87.5 % of the total free amino acid output. The amount of glutamic acid in glutamine, or as glutamic acid, is a major quantity, 50 to 100 mg per day, but is not included in these calculations. The remaining 13 amino acids together form only 12.5 % of the total output. Differences in the average values of these 13 acids are therefore not to be taken as significant.

A typical chromatogram as obtained with a 24-hr urine specimen of a normal adult is given in Fig. 2.

3.1.3. *Histidinuria in the Course of Normal Menstrual Cycle*

Soupart (S22) showed that the output of histidine in the urine varies during the normal menstrual cycle and reaches a maximum value between the 15th and 20th day. This coincides with a similar peak of excretion of phenolic steroids, suggesting a possible relation between histidine excretion and hormonal activity. In still unpublished observations, he finds similar correlations during pregnancy and lactation.

3.1.4. *Amino Aciduria in General and Histidinuria in Particular in the Course of Normal Pregnancy*

Color reactions for testing histidinuria are unreliable. The daily output of free histidine correctly measured varies as follows in normal adults (S22, S24): in men (11 subjects), 30–212 mg/24 hr, average, 140 mg; in nonpregnant women (27 subjects), 20–208 mg/24 hr, average, 115 mg.

In some of these cases, the figures given for histidine may be slightly on the high side, since 3-methylhistidine (T1) happens sometimes not to be completely separated from histidine (S22, S23).

Daily fluctuations are smaller in a given subject. Within the limits

TABLE 1

URINARY EXCRETION OF FREE AMINO ACIDS IN NORMAL HEALTHY MALES
(mg/24-hr collection)

Amino acid	Range[a,e,f]			Average values[a,e,f]		
	Stein[c] 6 cases	Evered[c] 3 cases	Soupart[b,d] 6 cases	Stein	Evered	Soupart
Ornithine	—	—	0–4[k]	—	—	1
Arginine	—	—	0–14	< 10	—	6
α-Aminoadipic acid	—	—	5–13	—	—	8
Methionine	—	5–9	5–11	< 10	6	7
Aspartic acid	—	—	3–29	< 10	—	8
β-Alanine	—	—	3–10	—	—	6
Valine	4–10	5–8	4–17	< 10	5	10
Lysine	7–48	7–17	0–14[k]	19	12	7
Cystine	10–21	5–15	3–33	< 10	9	14
Leucine	10–25	5–10	6–20	14	8	11
Isoleucine	10–30	5–22	8–24	18	13	15
Phenylalanine	9–31	11–18	8–15	18	14	13
Tyrosine	15–50	18–26	7–27	35	23	19
Threonine	15–50	17–37	2–35	28	28	17
β-Aminoisobutyric acid	—	9–18	6–37	—	13	22
Alanine	20–70[h]	17–37[h]	5–32	46	28	22
3-Methylhistidine	—	33–47[i]	35–87[k,l]	—	40	65
Serine	25–75	44–134[m]	27–65	43	93	42

TABLE 1 (*Continued*)

Amino acid	Range[a,e,f]			Average values[a,e,f]		
	Stein[c] 6 cases	Evered[c] 3 cases	Soupart[b,d] 6 cases	Stein	Evered	Soupart
Glutamine[g]	40–100	—	42–103	—	—	—
1-Methylhistidine	50–210	9–29	22–114[k]	180	22	73
Histidine	110–320[j]	59–130[j]	20–213[k]	216	97	138
Taurine	85–300	35–81	44–231	156	59	123
Glycine	70–200	69–148	53–189	132	109	104

[a] Values lower than 10 mg/24 hr are not to be taken as absolute.

[b] Not included in Soupart's data are citrulline, which is part of the glycine peak, and γ-aminobutyric acid, traces of which seem to be present in almost all cases but which has not been definitely identified by use of the usual criteria (Stein, 1953, S32). Tryptophan, not recovered quantitatively on Dowex 50, is not included.

[c] 100-cm column of Dowex 50-X8 or X12 (Moore & Stein, 1951, M14).

[d] 150-cm column of Dowex 50-X5, elution gradient of gradually increasing pH and concentration (Moore & Stein, 1954, M15).

[e] — implies that no range is given by authors.

[f] 0 implies that, at least in one of the cases investigated, no peak was found for this amino acid. The arithmetic mean was thus calculated including the number of cases where no peak was found.

[g] The chromatograms always indicate the excretion of considerable amounts of glutamine (and glutamic acid), but it is not possible to arrive at accurate values for glutamine (see text). The values for this amide are therefore to be considered as very approximate, indicating merely that it is found in urine in considerable amounts.

[h] Includes α-aminoadipic acid (100-cm column).

[i] Values obtained on 15-cm column, which has a poor resolving power in this range in the case of urine.

[j] Histidine peak obtained on 100-cm column contains 3-methylhistidine. These results, although minimal owing to incomplete recovery of the basic amino acids in experiments of this type, involve an overestimation for histidine.

[k] These are to be considered as maximal values because of elevated blank reading in the range from the NH_3 peak to the 3-methylhistidine peak (Moore & Stein, 1954, M15). For histidine, values obtained by an enzymic specific decarboxylation method are a little lower than those obtained on 150-cm column (Soupart, 1958, S23).

[l] May include some anserine and carnosine, if present in urine.

[m] Contains asparagine and glutamine, calculated as serine.

TABLE 2
URINARY EXCRETION OF FREE AMINO ACIDS IN NORMAL HEALTHY FEMALES
(mg/24-hr collection)

Amino acid	Evered (1 case)[c] Single value[a,e]	Soupart (9 cases)[b,d]	
		Range[a,f]	Average[a]
Ornithine	—	0–11[k]	2
Arginine	24	0–11	4
α-Aminoadipic acid	—	0–13	4
Methionine	5	3–12	5
Aspartic acid	—	2–11	4
β-Alanine	—	2–9	3
Valine	5	0–30	6
Lysine	8	0–16[k]	8
Cystine	5	0–13	6
Leucine	5	2–16	9
Isoleucine	5	5–20	10
Phenylalanine	9	6–41	13
Tyrosine	11	9–26	15
Threonine	11	5–33	23
β-Aminoisobutyric acid	5	10–52	29
Alanine	12[h]	9–44	24
3-Methylhistidine	21[i]	30–69[k,l]	48
Serine	30[m]	22–61	37
Glutamine[g]	—	43–88	—
1-Methylhistidine	10	26–155[k]	65
Histidine	130[j]	79–208[k]	128
Taurine	80	27–161	87
Glycine	24	67–312	142

For footnotes see Table 1.

indicated above certain women excrete histidine at a relatively high mean level (above 100 mg), others at a lower one (below 100 mg); hence an output of 130 mg, for instance, may be normal among certain individuals but abnormally high in others. When this specific level is not known, which is of course the usual case, it may be taken that a daily excretion of 210 mg is an upper limit. To be on the safe side a rounded figure of 250 mg may be adopted. In normal pregnancy, the output rises above 200 mg very early within the first week and may fluctuate between 200 and 600 mg. Among 39 cases, the average level was 334 mg/24 hr within the first 14 weeks of pregnancy. Hyperhistidinuria is therefore an early sign of pregnancy in normal women.

In regard to the urinary excretion of other free amino acids than histidine during normal pregnancy, the situation is as follows. There is hyperexcretion of glycine, glutamine (including glutamic acid), serine, phenylalanine, tyrosine, α-alanine, arginine, and to a lesser degree, isoleucine, leucine, α-aminoadipic acid, methionine, lysine, and β-alanine.[4]

3.2. Amino Aciduria in Children, in Infancy and Prematurity

In children above 2 years of age, the situation is not substantially different from adults as far as the relative amounts of amino acids excreted are concerned. In fact, only very few really satisfactory data have been obtained so far in urines from children above 2 years of age. We present therefore in Fig. 3 a typical chromatogram of a 24-hr specimen of urine of a normal child, 2 years old, as drawn from Vis' observations. Comparison of quantitative data concerning daily excretions are rendered difficult because of the differences in body size. The 24-hr basis alone is not suitable. It is necessary to compare data on the basis of percentage of total nitrogen excretion, or per kilogram of body weight, a procedure adopted by Jonxis and Huisman. We do not believe it suitable or safe to choose a comparison factor based on creatinine excretion.

In early infancy the situation is known to be quite different from that among older children. Simon noticed in 1911 that the newborn infant excreted 10 % of its total nitrogen output in the form of amino nitrogen, whereas in adults it drops to 2 % (S15). This was confirmed in later years (B4, C11, G5) and extended to the study of individual amino acid excretion by microbiological methods (S10, S11), by paper chromatography (S20, S21), and finally by ion exchange column chromatography (D28, D30) at the time Moore and Stein's first method of 1951 was still the only one available.

Dustin *et al.* (D28, D30) have studied the daily excretion of the 18 amino acids usually found in proteins; they have expressed their output in terms of α-amino nitrogen. By summing up these figures and by expressing the total amount so obtained in per cent of total nitrogen excretion of the 24-hr samples of urine, they found more than 4 % in a prematurely born infant and 2.9 % in a full-term infant 25 days of age, whereas the corresponding figure in the case of normal adults was less than 1 % (S32). This peculiar situation at birth, which disappears during the first months of life, is to be correlated with the still incomplete maturation of renal function in early infancy and consequently to a condition probably related to a temporary deficiency in the process of tubular reabsorption. Our insufficient knowledge of amino acid blood

[4] For more detailed information on amino aciduria in relation to hormonal influences and to pregnancy, see P. Soupart's Ph. D. Thesis ("L'aminoacidurie de la grossesse," Les Editions Acta Medica Belgica, Brussels, 1959; also *Ann. soc. roy. sci. méd. et nat. Bruxelles,* 1959, **12**, 33-185).

TABLE 3
FREE URINARY AMINO ACIDS OF HEALTHY ADULTS
(mM × 10^2/24 hour collection)[d]

Amino acid	Women			Men			General range for both		Average in % of total average
	Lower	Higher	Average	Lower	Higher	Average	sexes	Average	values[c]
Ornithine	0	8	1	0	3	0.5	0–8[a]	1.1	0.17
Arginine	1	6	2	2	8	3	1–8	2.6	0.41
α-Aminoadipic acid	0	8	3	3	8	5	0–8[a]	3.4	0.54
Methionine	2	8	3	2	7	5	2–8	3.6	0.57
Aspartic acid	1	8	3	1	22	6	1–22	4.1	0.65
β-Alanine	2	10	3	4	12	6	2–12	4.7	0.74
Valine	0	26	3	4	15	9	0–26[a]	7.2	1.13
Lysine	0	12	4	0	10	5	0–12[a]	4.4	0.69
½ Cystine	3	11	6	3	28	12	3–28	8.2	1.29
Leucine	1	12	7	2	15	8	1–15	7.4	1.17
Isoleucine	4	15	7	6	18	11	4–18	9.4	1.48
Phenylalanine	4	25	8	5	9	8	4–25	8.0	1.26
Tyrosine	5	15	9	4	15	11	4–15	9.4	1.48
Threonine	4	27	19	2	30	15	2–30	16.8	2.65
β-Aminoisobutyric acid	0	51	25	6	36	22	0–51[a]	25.2	3.97

[a] Ornithine was found in 4 out of 15 cases studied, α-aminoadipic acid in 13 out of 15, valine in 11 out of 13 (2 determinations lost), lysine in 10 out of 14 (1 determination abnormally high, not included), β-aminoisobutyric acid in 14 out of 15.

[b] Average of 27 female subjects and of 11 male subjects.

[c] Only very approximate values for glutamine (see footnotes of Tables 1 and 2, and also text); consequently the percentages given in the last column on the right are also approximate ones. They are calculated on the basis of an average output of 51.5 10^{-2} mM of glutamine. However, even if one supposes this latter figure to be definitely too low, say by some 50%, a recalculation of the percentages in the last column would be only insignificantly altered.

[d] Values expressed in one hundredths of a mM; 0.1 mM = ten 10^{-2} mM.

TABLE 3 (*Continued*)

Amino acid	Women			Men			General range for both sexes	Average	Average in % of total average values
	Lower	Higher	Average	Lower	Higher	Average			
Alanine	10	50	26	6	36	24	6–50	25.7	4.05
3-Methylhistidine	18	41	28	21	52	38	18–52	32.3	5.09
Serine	21	58	36	26	62	40	21–62	37.4	5.89
Glutamine[c]	30	53	—	29	77	—	29–77	—	8.11
1-Methylhistidine	16	93	43	13	67	43	13–93	43.3	6.82
Histidine	13	134	74[b]	13	137	90[b]	13–137	79.0	12.45
Taurine	22	129	70	35	185	99	22–185	81.2	12.80
Glycine	89	416	190	71	252	138	71–416	168.7	26.58

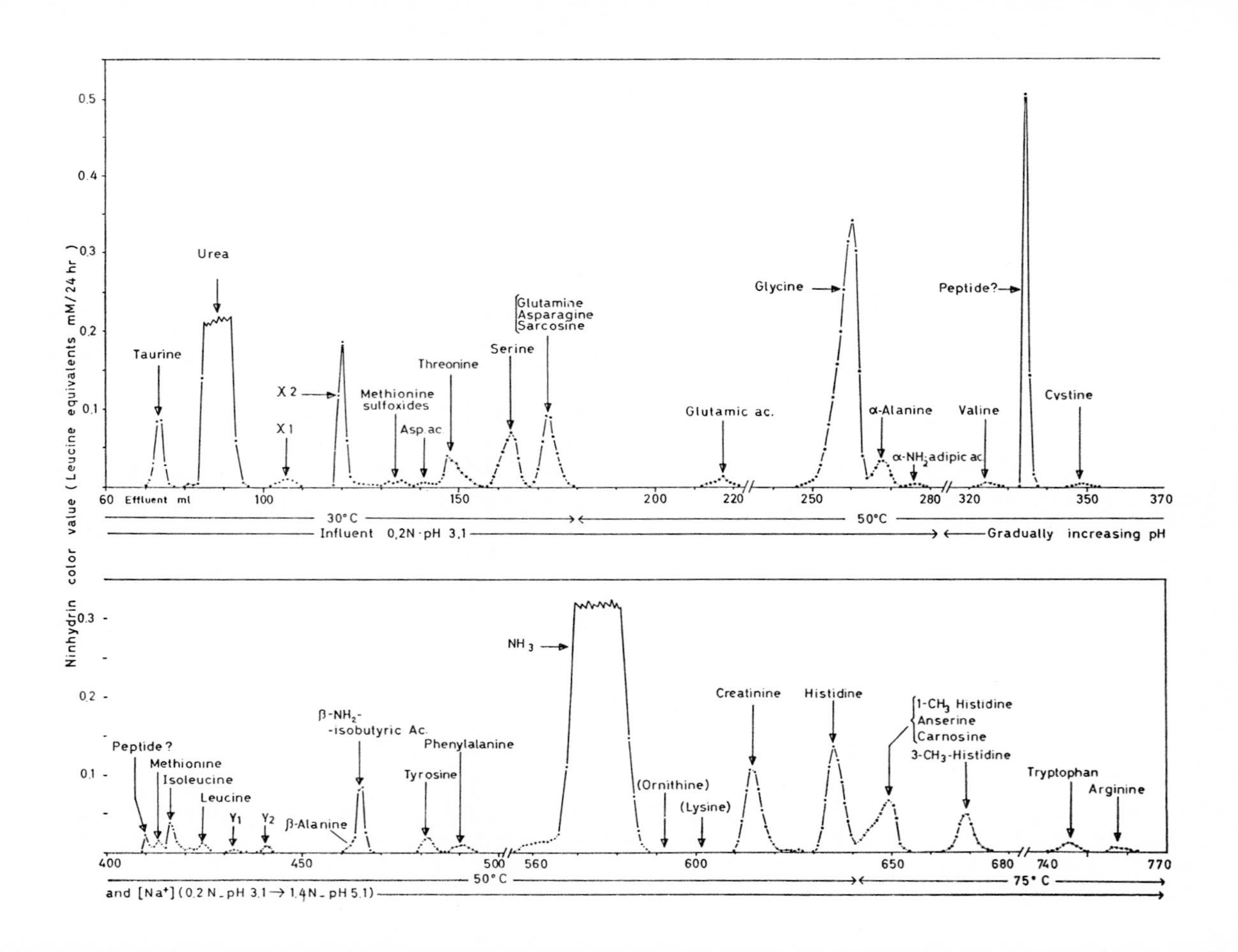
Ninhydrin color value (Leucine equivalents mM/24 hr)
0.1
0.2
0.3
0.4
0.5
Taurine
Urea
X 2
X 1
Methionine sulfoxides
Asp. ac
Threonine
Serine
Glutamine
Asparagine
Sarcosine
Glutamic ac.
Glycine
α-Alanine
α-NH$_2$adipic ac
Valine
Peptide?
Cystine
60 Effluent ml
100
150
200
220
250
280
320
350
370
30°C
50°C
Influent 0,2N · pH 3,1
Gradually increasing pH
Peptide?
Methionine
Isoleucine
Leucine
Y$_1$
Y$_2$
β-Alanine
β-NH$_2$-
-isobutyric Ac.
Tyrosine
Phenylalanine
NH$_3$
(Ornithine)
(Lysine)
Creatinine
Histidine
1-CH$_3$ Histidine
Anserine
Carnosine
3-CH$_3$-Histidine
Tryptophan
Arginine
400
450
500
560
600
650
680
740
770
50°C
75° C
and [Na$^+$] (0.2 N. pH 3.1 → 1.4N. pH 5.1)

levels in these conditions prevents us, so far, from having a better understanding of the situation.

3.3. Factors Influencing Amino Aciduria and the Mechanism of Renal Excretion

The filtration-reabsorption theory of renal function indicates that the two main factors which come into play are the amino acid blood levels and the tubular function (both reabsorption or eventually secretion). For early references to studies of filtration and tubular reabsorption of amino acids in the kidney, see S17.

Dent (D7, D8) has based his classification of pathological processes of amino aciduria on these two fundamental factors (see Section 4.1). Amino acid blood levels are influenced in turn by various other indirect factors. Exogenous supply of amino acids is a dietary factor which has been considered for instance in the case of low-protein diets (H12).

It is generally admitted that within the limits of normal variations in protein intake, diet does not influence appreciably the excretion of amino acids and the blood amino acid content (B14). Even on high-protein diets reaching 3–4 g of protein per kilogram of body weight, the total urinary nitrogen increases but the ratio of amino nitrogen to total nitrogen drops. Combined amino acids in the urine do, however, tend to be increased in those conditions (H29).

But many metabolic processes are likely to influence amino acid blood levels and urinary excretion both ways, some of which may be genetically controlled specifically or else may be influenced by a variety of body functions. Endocrine factors have for instance been taken into consideration in the case of histidinuria (S22). It seems, however, that in our present state of knowledge we are still altogether insufficiently informed about amino acid blood levels and on their possible fluctuations. As long as the data published in the literature on amino acidemia remain as scarce as they are at the present time, we feel that our knowledge of the

---

Fig. 2. Chromatogram of normal adult female subject (urinary amino acids, daily output at mid-cycle; 1,010 ml/24 hr; sample: 2 ml). Elution curve from Dowex 50-X5 of free amino acids and closely related substances. Peaks X_1, X_2 and Y_1, Y_2 are found regularly but correspond to unknown substances; they do not disappear after hydrolysis. The peak marked "peptide?" in the region near fraction 410 is usually present and disappears after hydrolysis. The one at fraction 325 also disappears after hydrolysis, but its presence was seen only once in a group of more than 50 different samples of urines of normal subjects, including cases of pregnancy (Soupart, footnote p. 223). Amino acids in brackets are usually found but were practically absent from the present sample of urine.

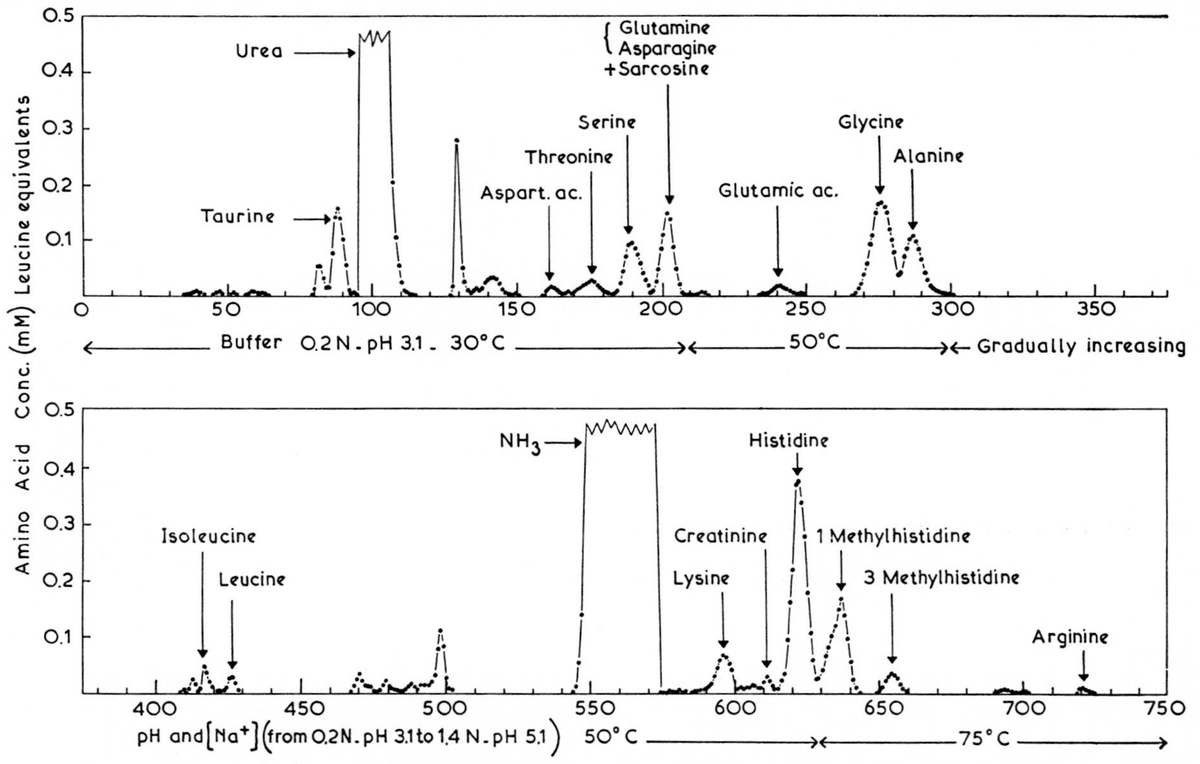

Fig. 3. Chromatogram of a 24-hr urine specimen of a normal child, 2 years old (D24) (sample: 1.5 ml; 240 ml/24 hr). In order to compare this with the case of a normal adult (Fig. 2), note that in Fig. 2 the sample amounted to 0.20 % of the total volume of the 24-hr specimen, whereas in the case of this child it was 0.625 %. With Dent's procedure on paper, the sample amounted to only 0.025 ml, and seven spots showed up. In order to detect the other amino acids present, the sample should be larger but then the spots would tend to overlap and the picture would be confusing.

blood clearance of amino acids will remain unsatisfactory (see also our comments on this matter in Section 1.3.2).

Another factor which influences the urinary excretion of a given amino acid is the overloading effect on the kidney function produced by another amino acid introduced into the body parenterally in relatively high doses. Several instances of such effects are referred to below (see Section 4.2.1, Cystinuria, for instance).

Dustin has also collected experimental data on these mutual influences of certain amino acids on others in regard to their excretion in the urine (still unpublished data). It is because of those mutual effects and of the very scanty information we have on them that clearance tests for amino acids, as observed in overloading conditions of their administration, seem difficult to interpret at the present time. Many publications concern such investigation under rather artificial conditions with regard to blood clearance of amino acids. There are data, however, which have been obtained without overloading the kidney artificially by parenteral injections of large amounts of an amino acid (C13) (also footnote p. 223).

4. Amino Aciduria in Disease

4.1. Dent's Classification

Dent's classification is the basis of most classifications of abnormal types of amino aciduria. He divides the pathological conditions into three categories (D9): (*A*) *"Overflow amino aciduria,"* in which the blood level of the amino acid concerned is definitely raised above normal [example: Section 4.2.1.1. (*a*)]; (*B*) *"Renal amino aciduria,"* in which case the blood level is either normal or below normal, and yet the urinary excretion is above normal [tubular reabsorption deficiency; example: Section 4.2.1.1. (*e*)]; (*C*) *"No threshold amino aciduria,"* in which Dent attributes the condition to an extrarenal disturbance of the metabolism of an amino acid with a high blood clearance, and in which, on account of this latter circumstance, the corresponding blood level may stay normal or is hardly increased.

Dent's proposal presupposes that our knowledge of amino acid blood levels and of clearances is sufficiently accurate to make it possible to classify easily all observed disturbances into one of those three categories. Once again, that is far from being the case in many instances; our present volume of information seems to be still far from what it should be, as far as quantitative data are concerned, to render this possible. We have found it preferable therefore, and for the time being, to adopt the following classification.

4.2. Provisional Classification of Pathological Disorders Adopted in the Present Chapter

4.2.1. *Congenital Amino Aciduria*

4.2.1.1. *Disturbances in the Metabolism of the Amino Acid(s) Concerned*

(*a*) *Phenylpyruvic oligophrenia.* There are four review articles on the subject (J4, K4, L2, S9). The condition was first described by Fölling (F9). The patients have fair hair and blue eyes (little melanin pigments). The biochemical lesion may be seen in subjects with normal IQ (C15, L9). The concentration of phenylalanine in the blood and in the cerebrospinal fluid is high (B23, F10, J2, J5). There is a high urinary excretion of phenylalanine, of phenylpyruvic acid (F9, J2), of phenyllactic acid (Z1), and of phenylacetic acid, chiefly in the form of phenylacetylglutamine (W13). The condition is attributed by Jervis to a catabolic block of phenylalanine; its oxidation to tyrosine is hampered (inhibition of phenylalanine oxidase; J3). This block is not complete, one-tenth of the phenylalanine being still convertible into tyrosine (U1). The urine of these patients also contains *p*-hydroxyphenylacetic, *p*-hydroxyphenyllactic and *o*-hydroxyphenylacetic acids (A10). On the basis of these findings, Boscott and Bickel (B36) have suggested that the metabolism of phenylalanine is blocked in more than one stage of the process. Other researchers interpret the findings differently (B28, D1, D2). Studies of the excretion of indole derivatives, namely indoleacetic, indolelactic and 5-hydroxyindoleacetic acids have suggested that the metabolism of tryptophan is also disturbed (A8).

The large quantities of L-phenylalanine accumulated in the tissues have been held responsible for the inhibition of tyrosine catabolism, hence the reduced production of melanins in comparison with normal individuals (W13). Phenylpyruvic acid accumulation has also been claimed to be responsible for that inhibition (B28). Phenylalanine clearance is considered to stay normal (E5, J7). Phenylalanine-poor diets have produced slight temporary improvements in the condition (A9, B18, B25, B31, H19, H24, P1). It has also been claimed that phenylketonuria is a condition that can be improved by adding glutamine or glutamic acid, asparagine, or glycine to the diet (H1, M10). The patients are all homozygotes, whereas the parents are heterozygotes; the latter can be detected by an overloading test with phenylalanine (B17, H23).

(*b*) *Tyrosinosis.* Medes (M8), in 1932, described the case of a man suffering from myasthenia gravis, who excreted daily 1.6 g of *p*-hydroxyphenylpyruvic acid. She noticed that the intake of phenylalanine produced an increase in the output of tyrosine, and of *p*-hydroxyphenylpyruvic and *p*-hydroxyphenyllactic acids, whereas it did not produce an increased excretion of 2,5-dihydroxyphenylpyruvic acid nor of homogentisic acid. Since this is the only such case published so far, the reality of an authentic congenital biochemical lesion has been questioned (S9), considering that it is well established now that the excretion of *p*-hydroxyphenylpyruvic acid may be abnormally increased in certain noncongenital conditions of hepatic deficiency (F3, F4, G6), of malignant tumors, of infections (G6), and also in vitamin C deficiency (L6, M20).

(*c*) *Alcaptonuria.* It is not well established whether alcaptonuric urines show evidence of abnormal amino acid excretion. Certain research workers claim that they do not (B12, B23, L1, S12), whereas Careddu *et al.* find that there is an increased excretion of tyrosine, of phenylalanine, of threonine, and of valine (C4). In our laboratory Vis has studied a case of alcaptonuria on Dowex 50-X5 columns. He found a very definite increase in tyrosine and in phenylalanine excretions; he located a peak of a still unidentified nonhydrolyzable constituent yielding a yellow color on reaction with ninhydrin (still unpublished observations).

(*d*) *Hartnup's syndrome* is a rare manifestation of a "familial" character described by Baron *et al.* (B5): cerebellar ataxia associated with pellagroid dermatosis, excretion of indoxylacetic acid and indican (B5, J1), and an abundant amino aciduria probably of renal origin (E5). It is suggested that the metabolic block affecting tryptophan interferes with its conversion into nicotinic acid, and tryptophan is thus excreted as such or in the form of indole derivatives.

(*e*) *De Toni-Debré-Fanconi syndrome.* This is a condition described in a series of three initial publications by pediatricians: de Toni in 1933 (D21), Debré *et al.* in 1934 (D3), and Fanconi in 1936 (F1). In certain cases of dwarfism associated with vitamin-resistant hypophosphatemic rachitis, renal diabetes (renal glycosuria and phosphaturia), there is an excessive urinary output of organic acids. These were ultimately recognized to be amino acids (M3). De Toni in 1956 surveyed the historical development of our knowledge of this syndrome (D22). Today the complete picture includes dwarfism, polydypsia, polyuria, late vitamin D-resistant rachitis, hypophosphatemia, normal or high blood calcium level, low blood potassium level, glycosuria, slight albuminuria, high uric acid excretion, and an abundant amino aciduria (R4). The syndrome finds its origin in a disturbance of the proximal tubules. Bickel *et al.* (B27) have claimed that all cases of this syndrome are in fact manifestations of cystinosis. De Toni pointed to the confusion this view has brought about (D22) and the matter now seems to be definitely settled. The etiology of the condition is not clear. The influence of heavy metals and of lysol has been considered (see Section 4.2.2.3), as well as that of maleic acid (H9) of glycogenosis (F2), and of the toxic effect of vitamin D derivatives (V1). The syndrome has also been considered to be an eventual complication of lipoidic nephrosis (H15, T3, W14). The clinical picture is not always a complete one. It may be reduced to renal glycosuria only, or to phosphaturia alone, or a combination of one of these two manifestations with excessive amino aciduria (J15, L12, S14). In the latter case, a familial character of the manifestation has been considered (E5, L12, S14). Other partial manifestations of the syndrome have also been described (D23).

Wallis and Engle (W2) have reported cases among adults ("Milkman syndrome"), the condition being eventually attributed to the toxicity of heavy metals or of Bence Jones proteins (E3). According to Dent and Harris (D11) the condition is transmitted as a recessive gene.

This hyperamino aciduria is a general one involving most of the amino acids. From his urine and blood paper chromatographic findings, Dent considers the condition to be one of tubular deficiency of the kidney and correlates the findings with those concerning the glycosuria in the same subject (D7). According to Evered amino acid clearances are increased in two cases of this syndrome and in one case of cystinosis (E5). He also recognized the excretion of citrulline.

Dubois *et al.* (D24) have made an intensive study of a case of de Toni-Debré-Fanconi syndrome in a 2-year-old child in which there were no cystine deposits; they also found an abundant excretion of citrulline, but no correlation between glycosuria and amino aciduria in the course of a glucose tolerance test (Fig. 4).

It has been claimed that calcium gluconate infusions reduce both phosphate and amino acid excretion (R4, R5). This has been only partly confirmed (F20) and needs to be further investigated.

(*f*) *Cystinosis.* This condition begins by presenting itself in a form that suggests a de Toni-Debré-Fanconi syndrome; consequently many clinicians consider both as being one and the same disease, sometimes referred to as *Lignac-Fanconi disease* (B27), since Lignac was the first to have described cystinosis (L7). In fact, Abderhalden had published, as far back as 1903, a post-mortem observation of a 2-year-old child in which he found important deposits of cystine in the liver and spleen (A1). Several general reviews of the literature on this subject have already been published (B27, F18, H14, W15), although many of them are incomplete. The specific feature of the condition is a deposit of

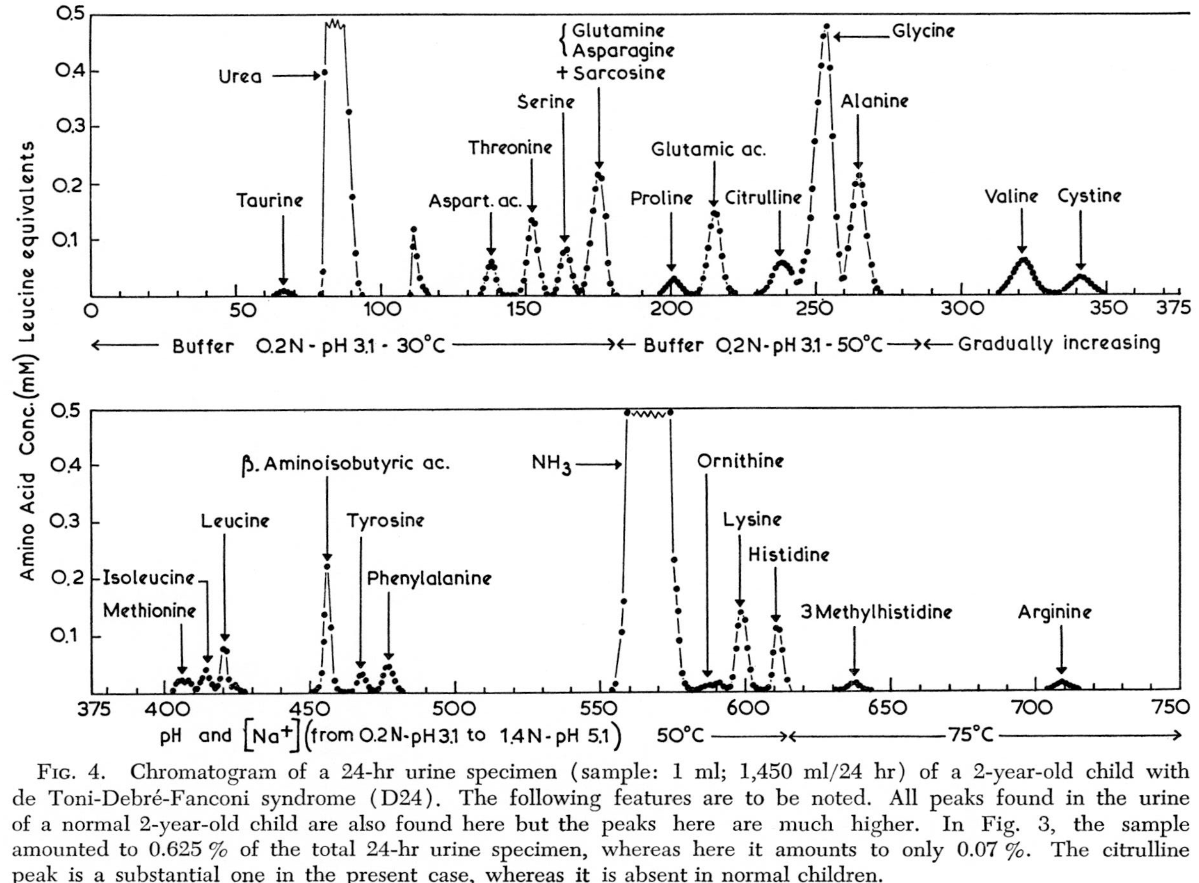

FIG. 4. Chromatogram of a 24-hr urine specimen (sample: 1 ml; 1,450 ml/24 hr) of a 2-year-old child with de Toni-Debré-Fanconi syndrome (D24). The following features are to be noted. All peaks found in the urine of a normal 2-year-old child are also found here but the peaks here are much higher. In Fig. 3, the sample amounted to 0.625 % of the total 24-hr urine specimen, whereas here it amounts to only 0.07 %. The citrulline peak is a substantial one in the present case, whereas it is absent in normal children.

cystine crystals in the reticuloendothelial tissue of the liver, spleen, and bone marrow, in the cornea, and in the kidney, chiefly in the proximal renal tubules. In acute cases, the de Toni-Debré-Fanconi syndrome appears with its complete symptomatology as well as cystinosis. In chronic cases, however, the renal diabetes may be less apparent and glycosuria may even be absent (C7, D20, D22, H20, L7, W8); there may be pentosuria instead (L7); in other cases the rachitic symptoms may be absent (K5). Satisfactory observations on the amino acid urinary excretion in cystinosis are scarce. Chromatography on starch columns has shown the absence of taurine from urine, a small content of conjugated glycine, but a high content of glutamic and aspartic acids after hydrolysis (W15). Bickel and Hickmans (B22) claim that the plasma levels of these amino acids are high; they believe that the condition is related to a disturbance other than renal (B22) and that their clearance is also on the high side (see also D20).

The condition develops progressively and becomes detectable only after several months (B20). That tyrosine crystals are also deposited as well as cystine was not confirmed (B1, F19).

Cystinosis can be produced experimentally in dogs by administration of 0.60 to 0.75 g of cystine per kilogram, and this produces a generalized hyperamino aciduria as well as albuminuria and finally mellituria in the premortal stage of the intoxication (S27, S29). This experimental cystinosis is attributed to a toxic effect of cystine at first on the renal functions and ultimately also on other tissues (S28, S30). From the genetic point of view, little is known. Some researchers believe cystinosis to be related to cystinuria (A1, H14, H16, P7).

(*g*) *Classic cystinuria* is a condition in which four amino acids are excreted in abnormal amounts in the urine: cystine, lysine, arginine, and ornithine. It may be, but is not necessarily, accompanied by cystine calculosis. This particular situation of hyper-amino aciduria affecting specifically four amino acids was not recognized at first. Originally it was the excessive output of cystine only that was detected because it is almost insoluble in slightly acid aqueous media and therefore is easily recognized as producing a urinary sediment and eventually urinary calculosis (see also D13).

Excretion of cystine in abnormal amounts can occur in many other cases: for instance in cystinosis, the characteristic feature of which is the deposit of cystine crystals in the body tissues, although cystinosis is not always accompanied by excessive excretion of cystine alone in the urine. In de Toni-Debré-Fanconi syndrome there is also an increased cystine output, but this is part of a generalized hyperamino aciduria such as occurs also in hepatic hyperamino aciduria. We insist therefore that the condition we refer to as classic cystinuria has the specific feature of a high excretion of cystine, arginine, ornithine, and lysine, whether or not it is accompanied by cystine calculosis.

Cystinuria has been known for a long time (A2, H18, Y1), but it was Dent and Rose (D12) who noticed, in 15 patients suffering from cystine calculosis, that cystine was not the only amino acid abnormally abundantly excreted by the kidney. They observed that arginine and lysine were also excreted in abnormally large amounts, and Stein (S31) reported this to be true of ornithine as well. Bickel refers to the condition by using the expression lysine cystinuria, but this is in fact misleading because it infers that these are the only two amino acids concerned, whereas that is not the case. Taurine excretion, on the other hand, is low (S31). The cystine plasma level is either normal or low (D19, F11).

Dent *et al.* have tested the effect of ingestion of cystine, cysteine, or methionine on the blood level and urine output of cystine, using microbiological and polarographic methods; their findings are considered to confirm the renal mechanism producing cystinuria. Only the administration of cysteine can raise the cystine level in the blood and its output in the urine, both in normal subjects and in cystinuric patients; cystine clearances in the latter are 20–30 times more important than in the normal subject (D18, D19).

The competitive mechanism of cystine, lysine, arginine, and ornithine excretion in normal subjects has been stressed by testing the effect of intravenous injection of 5 g of lysine (R3). Small variations in cystine excretion are accompanied by corresponding fluctuations in lysine output and vice versa, whereas important variations in cystine or lysine excretion are necessary to influence the arginine output (H5, H7, H8). The genetics of the condition are complex. The gene is presumably recessive and patients with calculosis are presumably homozygotes (D10, H6). Some of the offspring of heterozygotes show no anomalies (recessive cystinuria), whereas others are excreters of cystine and lysine only, their output of arginine and ornithine being presumably normal (incomplete recessive cystinuria; H5, H7, H8). Pfändler and Berger (P7) have studied two families in which there were 3 cases of cystinosis, 8 cases of cystinuria associated with lysinuria, and 19 cases of more generalized hyperamino aciduria; they believe therefore that these various conditions are closely related. This same point is also discussed by Hooft and Herpol (H14).

(*h*) β-*Aminoisobutyric acid excreters.* The presence of this amino acid in urine was first observed by Crumpler *et al.* (C21) in 1951. It has been held to be related to the metabolism of nucleic acids, and in particular to pyrimidine metabolism (F5). It has been recognized to be present in minute amounts in most human urines, but certain apparently normal subjects excrete large amounts of it; of 345 subjects in Great Britain some 10 % are in that sense β-aminoisobutyric acid excreters (H3). This observation was made by paper chromatography (see also Section 3.1 for more accurate quantitative data on the majority of normals).

The paper chromatographic findings were obtained by judging the relative intensity of the β-aminoisobutyric acid and alanine spots. The high excretion of β-aminoisobutyric acid as seen in one out of 10 normal subjects is considered a hereditary character, monogenic and recessive (G2, H3). Similar observations were made in Italy (C2). In Gambia (Central African tribes) 25 % of the individuals surveyed were β-aminoisobutyric acid excreters and, here too, this character is hereditary and recessive (L8). Among Asians the percentage of β-aminoisobutyric acid excreters is higher still (D5); it is independent of age, sex, and diet, but it is considered to be under the control of more than one gene. The distribution curve is bimodal; among Mongoloid subjects 60 % are β-aminoisobutyric acid excreters, whereas among the Negroid type of individuals the frequency is of the order of 40 %, and this percentage drops to 10 % among Caucasians (G3). Originally it was considered to be a renal type of amino aciduria (E5), but this has become doubtful; the question remains unsettled; Dent includes it rather in his third category of amino aciduria processes ("no threshold" category), the threshold being very low (D9), if there is one.

4.2.1.2. *Other Metabolic Disturbances Accompanied by Amino Aciduria*

(*a*) *Wilson's disease.* This condition has been described as a congenital biochemical lesion concerning copper metabolism and characterized by: (*1*) the absence or marked reduction of ceruloplasmin (S1) with either a reduction in the total copper content of the plasma or a relative increase in the albumin-bound copper fraction (C6); (*2*) a pronounced copper excretion in the urine (B7, M5); (*3*) a deposit of copper in various tissues such as the central nervous system, the cornea, the liver, and the kidney; and (*4*) perhaps also an excessive intestinal absorption of copper (B40, C6, M6).

The hyperamino aciduria which very often accompanies this condition was first described by Uzman and Denny-Brown (U3); it has been confirmed since (B26, C17, P8), although it is, however, not always observed; it has been considered to be of renal origin (C17) because the α-aminonitrogen level in the plasma is normal or subnormal (B26, C17).

A thorough investigation of this type of amino aciduria has led to a more accurate and more complete picture of it (S34). Taurine excretion is low; proline, citrulline, and cystine excretion may be abnormally high, and when this is the case there does not seem to be a relation between the cystine output, which may be considerable, and that of lysine. Moreover, whereas the intake of a meal has but little influence on amino aciduria in normals, it has a definite effect on the output of amino acids in the case of Wilson's disease; finally the amino acid fasting levels in plasma are at the lower limit of normal.

The cause of this amino aciduria is still undetermined. It has been reported that copper excretion may be increased under the influence of ingestion of large amounts of glycine or alanine (M7). This however has not been confirmed (I1, W4), but it does seem as if the ingestion of protein tends to increase the excretion of copper and of amino acids (B8, I1). BAL (British Anti-Lewisite) treatment which induces negative copper balances by stimulating copper excretion does not seem capable of influencing significantly the output of α-amino nitrogen (D6, U2, W5). Two theories of pathogenesis have been advanced to explain the condition. The one invokes the lack of synthesis of ceruloplasmin and the assumption that the free copper fraction develops a heavy metal toxic effect on the proximal renal tubules (B6); the situation is presumably comparable to a Fanconi syndrome, considering the association of amino aciduria, glycosuria, phosphaturia, and excessive uric acid excretion, and osteomalacia; clearance estimates in 11 cases point presumably to a deficient tubular reabsorption of those substances (B9). The other theory is based on the assumption of an abnormal protein synthesis (U4), the peptides so formed having a special affinity for copper which is presumably eliminated in the urine in some sort of organic compound (U2); a cupric peptide can be isolated from the liver in the case of Wilson's disease, whereas normal livers fail to show evidence of the presence of such a compound. Walshe (W5), however, has severely criticized this theory.

(*b*) *Galactosemia* is a condition due to a biochemical lesion preventing the conversion of galactose-1-phosphate into glucose-1-phosphate (S13). The congenital enzyme defect concerns the uridyl transferase; it has been located in the erythrocytes (K1) and in the liver (A6). Galactose accumulates in the tissues and is related there to various disturbances, such as cirrhosis, cataract, mental deficiency, or renal tubular deficiency. The condition may be either an acute or a chronic one. Holzel *et al.* (H13) were the first to discover that hyperamino aciduria completes the picture. The α-aminonitrogen plasma level was reported to be normal (H25, K8). The hyperamino aciduria was confirmed by paper chromatography (B21, C18, F12, H26). It does not occur constantly (A11, H26). On galactose-free diets, galactose disappears from the urine, and at the same time hyperamino aciduria usually retrocedes also; they both reappear a week or ten days after galactose consumption (K9). The α-amino nitrogen excretion in the urine has been reported not to vary during the course of a galactose tolerance test (B24). On the basis of α-amino nitrogen clearance determinations, the mechanism of this hyperamino aciduria is claimed to be of the renal type (C22). It seems however that a hepatic factor is also involved (B10, B21), and in our laboratory, ion exchange column chromatography of blood plasma and urine specimens from an acute case of galactosuria with marked hepatic disturbance, revealed that there were, along with extremely high plasma amino acid levels, a high urinary excretion of taurine, β-aminoisobutyric acid, and γ-aminobutyric acid. This hyperamino aciduria retroceded in the case of this patient when he was put on a galactose-free diet (V3).

(*c*) *Hypophosphatasia.* A condition characterized by the absence or at least a marked diminution of plasma and tissue alkaline phosphatase activity, producing a vitamin-resistant rachitis (R1) with normal or hypercalcemia and normal phosphate level in the plasma (F15) and a high urinary excretion of phosphoethanolamine (F17, M1, S2). The concentration of this ester is also high in the plasma (F16). On perfusion of the ester, a patient suffering from hypophosphatasia excreted it in the same way as a normal subject does (M2).

(*d*) *Lowe's syndrome* is a condition which was first described in 1952 (L11). It is characterized by glaucoma, mental deficiency, thermoregulation disturbance, abundance of subcutaneous fat, acidosis, lack of ammonia formation in the kidney, rachitis, increased output of organic acids, hyperamino aciduria which is only partly responsible for the high excretion of organic acids, normal glycemia, and normal amino acid blood levels.

Bickel and Thursby-Pelham (B24) have observed a similar situation in a child with cataract, dwarfism, rachitism, hypotonia, increased output of organic acids, hyperamino aciduria, and high plasma levels of glutamine and glutamic acid, whereas the phosphate level in the blood was low, the alkali reserve was slightly below normal, and ammonia production in the kidney was normal. Debré *et al.* (D4) have described a similar case with high amino aciduria, and a high output of glutamic acid and pyruvic acid.

Monnet *et al.* (M12) report that among 15 children below one year of age, suffering from congenital glaucoma, one developed a typical Lowe syndrome and four others presented hyperamino aciduria as the only extra pathological feature. In three other cases of this disease (two brothers and one third cousin), one of which reached adult age, there was a definite hyperamino aciduria, namely, a large excretion of glutamic acid by the two brothers (G4).

(*e*) *Muscular dystrophies.* In 28 cases of progressive muscular dystrophy, one case of dermatomyositis, and one case of congenital amyotonia, a hyperamino aciduria was recorded by two-dimensional paper chromatography (A5), mainly concerning taurine, arginine, leucine, threonine, and valine (H32). These observations were confirmed in other cases (S4), although not in all patients (B23).

(*f*) *Miscellaneous diseases.* Hyperamino aciduria has been reported to occur in several other congenital metabolic disturbances, but in only isolated and rare cases and, unfortunately, with altogether too scanty information, so that we are still far from knowing enough of the pathological conditions concerned to be able to comment on them usefully. They include the following: gargoylism (S3); a case of coproporphyrinuria accompanied by symptoms of rachitis and of ariboflavinosis (B15); a case of osteosclerosis (B34) in which the hyperamino aciduria was presumably associated with a high blood level of cystine and valine; a case of idiopathic hypoproteinemia (D31); fatal cases of progressive familial infantile cerebral dysfunction ("maple sugar urine disease"; M11) in which the abnormal urinary constituent was presumably a peptide formed of valine, leucine, and isoleucine (W11); cases of mental deficiency (A4, T4); and a case of infantile congenital cirrhosis (B2).

4.2.2. *Acquired Amino Aciduria*

4.2.2.1. *In Hepatic Disturbances.* It has been well known for a long time that liver function deficiencies influence amino nitrogen metabolism. They were only noticeable, however, in serious hepatic damage. In the case of relatively benign forms of virus hepatitis, however, blood levels and renal excretion of amino acids may be noticeably disturbed. Some of these alterations were identified with the aid of microbiological methods (F14) which showed that the output of cystine, histidine, or tryptophan may be sometimes abnormally high and sometimes abnormally low. Among 25 patients suffering from subacute cirrhosis, amino aciduria was found to be abnormally high in some 40 % of the cases and abnormally low in 20 % of the others (D27). In minor hepatic deficiencies amino aciduria was normal (microbiological determinations), but in more severe cases high blood levels occurred, particularly with regard to methionine, lysine, tyrosine, and tryptophan (S8), and in an acute case of yellow atrophy of the liver the output of these same amino acids in urine was found to be markedly increased. Kinsell *et al.* (K6) recognized that methionine, given parenterally to patients suffering from hepatic deficiency, disappears much more slowly from the bloodstream than it does in the case of normals; they noticed that plasma methionine fasting levels are increased only in cases of severe liver damage. Among 20 cirrhotic patients with severe liver deficiency, the output of methionine was definitely increased in more than one-half of them (G1; microbiological determinations); tryptophan excretion is also frequently increased, whereas for isoleucine and histidine the output is rather decreased if not normal; the situation is not influenced by diet, but it is related to the severity of the disease. Paper chromatography was used by Cachin *et al.* (C1); they stress the fact that the output of tyrosine is increased. In an extensive paper chromatographic study concerning 119 patients with severe liver damage, Walshe (W3) confirmed the findings he had already reported with Dent (D14), namely, that there is an excessive excretion of cystine, β-aminoisobutyric acid, 1-methylhistidine, and taurine, and frequently also of methionine, tyrosine, and phenylalanine, whereas in milder conditions, e.g. obstructive jaundice, no hyperamino aciduria is noticed. Pronounced output of ethanolamine is observed during regenerative phases or in the course of abnormal proliferation of hepatic cells (hepatoma) (D15, D16).

In reporting more extensive investigations on cystine excretion in cirrhosis, involving polarographic measurements, Walshe and Senior stress the point that abnormal amounts of cystine excreted in the urine vary with the severity of the cirrhosis, and they claim that renal clearances for that amino acid are normal (W6). In 18 children suffering from virus hepatitis, the blood level of α-amino nitrogen was increased in two-thirds of the cases (H22), and the hyperamino aciduria concerned essentially leucine, isoleucine, lysine, and methionine; a high taurine excretion was not mentioned and only small amounts of cystine were recorded in these paper chromatographic studies; moreover, no mention was made of β-aminoisobutyric acid. In 12 cases of hepatic coma, paper chromatographic observations revealed a markedly increased output of glutamic acid, glutamine, and tyrosine in 9, of cystine in 8, and of methionine in 7; whereas glycine was excreted in abnormally high amounts in only two (W16). Among 19 other children suffering from hepatitis, pronounced increased outputs of glycine, α-alanine, and taurine were noticed, with no abnormal levels of these amino acids in the blood, except for taurine in severe cases (H11). Bickel and Souchon insist on the high excretion of γ-aminobutyric acid, of methionine, and of ethanolamine (B23).

Müting and Wortmann (M23) stress the fact that the excretion of methionine is abundant and that its output is further increased under the influence of methionine ingestion, whereas in normals this does not happen after administration of that amino acid. Using Moore and Stein's first column chromatographic methods (M14), Iber *et al.* (I2) have studied the amino acid blood plasma levels in seven patients suffering from liver deficiency, five of them in coma. They observed essentially increased levels for methionine and tyrosine, although in two of the patients the methionine level was actually low; the phenylalanine level was high in 4 of those who were in coma. They attribute the hyperamino acidemia to a deficient hepatic consumption of the amino acids concerned and also to an increased catabolic process of muscular proteins. Injections of glutamate increase the amino acidemia, chiefly with regard to α-alanine and glutamic acid. The amino aciduria was not studied in these patients.

It is obvious that our knowledge will improve considerably when more complete investigations will have been made with Moore and Stein's most recent methods. In our laboratory, a case of hepatic coma was investigated in this way by Dubois *et al.* (D24) (see Fig. 5). It will be essential to accumulate more specific, detailed, and reliable quantitative data both on amino acid blood levels and urinary excretion. So far, it seems as if high excretions coincide with high clearance levels (taurine, β-aminoisobutyric acid, ethanolamine, methylhistidine), whereas amino acids having high plasma levels show low renal clearances (tyrosine, methionine, phenylalanine) (D17). The study of either blood or urine alone is liable therefore to lead to wrong interpretations. Interesting information would also be found in accurate determinations in the tissues; some experimental data have already been obtained after hepatectomy (F7). Obviously, the most recent column chromatographic methods are needed to permit substantial progress in our knowledge of the subject.

4.2.2.2. *In Nutritional Disturbances*

(*a*) *Protein deficiency.* Two different conditions are to be taken into consideration.

(*1*) *Low-protein and low-calorie intake.* This is the case of *starvation* leading to severe emaciation and marasmus. Dietary fluctuations in protein intake have been considered in Section 3.3. Low-protein diets

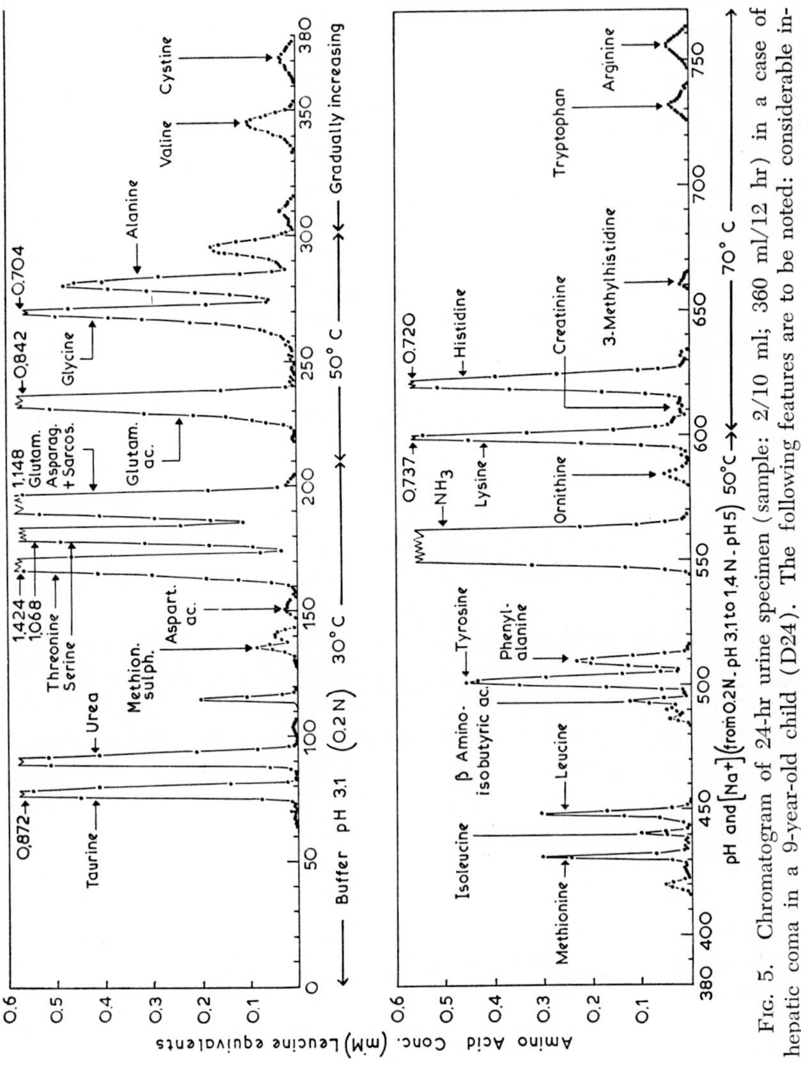

FIG. 5. Chromatogram of 24-hr urine specimen (sample: 2/10 ml; 360 ml/12 hr) in a case of hepatic coma in a 9-year-old child (D24). The following features are to be noted: considerable increase in output of substances corresponding to all peaks; presence of α-aminoadipic acid, α-amino-butyric acid (following the α-alanine peak), and tryptophan.

have but little effect on amino aciduria (H12). Normal subjects starved for a few days show a slight diminution of their amino acid blood plasma levels and renal excretion, except that leucine and valine blood levels may be abnormally high (C8, H29). In severe cases of starvation accompanied by pancreatic and other glandular fibrotic degeneration, hyperamino aciduria is a usual finding (S19). Increased excretion of phosphoethanolamine has also been observed in celiac disease (F6).

It is only when the renal tubular function becomes disturbed that hyperamino aciduria appears, in comparison to normal subjects, under the influence of protein overloading in the diet (S34). Dubois *et al.* attribute also to the same renal tubular deficiency the temporary rise in amino aciduria observed in the first days of high-protein dietary treatment of kwashiorkor cases (D25, D26), a situation which has also been seen by others (C9, M4). It has been attributed both to hepatic and renal deficiencies (C3).

(2) *Low-protein and relatively high-calorie intake.* An example is the abnormally high carbohydrate consumption in comparison to protein consumption of kwashiorkor patients. Cheung *et al.* (C9) were the first to study amino acid excretion in this condition, and in three cases, two Mexican and one Porto Rican, they used Moore and Stein's 1951 procedure. They express their findings by referring to the amount of creatinine excreted in a given time. Unfortunately the 24-hr urine volumes and the body weights of the patients are not indicated, and it is therefore impossible to convert their findings for comparison with other data in the literature. They claim that during the acute phase of this condition there is no hyperamino aciduria but that this does occur in later stages of recovery. In untreated cases, isoleucine is presumably excreted in higher proportions than leucine, and the same is described for phenylalanine in comparison to tyrosine, a situation they consider not to exist in normal children. In one instance, they note a low output of threonine.

A similar condition of malnutrition has been described in Europe ("Mehlnährschade"), and in Italy it has been studied from the point of view of amino acid excretion in three patients (M4).

In our laboratory, nine cases of colored children from Central Africa, suffering from typical kwashiorkor, have been examined, using Moore and Stein's 1954 method. Abnormal amino aciduria was observed in most cases, both before and in the early stage of protein treatment, by comparison with normal subjects of the same region. Taurine and β-amino-isobutyric acid were abundantly excreted (Fig. 6). Under protein treatment the drop in amino acid output is a most striking feature of the

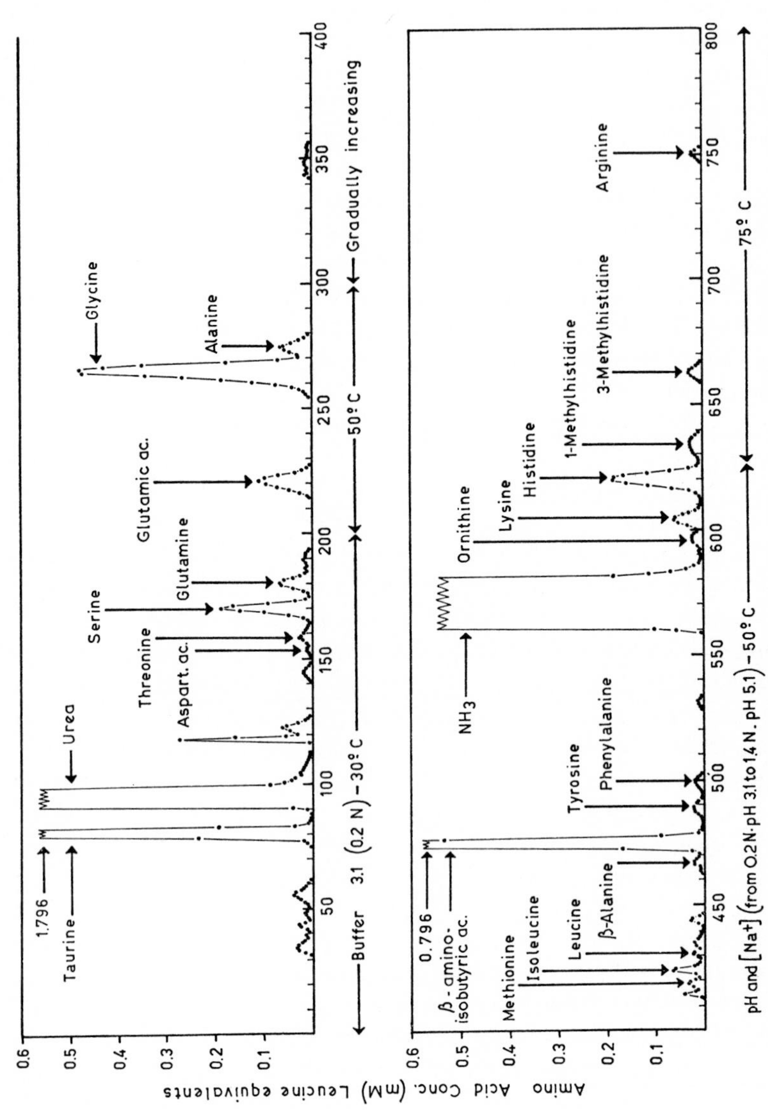

FIG. 6. Chromatogram of 24-hr urine specimen (sample: 1 ml; 135 ml/24 hr) of a colored child, 4 years old (Belgian Congo), suffering from kwashiorkor (D25). The following features are to be noted: there is a considerable excretion of taurine and of β-aminoisobutyric acid.

condition. Normal amino aciduria is restored in 10 to 15 days in patients responding favorably to the adjustment of the diet. In the case of β-aminoisobutyric acid, the return to normal is the last to take place. In three of the nine cases, amino acid blood levels were also determined. Before treatment they were normal or below normal and tryptophan was absent in the three cases. The situation strongly suggests that the condition is related to hepatic damage (fatty livers as recognized by biopsy samples obtained from these little patients). That a renal factor is also involved is most probable (D25, D26).

(*b*) *Vitamin C deficiency.* In prematurely born infants showing evidence of vitamin C deficiency, and fed on cow's milk diets, there is a condition of tyrosyluria; there is, namely, an increased excretion of *p*-hydroxyphenylpyruvic and *p*-hydroxyphenyllactic acids. This condition disappears in 48 hours under vitamin C treatment, a result which is also obtained with ACTH administration without the intervention of the adrenocortical relay (L4, L5, L6). There is an abnormally high α-amino nitrogen excretion in the urine of vitamin C-deficient patients, and also of glycine conjugates (J14).

Evidence for vitamin C deficiency in prematurely born infants showing hyperamino aciduria has been reported by Dustin and Bigwood (D29). In addition to the increased excretion of certain amino acids reported in common rachitis, there is also an increased output of tyrosine and phenylalanine in scurvy (J11). The blood levels of amino acids are normal (H28). The excessive output of tyrosine and phenylalanine is considered to be related to a renal factor; it is not a competitive effect due to increased excretion of serine and threonine (J13). In addition, therefore, to the increased output of certain amino acids reported in common rachitis (*vide infra*) there is also, in scurvy, an increased excretion of tyrosine and phenylalanine.

(*c*) *Common rachitis.* That the excretion of α-amino acids is increased in vitamin D deficiency is a well known fact (B13, B29, H21, H31, R6; see also Fig. 7). Normal amino aciduria is restored after three to four weeks treatment (H31). The hyperamino aciduria concerns essentially serine, threonine, histidine, lysine, and glutamic acid chiefly in its conjugated form (J6, J9); proline and hydroxyproline are also involved in the form of their conjugates (B29). This amino aciduria is a constant feature of the disease; only certain amino acids are concerned; their clearance is increased; the normal situation is always restored by proper treatment (J6, J9, J10, J12). High output of histidine through the renal tubules of a rachitic patient induces increased excretion of other amino acids, such as threonine, serine, and tyrosine (J6). There is no competitive effect

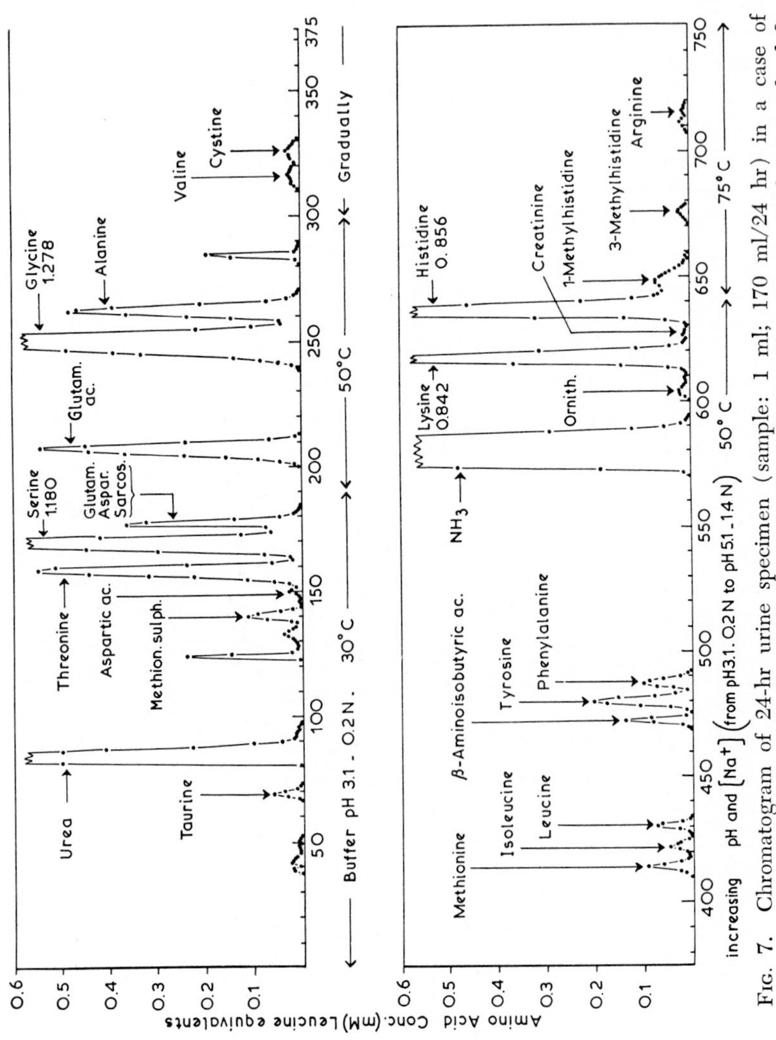

FIG. 7. Chromatogram of 24-hr urine specimen (sample: 1 ml; 170 ml/24 hr) in a case of common rachitis in a 9-month-old child (D24). The following features are to be noticed: definitely increased excretion of threonine, serine, glycine, and histidine.

between phosphaturia and amino aciduria. Plasma levels of the amino acids involved are normal (H28, J9).

According to Jonxis and Huisman (J6, J9, J12), hyperamino aciduria may be observed among the other members of the family of a patient with common rachitis and may sometimes (but not always) be corrected by vitamin D administration. There are presumably different grades of hereditary vitamin sensitiveness of the renal tubular function to vitamin D. The regulation does not take place through the mediation of the parathyroid hormone.

(*d*) *Vitamin B_{12} deficiency.* Hyperamino aciduria concerning essentially taurine and β-aminoisobutyric acid has been reported in Addison-Biermer's pernicious anemia (W7). In cases of acquired megaloblastic anemia, the excretion of amino acids stays normal (K3).

4.2.2.3. *In Intoxications*

(*a*) *Heavy metal intoxications.* Toxic heavy metals usually produce hyperamino aciduria of renal origin. The condition is either endogenous (Wilson's disease) or exogenous. The toxic effect is usually located at the proximal tubules and may therefore produce a Fanconi syndrome.

Lead produces a generalized hyperamino aciduria which may be accompanied by glycosuria (W12), hyperphosphaturia, and rickets (C12). Clarkson and Kench (C14) have studied cases of lead, mercury, uranium, and cadmium poisoning; they all produced hyperamino aciduria, particularly in the two latter cases. No blood level determinations have so far been made.

(*b*) *Maleic acid intoxication.* Experimental de Toni-Debré-Fanconi syndrome has been temporarily produced in rats (H9) by maleic acid injections. It is assumed that by forming complex products with SH-compounds such as glutathione or cysteine, the succinic acid dehydrogenase activity is inhibited (H17, M18, M19), disturbing thereby the Krebs Cycle. It has been claimed that it is by this mechanism that most of the toxic effects on the proximal renal tubules take place (L10).

(*c*) *Other intoxications.* Hyperamino aciduria occurs in various other cases of intoxication of either endogenous or exogenous origin, due, e.g., to hemoglobin (P3), lysol (S26), oxalic acid (E2), or *p*-aminosalicylic acid (J8).

5. General Conclusion

Abnormal patterns of amino acid excretion in the urine have been described in a wide variety of pathological conditions. So far, their description has led very frequently to quite confusing pictures when all published data and interpretations are confronted in an over-all survey of the literature.

In some of the most outstanding contributions to the subject, an attempt has been made to classify the observations in a logical set of categories, with the object of distinguishing among them various pathogenic mechanisms.

In our present state of knowledge, however, the information required to justify the classification has not yet reached the stage where such an attempt can be made satisfactorily.

Paper chromatography has been used as a major method of research because of its simplicity; it seems evident that our knowledge would have remained imperfect, had not new methods of investigation become available.

Adequate column chromatographic methods are now opening the way toward the required progress, allowing for more complete and more accurate quantitative data on amino acid blood levels as well as on amino acid renal excretion.

Present difficulties are also due to our lack of knowledge on the situation both in blood and in urine of normal individuals of various age groups.

When sufficient progress has been achieved in those directions, it will become possible to have a better understanding of what takes place in many pathological conditions. Some information has already been gathered along those lines and is reported in the present chapter.

References

References which became available at the time of proof corrections, are indicated in the text.

A1. Abderhalden, E., Familiale Cystindiathese. *Z. physiol. Chem.* **38**, 557-561 (1903).

A2. Ackermann, D., and Kutscher,, F., Über das Vorkommen von Lysin im Harn bei Cystinurie. *Z. Biol.* **57**, 355-359 (1911).

A3. Albanese, A. A., and Irby, V., Determination of urinary amino nitrogen by the copper method. *J. Biol. Chem.* **153**, 583-588 (1944).

A4. Allan, J. D., Cusworth, D. C., Dent, C. E., and Wilson, V. K., A disease, probably hereditary, characterised by severe mental deficiency and a constant gross abnormality of amino acid metabolism. *Lancet* **i**, 182-187 (1958).

A5. Ames, S. R., and Risley, H. A., Aminoaciduria in progressive muscular dystrophy. *Proc. Soc. Exptl. Biol. Med.* **68**, 131-135 (1948).

A6. Anderson, E. P., Kalckar, H. M., and Isselbacher, K. J., Defect in uptake of galactose-l-phosphate into liver nucleotides in congenital galactosemia. *Science* **125**, 113-114 (1957).

A7. Antener, I., Die kolorimetrische Bestimmung der Aminoacidurie. *Schweiz. Med. Wochschr.* **83**, 425-427 (1953).

A8. Armstrong, M. D., and Robinson, K. S., On the excretion of indole derivatives in phenylketonuria. *Arch. Biochem. Biophys.* **52**, 287-288 (1954).

A9. Armstrong, M. D., and Tyler, F. H., Studies on phenylketonuria. I. Restricted phenylalanine intake in phenylketonuria. *J. Clin. Invest.* **34**, 565-580 (1955).

A10. Armstrong, M. D., Shaw, K. N. F., and Robinson, K. S., Studies on phenylketonuria. II. The excretion of o-hydroxyphenylacetic acid in phenylketonuria. *J. Biol. Chem.* **213**, 797-804 (1955).

A11. Arthurton, M. W., and Meade, B. W., Congenital galactosaemia. *Brit. Med. J.* **ii**, 618-620 (1954).

A12. Awapara, J., and Sato, Y., Paper chromatography of urinary amino acids. *Clin. Chim. Acta* **1**, 75-79 (1956).

B1. Baar, H. S., and Baar, S., The nature of the crystalline material in tissues of Lignac-Fanconi's disease. *Acta Paediat.* **45**, 551-557 (1956).

B2. Baber, M. D., A case of congenital cirrhosis of the liver with renal tubular defects akin to those in the Fanconi syndrome. *Arch. Disease Childhood* **31**, 335-339 (1956).

B3. Balikov, B., and Castello, R. A., Urinary peptides. Method for assay. *Clin. Chem.* **2**, 83-90 (1956).

B4. Barlow, A., and McCance, R. A., Nitrogen partition in newborn infants' urine. *Arch. Disease Childhood* **23**, 225-230 (1948).

B5. Baron, D. N., Dent, C. E., Harris, H., Hart, E. W., and Jepson, J. B., Hereditary pellagra-like skin rash with temporary cerebellar ataxia, constant renal aminoaciduria, and other bizarre biochemical features. *Lancet* **ii**, 421-428 (1956).

B6. Bearn, A. G., Wilson's disease; an inborn error of metabolism with multiple manifestations. *Am. J. Med.* **22**, 747-757 (1957).

B7. Bearn, A. G., and Kunkel, H. G., Biochemical abnormalities in Wilson's disease. *J. Clin. Invest.* **31**, 616 (1952).

B8. Bearn, A. G., and Kunkel, H. G., Abnormalities of copper metabolism in Wilson's disease and their relationship to the aminoaciduria. *J. Clin. Invest.* **33**, 400-409 (1954).

B9. Bearn, A. G., Yü, T. F., and Gutman, A. B., Renal function in Wilson's disease. *J. Clin. Invest.* **36**, 1107-1114 (1957).

B10. Berger, H., La pathogénèse de la galactosémie. *J. génét. hum.* **4**, 7-22 (1955).

B11. Berger, H., Die Bestimmung der im Harn ausgeschiedenen Aminosäuren und ihre Bedeutung in der Klinik. *Schweiz. Med. Wochschr.* **86**, 711-714 (1956).

B12. Berger, H., Hereditäre chronische Hyperaminoacidurien. *Modern Problems Paediat.* **3**, 238-266 (1957).

B13. Berger, H., and Stalder, G., Zur Frage der Aminoacidurie bei der Vitamin D Mangelrachitis der Säuglinge. *Ann. Paediat.* **186**, 163-175 (1956).

B14. Berger, H., and Antener, I., Influence of protein content of the food upon the excretion of free α-aminonitrogen in the urine as a measure of aminoaciduria. *Intern. Z. Vitaminforsch.* **28**, 17-32 (1957).

B15. Berger, H., Weber, J. R., Antener, I., and Pfändler, U., Schwere Ariboflavinose, Spätrachitis und Amindiabetes bei chronischer erblicher Koproporphyrie. *Intern. Z. Vitaminforsch.* **26**, 96-129 (1955).

B16. Berry, H. K., Paper chromatographic method for estimation of phenylalanine. *Proc. Soc. Exptl. Biol. Med.* **95**, 71-73 (1957).

B17. Berry, H. K., Sutherland, B., and Guest, G. M., Phenylalanine tolerance tests on relatives of phenylketonuric children. *Am. J. Human Genet.* **9**, 310-316 (1957).

B18. Berry, H. K., Sutherland, B., Guest, G. M., and Umbarger, B., Chemical and clinical observations during treatment of children with phenylketonuria. *Pediatrics* **21**, 929-940 (1958).

B19. Bhattacharya, K. R., Datta, J., and Roy, D. K., Paper chromatographic method for the estimation of histidine. *Ann. Biochem. and Exptl. Med. (Calcutta)* **17**, 1-4 (1957).

B20. Bickel, H., Die Entwicklung der biochemischen Läsion bei der Lignac-Fanconischen Krankheit. *Helv. Paediat. Acta* **10**, 259-268 (1955).

B21. Bickel, H., and Hickmans, E. M., Paper chromatographic investigations on the urine of patients with galactosaemia. *Arch. Disease Childhood* **27**, 348-350 (1952).

B22. Bickel, H., and Hickmans, E. M., Some biochemical aspects of Lignac Fanconi disease. *Acta Paediat.* **42**, Suppl. 90, 137-170 (1952).

B23. Bickel, H., and Souchon, F., Die Papierchromatographie in der Kinderheilkunde. *Arch. Kinderh. Beih.* **31**, 1-151 (1955).

B24. Bickel, H., and Thursby-Pelham, D. C., Hyperaminoaciduria in Lignac-Fanconi disease, in galactosaemia and in an obscure syndrome. *Arch. Disease Childhood* **29**, 224-231 (1954).

B25. Bickel, H., Gerrard, J. W., and Hickmans, E. M., The influence of phenylalanine intake on the biochemistry and behaviour of a phenylketonuric child. *Acta Paediat.* **43**, 64-77 (1954).

B26. Bickel, H., Neale, F. C., and Hall, G., A clinical and biochemical study of hepatolenticular degeneration (Wilson's disease). *Quart. J. Med.* **26**, 527-558 (1957).

B27. Bickel, H., Smallwood, W. C., Smellie, J. M., Baar, H. S., and Hickmans, E. M., Cystine storage disease with aminoaciduria and dwarfism (Lignac Fanconi disease). *Acta Paediat.* **42**, Suppl. 90, 9 (1952).

B28. Bickis, I. J., Kennedy, J. P., and Quastel, J. H., Phenylalanine inhibition of tyrosine metabolism in the liver. *Nature* **179**, 1124-1126 (1957).

B29. Biserte, G., Breton, A., and Fontaine, G., Etude analytique des acides aminés urinaires dans le rachitisme commun de l'enfance. *Arch. franç. pédiat.* **12**, 988-997 (1955).

B30. Biserte, G., Paysant, P., and Schoonaert, T., Séparation des acides aminés par électrophorèse et chromatographie sur papier (Mise au point technique). *J. Chromatography* in press.

B31. Blainey, J. D., and Gulliford, R., Phenylalanine-restricted diets in the treatment of phenylketonuria. *Arch. Disease Childhood* **31**, 452-566 (1956).

B32. Blass, J., Sur une technique simple de chromatographie sur papier des acides aminés urinaires pour usage courant en clinique. *Ann. biol. clin. (Paris)* **13**, 56-59 (1955).

B33. Block, R. J., Durrum, E. L., and Zweig, G., "A Manual of Paper Chromatography and Paper Electrophoresis," 2nd rev. ed. Academic Press, New York, 1958.

B34. Boehncke, H., Aminoacidurie bei Cystinspeicher und Marmorknochen Krankheit. *Kinderärtzl. Praxis.* **21**, 232-234 (1952).

B35. Boissonas, R. A., and Lo Bianco, S., A new method for qualitative and quantitative study of aminaciduria by paper chromatography. *Experientia* **8**, 425 (1952).

B36. Boscott, R. J., and Bickel, H., Detection of some new abnormal metabolites in urine of phenylketonuria. *Scand. J. Clin. & Lab. Invest.* **5**, 380-382 (1953).

B37. Boulanger, P., and Biserte, G., Les acides aminés libres et combinés de l'urine humaine. *Exposés ann. biochim. méd.* **18**, 123-164 (1956).

B38. Boulanger, P., Biserte, G., Paysant, P., and Schoonaert, T., Etude sur les acides aminés urinaires. Résultats qualitatifs et quantitatifs nouveaux obtenus par électrophorèse et chromatographie sur papier. *Clin. Chim. Acta* in preparation.

B39. Braun, P., Kisfaludy, S., and Dubsky, M., Quantitative analysis of free aminoacids in normal and pathological human serum and urine. *Acta Med. Acad. Sci. Hung.* **7**, 147-159 (1955).

B40. Bush, J. A., Mahoney, J. P., Markowitz, H., Gubler, C. J., Cartwright, G. E., and Wintrobe, M. M., Studies on copper metabolism. XVI. Radioactive copper studies in normal subjects and in patients with hepatolenticular degeneration. *J. Clin. Invest.* **34**, 1766-1778 (1955).

B41. Busson, F., and Guth, P., Nouvelle technique d'analyse chromatographique des acides aminés libres du sérum sanguin. *Méd. trop.* **16**, 833-835 (1956).

C1. Cachin, M., Durlach, J., and Blass, J., Les acides aminés du sérum sanguin en pathologie hépatique. *Semaine hôp.* **28**, 3231-3238 (1952).

C2. Calchi-Novati, C., Ceppellini, R., Biancho, I., Silvestroni, E., and Harris, H., β-aminoisobutyric acid excretion in urine. A family study in an Italian population. *Ann. Eugenics* **18**, 335-336 (1954).

C3. Careddu, P., Osservazioni cromatografiche sull'aminoacidemia e sulla aminoaciduria nella distrofia del lattante. *Minerva Pediat.* **5**, 253-257 (1953).

C4. Careddu, P., Pugioni, J., and Vullo, C., Osservazioni cromatografiche sulla amino-aciduria basale e da carico nell' alcaptonuria. *Ann. ital. pediat.* **7**, 299 (1954).

C5. Carsten, M. E., A procedure for the ion exchange chromatography of the aminoacids of urine. *J. Am. Chem. Soc.* **74**, 5954-5959 (1952).

C6. Cartwright, G. E., Hodges, R. E., Gubler, C. J., Mahoney, J. P., Daum, K., Wintrobe, M. M., and Bean, W. B., Studies on copper metabolism. XIII. Hepatolenticular degeneration. *J. Clin. Invest.* **33**, 1487-1501 (1954).

C7. Caussade, L., Neimann, N., Pierson, B., and Fernier, Y., Considérations sur la cystinose. *Presse méd.* **62**, 646-650 (1954).

C8. Charkey, L. W., Kano, A. K., and Hougham, D. F., Effects of fasting on blood nonprotein aminoacids in humans. *J. Nutrition* **55**, 469-480 (1955).

C9. Cheung, M. W., Fowler, D. I., Morton, P. M., Snyderman, S. E., and Holt, L. E., Jr., Osservazioni sul metabolismo di aminoacidi nel kwashiorkor. *Pediat. intern.* **6**, 91-100 (1956).

C10. Cheval, M., and Hans, M. J., La determination de l'histidine dans l'urine et son application comme méthode de diagnostic de la grossesse et comme contrôle de l'évolution de la gestation. *Bruxelles-Méd.* **35**, 1685-1708, 1740-1750 (1955).

C11. Childs, B., Urinary excretion of free alpha-amino acid nitrogen by normal infants and children. *Proc. Soc. Exptl. Biol. Med.* **81**, 225-226 (1952).

C12. Chisolm, J. J., Jr., Harrison, H. C., Eberlein, W. R., Harrison, H. E., and Mohr, F. C., Aminoaciduria, hypophosphatemia and rickets in lead poisoning. *Am. J. Diseases Children* **89**, 159-168 (1955).

C13. Christensen, P. J., Date, J. W., Schoenheyder, F., and Volqvartz, K., Amino-acids in blood plasma and urine during pregnancy. *Scand. J. Clin. & Lab. Invest.* **9**, 54-61 (1957).

C14. Clarkson, T. W., and Kench, J. E., Urinary excretion of amino acids by men absorbing heavy metals. *Biochem. J.* **62**, 361-372 (1956).

C15. Coates, S., Norman, A. P., and Woolf, L. I., Phenylketonuria with normal intelligence and Gover's muscular dystrophy. *Arch. Diseases Childhood* **32**, 313-317 (1957).

C16. Consden, R., Gordon, A. H., and Martin, A. J. P., The identification of lower peptides in complex mixtures. *Biochem. J.* **41**, 590-596 (1947).

C17. Cooper, A. M., Eckhardt, R. D., Faloon, W. W., and Davidson, C. S., Investigation of the amino aciduria in Wilson's disease (hepatolenticular degeneration): demonstration of a defect in renal function. *J. Clin. Invest.* **29**, 265-278 (1950).

C18. Cox, P. J. N., and Pugh, R. J. P., Galactosaemia. *Brit. Med. J.* **ii**, 613-618 (1954).

C19. Cramer, F., "Papierchromatographie," 4th ed. Verlag Chemie, Weinheim Bergstr., 1958.

C20. Crokaert, R., "Contribution à l'étude de la β-alanine et de ses composés dans les milieux biologiques." Editions Acta Med. Belg., Brussels, 1953; See also *Ann. soc. roy. sci. méd. et nat.* (*Bruxelles*) **6**, 157-254 (1953).

C21. Crumpler, H. R., Dent, C. E., Harris, H., and Westall, R. G., β-aminoisobutyric acid (α-methyl-β-alanine): a new amino-acid obtained from human urine. *Nature* **167**, 307-308 (1951).

C22. Cusworth, D. C., Dent, C. E., and Flynn, F. V., The amino-aciduria in galactosemia. *Arch. Disease Childhood* **30**, 150-154 (1955).

D1. Dalgliesh, C. E., Unspecific hydroxylation reactions in the metabolism of aromatic aminoacids. *Biochem. J.* **58**, Proc. xiv (1954).

D2. Dancis, J., and Balis, M. E., A possible mechanism for the disturbance in tyrosine metabolism in phenylpyruvic oligophrenia. *Pediatrics* **15**, 63-67 (1955).

D3. Debré, R., Marie, J., Cleretat, F., and Messimy, R., Rachitisme tardif coexistant avec une néphrite chronique et une glycosurie. *Archives de médecine des enfants* **37**, 597-606 (1934).

D4. Debré, R., Royer, P., Lestradet, H., and Straub, W., L'insuffisance tubulaire congénitale avec arriération mentale, cataracte et glaucome. *Arch. franç. pédiat.* **12**, 337-348 (1955).

D5. de Groucny, J., and Eldon-Sutton, H., Genetic study of β-aminoisobutyric acid excretion. *Am. J. Human Genet.* **9**, 76-80 (1957).

D6. Denny-Brown, D., and Porter, H., The effect of BAL (2,3-dimercaptopropanol) on hepatolenticular degeneration (Wilson's disease). *New Engl. J. Med.* **245**, 917-925 (1951).

D7. Dent, C. E., The amino-aciduria in Fanconi syndrome. A study making extensive use of techniques based on paper partition chromatography. *Biochem. J.* **41**, 240-253 (1947).

D8. Dent, C. E., A study of the behaviour of some sixty amino-acids and other ninhydrin-reacting substances on phenol-"collidine" filter-paper chromatograms, with notes as to the occurrence of some of them in biological fluids. *Biochem. J.* **43**, 169-80 (1948).

D9. Dent, C. E., Clinical applications of amino acid chromatography. *Scand. J. Clin. & Lab. Invest.* **10**, Suppl. 31, 122-127 (1957).

D10. Dent, C. E., and Harris, H., The genetics of "cystinuria." *Ann. Eugenics* **16**, 60-87 (1951).

D11. Dent, C. E., and Harris, H., Hereditary forms of rickets and osteomalacia. *J. Bone and Joint Surg.* **38B**, 204-226 (1956).

D12. Dent, C. E., and Rose, G. A., Amino acid metabolism in cystinuria. *Quart. J. Med.* **20**, 205-219 (1951).

D13. Dent, C. E., and Senior, B., Studies on the treatment of cystinuria. *Brit. J. Urol.* **27**, 317-332 (1955).

D14. Dent, C. E., and Walshe, J. M., Amino acid metabolism in liver disease. *In* "Liver Disease," Ciba Foundation Symposium, pp. 22-31. Churchill, London, 1951.

D15. Dent, C. E., and Walshe, J. M., Primary carcinoma of liver: description of case with ethanolaminuria, new and obscure metabolic defect. *Brit. J. Cancer* **7**, 166-180 (1953).

D16. Dent, C. E., and Walshe, J. M., Ethanolaminuria. A new and obscure metabolic defect. *Metabolism, Clin. and Exptl.* **2**, 474 (1953).

D17. Dent, C. E., and Walshe, J. M., Amino-acid metabolism. *Brit. Med. Bull.* **10**, 247-250 (1954).

D18. Dent, C. E., Heathcote, J. G., and Joron, G. E., The pathogenesis of cystinuria. I. Chromatographic and microbiological studies of the metabolism of sulfur containing amino acids. *J. Clin. Invest.* **33**, 1210-1215 (1954).

D19. Dent, C. E., Senior, B., and Walshe, J. M., The pathogenesis of cystinuria. II. Polarographic studies of the metabolism of sulfur containing amino acids. *J. Clin. Invest.* **33**, 1216-1226 (1954).

D20. Dern, P. L., Aminoaciduria with cystinosis: case report with determination of urinary aminoacids and ocular cystine. *Ann. Internal Med.* **46**, 138-144 (1957).

D21. de Toni, G., Remarks on relations between renal rickets (renal dwarfism) and renal diabetes. *Acta Paediat.* **16**, 479-484 (1933).

D22. de Toni, G., Renal rickets with phospho-glucoamino renal diabetes (de Toni-Debré-Fanconi Syndrome). *Ann. Paediat.* **187**, 42-80 (1956).

D23. de Toni, G., and Durand, P., Guarigione clinica del diabete renale e del rachitismo renale in un grave caso di sindrome di de Toni-Debré-Fanconi giunto all'età adulta con persistenza del nanismo e della iperaminoaciduria. *Minerva Pediat.* **7**, 1053-60 (1955).

D24. Dubois, R., Vis, H., and Loeb, H., Etude chromatographique d'un cas de diabète gluco-amino-phosphaté. *Ann. Paediat.* **191**, 1-15 (1958).

D25. Dubois, R., Vis, H., Loeb, H., and Vincent, M., Note sur l'aminoacidurie du kwashiorkor. *Atti 25. Congr. ital. pediat.* **3** (1957).

D26. Dubois, R., Vis, H., Loeb, H. and Vincent, M., Alterations de l'amino acidurie observées dans des cas de Kwashiorkor. Etude par chromatographie sur colonne d'échangeurs d'ions. *Helv. Paediat. Acta* **14**, 13-43 (1959).

D27. Dunn, M. S., Akawie, S., Yeh, H. L., and Martin, H., Urinary excretion of aminoacids in liver disease. *J. Clin. Invest.* **29**, 302-312 (1950).

D28. Dustin, J. P., De l'élimination des acides aminés dans les urines de nourrisson. *Exptl. Med. and Surg.* **12**, 233-237 (1954).

D29. Dustin, J. P., and Bigwood, E. J., Aminoaciduria in infancy and ascorbic acid deficiency. *Proc. Nutrition Soc.* **12**, 293-300 (1953).

D30. Dustin, J. P., Moore, S., and Bigwood, E. J., Chromatographic studies on the excretion of aminoacids in early infancy. *Metabolism* **4**, 75-79 (1955).

D31. Dvorák, J., Kubový, A., Zázvorka, Z., and Hons, J., Ein Fall von idiopathischer Hypoalbuminämie mit Aminoacidurie beim Säugling. *Helv. Paediat. Acta* **9**, 258-263 (1954).

E1. Eckhardt, R. D., and Davidson, C. S., Urinary excretion of Aminoacids by a normal adult receiving diets of varied protein content. *J. Biol. Chem.* **177**, 687-695 (1949).

E2. Emalie-Smith, D., Hohnstone, J. H., Thomas, M. B., and Lowe, K. G., Aminoaciduria in acute tubular necrosis. *Clin. Sci.* **15**, 171-176 (1956).

E3. Engle, R. L., and Wallis, L. A., Multiple myeloma and the adult Fanconi Syndrome. *Am. J. Med.* **22**, 5-23 (1957).

E4. Evered, D. F., "A study of amino acids in biological fluids using ion-exchange resins." Ph.D. thesis, London University (1954).

E5. Evered, D. F., Excretion of aminoacids by the human. A quantitative study with ion-exchange chromatography. *Biochem. J.* **62**, 416-427 (1956).

F1. Fanconi, G., Der frühinfantile nephrotisch-glykosurische Zwergwuchs mit hypophosphatämischer Rachitis. *Jahrb. Kinderheilk.* **147**, 299-338 (1936).

F2. Fanconi, G., and Bickel, H., Die chronische Aminoacidurie (Aminosäurediabetes oder nephrotisch-glukosurischer Zwergwuchs) bei der Glykogenose und der Cystinkrankheit. *Helv. Paediat. Acta* **4**, 359-396 (1949).

F3. Felix, K., Leonhardi, G., and Glasenapp, I. von, Über die Ausscheidung von p-Oxyphenylbrenztraubensäure, Brenztraubensäure, Phenol und Acetessigsäure im Harn nach Verabreichung von p-Oxyphenylbrenztraubensäure an gesunde und Leberkranke. *Z. physiol. Chem.* **287**, 133-141 (1951).

F4. Felix, K., Leonhardi, G., and Glasenapp, I. von, Über Tyrosinosis. *Z. physiol. Chem.* **287**, 141-147 (1951).

F5. Fink, K., Henderson, R. B., and Fink, R. M., β-Aminoisobutyric acid: a possible factor in pyrimidine metabolism. *Proc. Soc. Exptl. Biol. Med.* **78**, 135-141 (1951).

F6. Fischer, O. D., and Neill, D. W., Excretion of ethanolamine phosphoric acid in coeliac disease (Preliminary communication). *Lancet* i, 334-335 (1955).

F7. Flock, E. V., Mann, F. C., and Bollman, J. L., Free aminoacids in plasma and muscle following total removal of the liver. *J. Biol. Chem.* **192**, 293-300 (1951).

F8. Folin, O., A colorimetric determination of the aminoacid nitrogen in normal urine. *J. Biol. Chem.* **51**, 393-394 (1922).

F9. Fölling, A., Über Ausscheidung von Phenylbrenztraubensäure in den Harn als Stoffwechselanomalie in Verbindung mit Imbezillität. *Z. physiol. Chem.* **227**, 169-176 (1934).

F10. Fölling, A., Closs, K., and Gammes, T., Vorläufige Schlussfolgerungen aus Belastungsversuchen mit Phenylalanin an Menschen und Tieren. *Z. physiol. Chem.* **256**, 1-14 (1938).

F11. Fowler, D. I., Harris, H., and Warren, F. L., Plasma cystine levels in cystinuria. *Lancet* **i**, 544-545 (1952).

F12. Fox, E. G., Fyfe, W. M., and Mollison, A. W., Galactose diabetes. *Brit. Med. J.* **i**, 245-247 (1954).

F13. Frank, H., and Petersen, H., Eine Methode zur quantitativen Bestimmung von Histidin in Papierchromatogrammen. *Z. physiol. Chem.* **299**, 1-5 (1955).

F14. Frankl, W., Martin, H., and Dunn, M. S., The apparent concentration of free tryptophan, histidine and cystine in pathological human urine measured microbiologically. *Arch. Biochem.* **13**, 103-112 (1947).

F15. Fraser, D., Hypophosphatasia. *Am. J. Med.* **22**, 730-746 (1957).

F16. Fraser, D., and Yendt, E. R., Metabolic abnormalities in hypophosphatasia. *Am. J. Diseases Children* **90**, 552 (1955).

F17. Fraser, D., Yendt, E. R., and Christie, F. H. E., Metabolic abnormalities in hypophosphatasia. *Lancet* **i**, 286 (1955).

F18. Freudenberg, E., Cystinosis, Cystine disease (Lignac's disease) in children. *Advances in Pediat.* **4**, 265-292 (1949).

F19. Freudenberg, E., Weitere Beobachtungen zur Frage der "Cystinosis." *Ann. Paediat.* **182**, 85-106 (1954).

F20. Frezal, J., Lestradet, H., Jacob, P., and Lortholary, P., Modification transitoire de l'aminoacidurie, au cours d'une perfusion calcique, dans le syndrome de Toni-Debré-Fanconi. *Rev. franç. études clin. et biol.* **3**, 626-630 (1958).

G1. Gabuzda, G. J., Jr., Eckhardt, R. D., and Davidson, C. S., Urinary excretion of aminoacids in patients with cirrhosis of the liver and in normal adults. *J. Clin. Invest.* **31**, 1015-22 (1952).

G2. Gartler, S. M., A family study of urinary β-aminoisobutyric acid excretion. *Am. J. Human Genet.* **8**, 120-126 (1956).

G3. Gartler, S. M., Firschein, I. L., and Kraus, B. S., Genetic and racial variation of β-aminoisobutyric acid excretion. *Am. J. Human Genet.* **9**, 200-207 (1957).

G4. Gérard-Lefebvre, Biserte, G., Woillez, M., Traisnel, M., Gosselin, J., and Combaud, A., Etude clinique, génétique et biologique du syndrome de Lowe-Bickel. *Pédiatrie* **12**, 527-534 (1957).

G5. Goebel, F., Ueber die Aminosäurefraktion im Säuglingsharn. *Z. Kinderheilk.* **34**, 94-141 (1922).

G6. Gros, H., and Kirnberger, E. J., Spontanausscheidung von p-Oxyphenylbrenztraubensäure im Harn. *Klin. Wochschr.* **32**, 115-118 (1954).

H1. Hamann, I., Enige onderzoekingen over de phenylalanine en tyrosine-stofwisseling bij een patientje met oligophrenia phenylpyruvica. *Maandschr. Kindergeneesk.* **24**, 28-36 (1956).

H2. Hamann, I., van der Schaaf, P. C., and Huisman, T. H. J., The Giri method in qualitative paper chromatographic analysis of amino acids in biological liquids. *Ned. Tijdschr. Geneesk.* **99**, II, 1512-1516 (1955).

H3. Harris, H., Family studies on urinary excretion of β-aminoisobutyric acid. *Ann. Eugenics* **18**, 43-49 (1953).

H4. Harris, H., Aminoaciduria in man. *Proc. 3rd Intern. Congr. Biochem., Brussels* pp. 467-474 (1955).

H5. Harris, H., and Robson, E. B., Cystinuria. *Am. J. Med.* **22**, 774-783 (1957).

H6. Harris, H., and Warren, F. L., Quantitative studies on urinary cystine in patients with cystine stone formation and their relatives. *Ann. Eugenics* **18**, 125-171 (1953).

H7. Harris, H., Mittwoch, U., Robson, E. B., and Warren, F. L., Excretion of aminoacids in cystinurics. *Biochem. J.* **57**, Proc. xxxiii (1954).

H8. Harris, H., Mittwoch, U., Robson, E. B., and Warren, F. L., Phenotypes and genotypes in cystinuria. *Ann. Human Genet.* **20**, 57-91 (1955).

H9. Harrison, H. E., and Harrison, H. C., Experimental production of renal glycosuria, phosphaturia, and aminoaciduria by injection of maleic acid. *Science* **120**, 606-608 (1954).

H10. Hawk, P. B., Oser, B. L., and Summerson, W. H., "Practical Physiological Chemistry," 13th ed. McGraw-Hill, New York, 1954.

H11. Heim, L., and Kaffanke, V., Die funktionsdiagnostische Bedeutung fortlaufender Aminosäurenbestimmungen in Blut und Urin bei kindlicher Hepatitis. *Ann. Paediat.* **185**, 348-361 (1955).

H12. Hier, S. W., Urinary excretion of individual aminoacids on normal and low protein diets. *Trans. N.Y. Acad. Sci.* **10**, 280-283 (1948).

H13. Holzel, A., Komrower, G. M., and Wilson, V. K., Amino-aciduria in galactosemia. *Brit. Med. J.* **i**, 194-195 (1952).

H14. Hooft, C., and Herpol, J., Cystinose et cystinurie. *Modern Problems Paediat.* **3**, 267-284 (1957).

H15. Hooft, C., and Vermassen, A., Syndrome de De Toni-Debré-Fanconi chez un enfant atteint de néphrose lipoidique. *Ann. Paediat.* **190**, 1-17 (1958).

H16. Hooft, C., Delbeke, M. J., and Herpol, J., Cystinose chronique associée à une hypothyroïdie probable. *Ann. Paediat.* **187**, 81-99 (1956).

H17. Hopkins, F. G., and Morgan, E. J., The influence of thiol-groups in the activity of dehydrogenases. *Biochem. J.* **32**, 611-620 (1938).

H18. Hoppe-Seyler, F. A., Lysinausscheidung im Harn bei Cystinurie. *Z. physiol. Chem.* **214**, 267-270 (1933).

H19. Horner, F. A., and Streamer, C. W., Effect of a phenylalanine-restricted diet on patients with phenylketonuria; clinical observations in three cases. *J. Am. Med. Assoc.* **161**, 1628-1630 (1956).

H20. Höttinger, A., Zur Cystindiathese. *Ann. Paediat.* **169**, 272-281 (1947).

H21. Höttinger, A., and Höttinger, G., Studien über Säure-Basen-Haushalt im kindlichen Organismus. *Monatschr. Kinderheilk* **52**, 204-238 (1932).

H22. Hsia, D. Y. Y., and Gellis, S. S., Amino-acid metabolism in infectious hepatitis. *J. Clin. Invest.* **33**, 1603-1609 (1954).

H23. Hsia, D. Y. Y., Driscoll, K., Troll, W., and Knox, W. E., Detection by phenylalanine tolerance tests of heterozygous carriers of phenylketonuria. *Nature* **178**, 1239-1240 (1956).

H24. Hsia, D. Y. Y., Knox, W. E., Quinn, K. V., and Paine, R. S., A one-year, controlled study of the effect of low-phenylalanine diet on phenylketonuria. *Pediatrics* **21**, 178-202 (1958).

H25. Hsia, D. Y. Y., Hsia, H. H., Green, S., Kay, M., and Gellis, S. S., Amino-aciduria in galactosemia. *Am. J. Diseases Children* **88**, 458-465 (1954).

H26. Hudson, F. P., Ireland, J. T., Ockenden, B. G., and White-Jones, R. H., Diagnosis and treatment of galactosaemia. *Brit. Med. J.* **i**, 242-245 (1954).

H27. Huisman, T. H. J., Een methode voor de quantitatieve bepaling van het gehalte van de verschillende aminozuren in urine. *Chem. Weekblad* **49**, 109-114 (1953).

H28. Huisman, T. H. J., The concentration of different aminoacids in the blood plasma in children suffering from rickets and scurvy. *Pediatrics* **14**, 245-253 (1954).

H29. Huisman, T. H. J., Amino acids in connection with nutrition of infants and children. *Voeding* **19**, 138-151 (1958).

H30. Huisman, T. H. J., and van der Schaaf, P. C., Een nieuwe quantitatieve chromatographische bepaling der verschillende aminozuren in eiwithydrolysaten volgens het principe der gradient-elutie-analyse. *Chem. Weekblad* **51**, 2-6 (1955).

H31. Huisman, T. H. J., Smith, P. A., and Jonxis, J. H. P., Rickets and amino-aciduria. *Lancet* **ii**, 1015-1017 (1952).

H32. Hurley, K. E., and Williams, R. J., Urinary aminoacids, creatinine and phosphate in muscular dystrophy. *Arch. Biochem. Biophys.* **54**, 384-391 (1955).

I1. Iber, F. L., Chalmers, C., and Uzman, L. L., Protein metabolism in hepato-lenticular degeneration. *Metabolism, Clin. and Exptl.* **6**, 388-396 (1957).

I2. Iber, F. L., Rosen, H., Levenson, S. M., and Chalmers, T. C., The plasma aminoacids in patients with liver failure. *J. Lab. Clin. Med.* **50**, 417-425 (1957).

J1. Jepson, J. B., Indolylacetyl-glutamine and other indole metabolites in Hartnup disease. *Biochem. J.* **64**, 14P (1956).

J2. Jervis, G. A., Metabolic investigations on a case of phenylpyruvic oligophrenia. *J. Biol. Chem.* **126**, 305-313 (1938).

J3. Jervis, G. A., Studies on phenylpyruvic oligophrenia. The position of the metabolic error. *J. Biol. Chem.* **169**, 651-656 (1947).

J4. Jervis, G. A., Phenylpyruvic oligophrenia (phenylketonuria). *Research Publs. Assoc. Research Nervous Mental Disease* **33**, 259-282 (1954).

J5. Jervis, G. A., Block, R. J., Bolling, D., and Kanze, E., Chemical and metabolic studies on phenylalanine. II. The phenylalanine content of the blood and spinal fluid in phenylpyruvic oligophrenia. *J. Biol. Chem.* **134**, 105-113 (1940).

J6. Jonxis, J. H. P., Amino-aciduria and rickets. *Helv. Paediat. Acta* **10**, 245-256 (1955).

J7. Jonxis, J. H. P., De uitscheiding van phenylalanine bij een tweetal patiëntjes met oligophrenia phenylpyruvica. *Maandschr. Kindergeneesk.* **25**, 272-279 (1957).

J8. Jonxis, J. H. P., Aminoacidurie. *Ergeb. inn. Med. u. Kinderheilk.* [N.S] **8**, 169-198 (1957).

J9. Jonxis, J. H. P., and Huisman, T. H. J., Amino-acidurie bij rachitische kinderen. *Voeding* **9**, 400-407 (1953).

J10. Jonxis, J. H. P., and Huisman, T. H. J., Amino-aciduria in rachitic children. *Lancet* **ii**, 428-431 (1953).

J11. Jonxis, J. H. P., and Huisman, T. H. J., Amino-aciduria and ascorbic acid deficiency. *Pediatrics* **14**, 238-244 (1954).

J12. Jonxis, J. H. P., and Huisman, T. H. J., The renal element in rachitic amino-aciduria. *Lancet* **ii**, 513-516 (1954).

J13. Jonxis, J. H. P., and Huisman, T. H. J., Onderzoekingen over de phenylalanine en tyrosinestofwisseling bij vitamine C deficiëntie in kinderen. *Maandschr. Kindergeneesk.* **25**, 130-138 (1957).

J14. Jonxis, J. H. P., and Wadman, S. K., De uitscheiding van aminozuren in de urine bij een patient met scorbuut. *Maandschr. Kindergeneesk.* **18**, 251-258 (1951).

J15. Juillard, E., and Piguet, C., Glucosurie et amino-acidurie familiale. *Ann. Paediat.* **183**, 257-270 (1954).

K1. Kalckar, H. M., Anderson, E. P., and Isselbacher, K. J., Galactosemia, a congenital defect in a nucleotide transferase. *Biochim. et Biophys. Acta* **20**, 262-268 (1956).

K2. Kapeller-Adler, R., Uber eine neue Methode zur quantitativen Histidinbestimmung und über deren Anwendbarkeit zur Untersuchung von biologischen Flüssigkeiten, insbesondere von Gravidenharnen. *Biochem* Z. **264**, 131-141 (1933).

K3. Keeley, K. J., and Politzer, W. M., Amino-aciduria in the megaloblastic anemias. *J. Clin. Pathol.* **9**, 142-143 (1956).

K4. Keup, W., Eine quantitative Mikromethode zur Bestimmung von Phenylbrenztraubensäure. *Biochem. Z.* **326**, 14-17 (1954).

K5. King, F. P., and Lochridge, E. P., Cystinosis (Cystine-storage disease): report of a case with biochemical isolation and quantitative determination of cystine in lymphnodes, spleen and liver. *Am. J. Diseases Children* **82**, 446-455 (1951).

K6. Kinsell, L. W., Harper, H. A., Bartow, H. C., Hutchin, M. E., and Hess, J. R., Studies in methionine and sulfur metabolism. I. The fate of intravenously administered methionine in normal individuals and in patients with liver damage. *J. Clin. Invest.* **27**, 677-688 (1948).

K7. Knox, W. E., and Hsia, D. Y. Y., Pathogenetic problems in Phenylketonuria. *Am. J. Med.* **22**, 687-702 (1957).

K8. Komrower, G. M., L'amino-acidurie dans la galactosémie. *Arch. franç. pédiat.* **10**, 185-190 (1953).

K9. Komrower, G. M., Schwarz, V., Holzel, A., and Golberg, L., A clinical and biochemical study of galactosaemia, a possible explanation of the nature of the biochemical lesion. *Arch. Disease Childhood* **31**, 254-264 (1956).

K10. Kropp, K., and Lang, K., A colorimetric method for the quantitative determination of phenylpyruvic acid in urine. *Klin. Wochschr.* **33**, 482-485 (1955).

L1. Ladu, B. N., Zannoni, V. G., Laster, L., and Seegmiller, J. E., The nature of the defects in tyrosine metabolism in alcaptonuria. *J. Biol. Chem.* **230**, 251-260 (1958).

L2. Lang, K., Die phenylpyruvische Oligophrenie. *Ergeb. inn. Med. u. Kinderheilk.* **6**, 78-99 (1955).

L3. Lederer, E., and Lederer, M., "Chromatography," 2nd ed. Elsevier, Amsterdam, 1957.

L4. Levine, S. Z., Het metabolisme der aromatische aminozuren bij het te vroeg geboren kind en de invloed, die ascorbinezuur en het adrenocorticotrope hormoon der hypophyse (ACTH) hierop uitoefenen. *Voeding* **12**, 306-309 (1951).

L5. Levine, S. Z., Marples, E., and Gordon, H. H., A defect in the metabolism of aromatic amino-acids in premature infants; the role of vitamin C. *Science* **90**, 620-621 (1939).

L6. Levine, S. Z., Marples, E., and Gordon, H. H., A defect in the metabolism of tyrosine and phenylalanine in premature infants: spontaneous occurrence and eradication by vitamin C. *J. Clin. Invest.* **20**, 209-219 (1941).

L7. Lignac, G. O. E., Gestoord cystine metabolism bij kinderen. *Ned. Tijdschr. Geneesk.* **68**, 2987-2996 (1924).

L8. Lindan, O., Paper chromatography of Gambian urines: amino-acid pattern and excretion of β-aminoisobutyric acid. Malnutrition in African mothers, infants and young children. *Rep. 2nd Inter-african (C.C.T.A.) Conf. on Nutrition, Gambia, 1952* pp. 195-196 (1954).

L9. Low, N. L., Armstrong, M. D., and Carlisle, J. W., Phenylketonuria, two unusual cases. *Lancet* **ii**, 917-918 (1956).

L10. Lowe, C. U., Congenital abnormalities of amino acid transport in renal tubules. *Pediatrics* **21**, 1039-1046 (1958).

L11. Lowe, C. U., Terrey, M., and MacLachlan, E. A., Organic aciduria decreased renal ammonia production, hydrophthalmos and mental retardation. A clinical entity. *Am. J. Diseases Children* **83**, 164-184 (1952).

L12. Luder, J., and Sheldon, W., A familial tubular absorption defect of glucose and amino-acids. *Arch. Disease Childhood* **30**, 160-164 (1955).

M1. McCance, R. A., Morrison, A. B., and Dent, C. E., The excretion of phosphoethanolamine and hypophosphatasia; preliminary communication. *Lancet* **i**, 131 (1955).

M2. McCance, R. A., Fairweather, D. V. I., Barrett, A. M., and Morrison, A. B., Genetic, clinical, biochemical and pathological features of hypophosphotasia based on the study of a family. *Quart. J. Med.* **20**, 523-537 (1956).

M3. McCune, D. J., Mason, H. H., and Clarke, H. T., Intractable hypophosphatemic rickets with renal glycosuria and acidosis (the Fanconi syndrome). Report of a case in which increased urinary organic acids were detected and identified, with a review of the literature. *Am. J. Diseases Children* **65**, 81-146 (1943).

M4. Maggioni, G., Modiano, G., Strigini, P., Signoretti, A., and Biaggi, G. Contributo alla conoscenza dell'aminoaciduria qualitativa e quantitativa nella distrofia da carenza proteica. *Atti 25. Congr. ital. pediat.* **2**, 100-105 (1957).

M5. Mandelbrote, B. M., Stanier, M. W., Thompson, R. H. S., and Thruston, M. N., Studies on copper metabolism in demyelinating disease of central nervous system. *Brain* **71**, 212-228 (1948).

M6. Matthews, W. B., The absorption and excretion of radiocopper in hepato lenticular degeneration (Wilson's disease). *J. Neurol. Neurosurg. Psychiat.* **17**, 242-246 (1954).

M7. Matthews, W. B., Milne, M. D., and Bell, M., The metabolic disorder in hepatolenticular degeneration. *Quart. J. Med.* **21**, 425-446 (1952).

M8. Medes, G., A new error of tyrosine metabolism: tyrosinosis. The intermediary metabolism of tyrosine and phenylalanine. *Biochem. J.* **26**, 917-940 (1932).

M9. Meister, A., "Biochemistry of the Amino Acids," pp. 397-400. Academic Press, New York, 1957.

M10. Meister, A., Udenfriend, S., and Bessman, S. P., Diminished phenylketonuria in phenylpyruvic oligophrenia after administration of L-glutamine, L-glutamate, or L-asparagine. *J. Clin. Invest.* **35**, 619-626 (1956).

M11. Menkes, J. H., Hurst, P. L., and Craig, J. M., New syndrome: progressive familial infantile cerebral dysfunction associated with unusual urinary substance. *Pediatrics* **14**, 462-466 (1954).

M12. Monnet, P., Matray, and Etienne, Glaucome congénital et néphropathie. *Pédiatrie* **10**, 617-622 (1955).

M13. Moore, S., Contribution to the discussion of H. Harris' report on "Aminoaciduria in man." *Proc. 3rd Intern. Congr. Biochem., Brussels* p. 475 (1955).

M14. Moore, S., and Stein, W. H., Chromatography of amino acids on sulfonated polystyrene resins. *J. Biol. Chem.* **192**, 663-681 (1951).

M15. Moore, S., and Stein, W. H., Procedures for the chromatographic determination of aminoacids on four per cent cross-linked sulfonated polystyrene resins. *J. Biol. Chem.* **211**, 893-906 (1954).

M16. Moore, S., and Stein, W. H., A modified ninhydrin reagent for the photometric determination of aminoacids and related compounds. *J. Biol. Chem.* **211**, 907-913 (1954).

M17. Moore, S., Spackman, D. H., and Stein, W. H., Chromatography of aminoacids on sulfonated polystyrene resins. *Anal. Chem.* **30**, 1185-1190 (1958).

M18. Morgan, E. J., and Friedmann, E., Interaction of maleic acid with thiol compounds. *Biochem. J.* **32**, 733-742 (1938).

M19. Morgan, E. J., and Friedmann, E., Maleic acid as inhibitor of enzyme reactions induced by —SH compounds. *Biochem. J.* **32**, 862-870 (1938).

M20. Morris, J. E., Harpur, E. R., and Goldbloom, A., The metabolism of L-tyrosine in infantile scurvy. *J. Clin. Invest.* **29**, 325-335 (1950).

M21. Moubasher, R., Estimation of α-aminoacids in pure solutions, in blood, and in urine with peri-naphthindan-2,3,4-trione hydrate. *J. Biol. Chem.* **175**, 187-193 (1948).

M22. Moubasher, R., and Sina, A., The gasometric determination of α-aminoacids by the peri-naphthindan-2,3,4-trione hydrate. Carbon dioxide method in pure solutions with remarks upon its use in blood and urine. *J. Biol. Chem.* **180**, 681-688 (1949).

M23. Müting, D., and Wortmann, V., Uber den Aminosäurenhaushalt bei Leberkrankheiten. *Deut. med. Wochschr.* **81**, 1853-1856 (1956).

N1. Nöller, H. G., Stelgens, P., and Kieser, H. J., Hochspannungselektrophoretische Untersuchungen am Urin bei einem Kinde mit Wilsonscher Erkrankung. *Monatsschr. Kinderheilk.* **105**, 343-346 (1957).

N2. Northrop, J. H., Convenient method for formol titration. *J. Gen. Physiol.* **9**, 767-769 (1926).

P1. Paine, R. S., Hsia, D. Y. Y., Hsia, H. H., and Driscoll, K., The dietary phenylalanine requirements and tolerances of phenylketonuric patients. *A.M.A. J. Diseases Children* **94**, 224-230 (1957).

P2. Page, E. W., Glendening, M. B., Dignam, W., and Harper, H. A., The causes of histidinuria in normal pregnancy. *Am. J. Obstet. Gynecol.* **68**, 110-118 (1954).

P3. Pare, C. M. B., and Sandler, M., Aminoaciduria in march hemoglobinuria. *Lancet* **i**, 702-704 (1954).

P4. Parry, T. E., Paper chromatography of fifty-six amino compounds using phenol and butanol-acetic acid as solvents, with illustrative chromatograms of normal and abnormal urines. *Clin. Chim. Acta* **2**, 115-125 (1957).

P5. Pentz, E. I., Davenport, C. H., Glover, W., and Smith, D. D., A test for the determination of taurine in urine. *J. Biol. Chem.* **228**, 433-445 (1957).

P6. Peters, J. P., and Van Slyke, D. D., "Quantitative Clinical Chemistry," Vol. II: Methods, pp. 398-400. Williams & Wilkins, Baltimore, 1932.

P7. Pfändler, U., and Berger, H., Zur Genetik der Cystinose (Cystinspeicherkrankheit) und ihre Beziehungen zur Cystinurie und Hyperaminoacidurie. *Ann. Paediat.* **187**, 1-41 (1956).

P8. Porter, H., Aminoacid excretion in degenerative diseases of the nervous system. *J. Lab. Clin. Med.* **34**, 1623-1626 (1949).

R1. Rathbun, J. C., Hypophosphatasia. A new developmental anomaly. *Am. J. Diseases Children* **75**, 822-831 (1948).

R2. Rider, A. A., and McCollum, E. V., Extraction of aminoacid containing substances from urine. *J. Lab. Clin. Med.* **45**, 215-218 (1955).

R3. Robson, E. B., and Rose, G. A., The effect of intravenous lysine on the renal clearances of cystine, arginine and ornithine in normal subjects, in patients with cystinuria and Fanconi syndrome and in their relatives. *Clin. Sci.* **16**, 75-93 (1957).

R4. Royer, P., and Prader, A., Les insuffisances congénitales du tubule rénal. Etude clinique et thérapeutique. *Rapport 21e Congr. Assoc. pédiatres de langue franç., Paris* pp. 259-357 (1957).

R5. Royer, P., Lestradet, H., and Gilly, R., Effet de la perfusion intraveineuse de gluconate de calcium sur l'hyperaminoacidurie du rachitisme commun et du rachitisme vitamino-résistant. *In.* "8th Intern. Congr. Paediat., Volume of Discussions," p. 268. Copenhagen, 1956.

R6. Royer, P., Spahr, A., and Gerbeaux, S., L'amino-acidurie du rachitisme commun et de la tétanie du nourrisson. *Semaine hôp.* **32**, 254-258 (1956).

S1. Scheinberg, I. H., and Gitlin, D., Deficiency of ceruloplasmin in patients with hepatolenticular degeneration (Wilson's disease). *Science* **116**, 484-485 (1952).

S2. Schlesinger, B. E., Luder, J., and Bodian, M., Rickets with alkaline phosphatase deficiency: an osteoblastic dysplasia. *Arch. Disease Childhood* **30**, 265-276 (1955).

S3. Schönenberg, H., Papierchromatographische Untersuchungen bei der Pfaundler-Hurlerschen Krankheit. *Monatsschr. Kinderheilk.* **102**, 404-407 (1954).

S4. Schönenberg, H., Papierchromatographische Untersuchungen bei der Dystrophia musculorum progressiva sowie anderen Myopathien. *Klin. Wochschr.* **33**, 513-515 (1955).

S5. Schönenberg, H., Kritische Bemerkungen zur klinischen Papierchromatographie der Aminosäuren. *Klin. Wochschr.* **34**, 269-272 (1956).

S6. Schram, E., and Bigwood, E. J., Fraction collector for chromatography. *Anal. Chem.* **25**, 1424 (1953).

S7. Schram, E., Dustin, J. P., Moore, S., and Bigwood, E. J., Application de la chromatographie sur échangeur d'ions à l'étude de la composition des aliments en acides aminés. *Anal. Chim. Acta* **9**, 149-162 (1953).

S8. Schreier, K., Untersuchungen über den Aminosäurenstoffwechsel bei Erkrankungen der Leber. *Med. Wochschr.* **76**, 868-872 (1951).

S9. Schreier, K., Die angeborenen Störungen im Phenylalaninstoffwechsel. *Modern Problems Paediat.* **3**, 285-307 (1957).

S10. Schreier, K., and Plückthun, H., Untersuchungen über den Gehalt an freien Aminosäuren im Serum und Urin. *Biochem. Z.* **320**, 447-465 (1950).

S11. Schreier, K., and Plückthun, H., Über die Aminosäuren; Übersicht über ihren Stoffwechsel und ihre physiologische Bedeutung: Bericht über Untersuchungen des Aminosäuregehaltes in Blut und Urin. *Z. Kinderheilk.* **68**, 480-511 (1950).

S12. Schreier, K., and Plückthun, H., Über die Alkaptonurie. Eine klinische und physiologisch-chemische Studie. *Kinderheilk.* **71**, 462-475 (1952).

S13. Schwarz, V., Golberg, L., Komrower, G. M., and Holzel, A., Metabolic observations on the erythrocytes from cases of galactosaemia. *Biochem. J.* **59**, Proc. xxii (1955).

S14. Seringe, P., Amiel, J. L., and Leluc, Diabète rénal pur. Diabète glucoaminé (Glycosurie et amino-acidurie rénales et familiales). *Semaine hôp.* (*Ann. pédiat.*) **33**, 1073-1081 (1957).

S15. Simon, S., Zur Stickstoffverteilung im Urin des Neugeborenen. *Z. Kinderheilk.* **2**, 1-17 (1911).

S16. Smirnova, L. G., and Chulkova, S., Paper chromatography and its application in the determination of urinary aminoacids. *Lab. Delo* **11**, 1-7 (1956).

S17. Smith, H. W., "The Kidney, Structure and Function in Health and Disease." Oxford University Press, London and New York, 1955.

S18. Sobel, C., Henry, R. J., Chiamori,, N., and Segalove, M., Determination of α-amino-nitrogen in urine. *Proc. Soc. Exptl. Biol. Med.* **95**, 808-813 (1957/

S19. Sochor, H., Pankreasinsuffizienz bei chronischer Aminoacidurie. *Oesterr. Z. Kinderheilk.* **7**, 264-273 (1952).

S20. Souchon, F., Papierchromatographische Untersuchungen der freien Aminosäuren im Säuglingsharn. *Z. ges. exptl. Med.* **118**, 219-229 (1952).

S21. Souchon, F., and Grunau, G., Zur Aminosäurenausscheidung bei Frühgeborenen. *Arch. Kinderheilk.* **144**, 143-152 (1952).

S22. Soupart, P., Contribution à l'étude de l'histidinurie au cours de la grossesse. Le débit urinaire d'histidine au cours du cycle menstruel normal. *Acta Clin. Belg.* **9**, 297-318, 319-324 (1954).

S23. Soupart, P., Le dosage de l'histidine dans l'urine par une méthode enzymatique spécifique et son application à l'étude de l'excrétion de l'histidine chez la femme normale et la femme enceinte. *Clin. Chim. Acta* **3**, 349-356 (1958).

S24. Soupart, P., Urinary excretion of free aminoacids in normal adult men and women. *Clin. Chim. Acta* **4**, 265-271 (1959).

S25. Spackman, D. H., Stein, W. H., and Moore, S., Automatic recording apparatus for use in the chromatography of aminoacids. *Anal. Chem.* **30**, 1190-1206 (1958).

S26. Spencer, A. G., and Franglen, G. T., Gross aminoaciduria following a lysol burn. *Lancet* **i**, 190-192 (1952).

S27. Stave, U., Aminosäuren-Verfütterung und Tubulusschaden. II. Die tubuläre Funktion nach Cystin-Verfütterung bei Kaninchen. *Kinderheilk.* **78**, 275-282 (1956).

S28. Stave, U., Über den Einfluss von Cystinfütterung auf Gewebs- Aminosäuren und α-Aminostickstoff im Nierenblut. *Kinderheilk.* **81**, 26-35 (1958).

S29. Stave, U., and Schlaak, E., Aminosäuren-Verfütterung und Tubulusschaden. I. Aminoacidurie nach Cystin-Verfütterung bei Kaninchen. *Kinderheilk.* **78**, 261-274 (1956).

S30. Stave, U., and Willenbockel, U., Aminosäuren-Verfütterung und Tubulusschaden. III. Die Aktivität der akalischen Nierenphosphatase nach Cystin-Verfütterung. *Kinderheilk.* **78**, 563-576 (1956).

S31. Stein, W. H., Excretion of aminoacids in cystinuria. *Proc. Soc. Exptl. Biol. Med.* **78**, 75-708 (1951).

S32. Stein, W. H., With the assistance of G. C. Carey, A chromatographic investigation of the amino acid constituents of normal urine. *J. Biol. Chem.* **201**, 45-58 (1953).

S33. Stein, W. H., and Moore, S., Electrolytic desalting of amino acids. Conversion of arginine to ornithine. *J. Biol. Chem.* **190**, 103-106 (1951).

S34. Stein, W. H., Bearn, A. G., and Moore, S., The aminoacid content of the blood and urine in Wilson's disease. *J. Clin. Invest.* **33**, 410-419 (1954).

T1. Tallan, H. H., Stein, W. H., and Moore, S., 3-Methylhistidine, a new aminoacid from human urine. *J. Biol. Chem.* **206**, 825-834 (1954).

T2. Tallan, H. H., Bella, S. T., Stein, W. H., and Moore, S., Tyrosine-O-sulphate as a constituent of normal human urine. *J. Biol. Chem.* **217**, 703-708 (1955).

T3. Tegelaers, W. H. H., and Tiddens, H. W., Nephrotic-glucosuric-aminoaciduric dwarfism and electrolyte metabolism. *Helv. Paediat. Acta* **10**, 269-278 (1955).

T4. Thelander, H. E., and Imagawa, R., Amino-aciduria, congenital defects, and mental retardation. A preliminary report. *J. Pediat.* **49**, 123-128 (1956).

T5. Turba, F., "Chromatographische Methoden in der Proteinchemie." Springer Verlag, Berlin, 1954.

U1. Udenfriend, S., and Bessman, S. P., The hydroxylation of phenylalanine and antipyrine in phenylpyruvic oligophrenia. *J. Biol. Chem.* **203**, 961-966 (1953).

U2. Uzman, L. L., The relation of urinary copper excretion to the aminoaciduria in Wilson's disease (hepatolenticular degeneration). *Am. J. Med. Sci.* **226**, 645-652 (1953).

U3. Uzman, L. L., and Denny-Brown, D., Aminoaciduria in hepatolenticular degeneration (Wilson's disease). *Am. J. Med. Sci.* **215**, 599-611 (1948).

U4. Uzman, L. L., Iber, F. L., Chalmers, C., and Knowlton, M., The mechanism of copper deposition in the liver in hepatolenticular degeneration (Wilson's disease). *Am. J. Med. Sci.* **231**, 511-518 (1956).

V1. Van Creveld, S., and Arons, P., Further experiences in a special case of renal osteoporosis with amino-aciduria, treated with dihydrotachysterol. *Ann. Paediat.* **182**, 191-202 (1954).

V2. van der Schaaf, P. C., and Huisman, T. H. J., Enkele verbeteringen in de chromatografische scheidingsmethode der aminozuren met behulp van de ionenuitwisselaar Dowex 50. *Chem. Weekblad* **50**, 273-275 (1954).

V3. Van Geffel, R., Devriendt, A., Dustin, J. P., Vis, H., and Loeb, H., La maladie du galactose. Considérations génétiques. Etude de l'aminoacidurie et de l'aminoacidémie. *Arch. franç. pédiat.* **15**, 1-27 (1959).

V4. Van Slyke, D. D., and Kirk, E., Comparison of gasometric, colorimetric, and titrimetric determinations of amino nitrogen in blood and urine. *J. Biol. Chem.* **102**, 651-682 (1933).

V5. Van Slyke, D. D., Dillon, R. T., MacFadyen, D. A., and Hamilton, P. B., Gasometric determination of carboxyl groups in free aminoacids. *J. Biol. Chem.* **141**, 627-669 (1941).

V6. Van Slyke, D. D., MacFadyen, D. A., and Hamilton, P. B., The gasometric determination of amino acids in urine by the ninhydrin-carbon dioxide method. *J. Biol. Chem.* **150**, 251-258 (1943).

V7. Vestergaard, P., and Leverett, R., Constancy of urinary creatinine excretion. *J. Lab. Clin. Med.* **51**, 211-218 (1958).

W1. Waksman, A., and Bigwood, E. J., Identification and quantitative chromatographic determination of γ-amino-n-butyric acid on ion exchange columns. *Ann. Acad. Sci. Fennicae Ser. A II,* **60**, 49-54 (1955).

W2. Wallis, L. A., and Engle, R. L., The adult Fanconi syndrome. II. Review of eighteen cases. *Am. J. Med.* **22**, 13-23 (1957).

W3. Walshe, J. M., Disturbances of aminoacid metabolism following liver injury; study by means of paper chromatography. *Quart. J. Med.* **22**, 483-505 (1953).

W4. Walshe, J. M., Penicillamine, a new oral therapy for Wilson's disease. *Am. J. Med.* **21**, 487-495 (1956).

W5. Walshe, J. M., Hepatolenticular degeneration (Wilson's disease). *Brit. Med. Bull.* **13**, 132-135 (1957).

W6. Walshe, J. M., and Senior, B., Disturbances of cystine-metabolism in liver disease. *J. Clin. Invest.* **34**, 302-310 (1955).

W7. Weaver, J. A., and Neill, D. W., Aminoaciduria in pernicious anaemia and subacute combined degeneration of cord. *Lancet* **i**, 1212-1213 (1954).

W8. Weber, H., Beitrag zur Frage der Nierenfunktionsstörung bei Cystinosis. *Helv. Paediat. Acta* **8**, 348-366 (1953).

W9. Werner, G., Eine Apparatur für Hochspannungs-Papierelektrophorese. *1st Eur. Congr. Clin. Chem.* p. 28 (1954).

W10. Westall, R. G., The aminoacids and other ampholytes of urine. 3. Unidentified substances excreted in normal human urine. *Biochem. J.* **60**, 247-255 (1955).

W11. Westall, R. G., Dancis, J., and Miller, S., Maple sugar urine disease.

Trans. Soc. Pediat. Research p. 110, preprints, *Am. J. Diseases Children* **94**, 571 (1957).

W12. Wilson, V. K., Thomson, M. L., and Dent, C. E., Aminoaciduria in lead poisoning. A case in childhood. *Lancet* **ii**, 66-68 (1953).

W13. Woolf, L. I., Excretion of conjugated phenylacetic acid in phenylketonuria. *Biochem. J.* **49**, Proc. ix-x (1951).

W14. Woolf, L. I., and Giles, H. M., Urinary excretion of amino acids and sugar in the nephrotic syndrome: chromatographic study. *Acta Paediat.* **45**, 489-500 (1956).

W15. Worthem, H. G., and Good, R. A., The de Toni-Fanconi syndrome with cystinosis. *A.M.A. J. Diseases Children* **95**, 653-688 (1958).

W16. Wu. C., Bollman, J. L., and Butt, H. R., Changes in free amino acids in plasma during hepatic coma. *J. Clin. Invest.* **34**, 845-849 (1955).

Y1. Yeh, H. L., Frankl, W., Dunn, M. S., Parker, P., Hughes, B., and György, P., Urinary excretion of amino acids by a cystinuric subject. *Am. J. Med. Sci.* **214**, 507-512 (1947).

Z1. Zeller, E. A., Isolierung von Phenylmilchsäure und Phenylbrenztraubensäure aus Harn bei Imbecillitas phenylpyruvica. *Helv. Chim. Acta* **26**, 1614-1618 (1943).

Z2. Zweig, G., and Hood, S. L., Microdesalter for qualitative paper chromatography of aminoacids. *Anal. Chem.* **29**, 438-441 (1957).

BILE PIGMENTS IN JAUNDICE

Barbara H. Billing

Department of Surgery, Postgraduate Medical School of London, London, England*

* Present address: Department of Medicine, Royal Free Hospital, London, England.

1. Introduction

The hypothesis that two types of bilirubin existed was first put forward in 1916 by Hijmans van den Bergh and Müller (H5). Numerous attempts were made to characterize the pigments, but it was not until 40 years later that the application of chromatographic and electrophoretic techniques provided the solution to the problem. In 1956, independent reports from Czechoslovakia (T1), the United States (S5), and England (B9) demonstrated that the main pathway of bilirubin excretion was by conjugation with glucuronic acid. In this way, bilirubin is converted from a fat-soluble pigment to a water-soluble one which is capable of giving a "direct" reaction in the van den Bergh test. The chief site of bilirubin glucuronide formation has been shown to be the liver microsomes (S8), with uridine diphosphate glucuronic acid as the glucuronyl donor in the transferase reaction. A more complete understanding of some of the defects in bilirubin metabolism which cause jaundice is now possible, and these will be discussed in this chapter.

2. Identification of "Direct Bilirubin"

2.1. "Indirect" and "Direct" Bilirubin

While estimating serum bilirubin, Hijmans van den Bergh and Müller (H5) observed that specimens from patients with hemolytic jaundice did not react promptly with diazotized sulfanilic acid, in acid solution, except in the presence of alcohol or some other accelerator substance. Bile, and urine and serum from patients with obstructive jaundice, reacted directly in aqueous solution. The two types of reactions were called indirect and direct, respectively. Explanations for these differences in behavior in the van den Bergh reaction have been numerous and often contradictory; they have been reviewed in detail by With in his monograph (W9). The possibility that they might be due to bilirubin bound to different compounds, such as conjugated bile salts or proteins, has now been disproved and subsequent sections will describe the work which confirms the hypothesis that there are two "types" of bilirubin, which are structurally different but chemically related. The different physiological and physicochemical characteristics of these pigments have been known for many years and are summarized in Table 1. It will be seen that the acceptance of a glucuronide structure for direct bilirubin can provide an explanation for these differences.

2.2. Chromatography of Bile Pigments

In 1953 Cole and Lathe (C5) subjected protein-free filtrates of icteric serum to reverse phase partition chromatography. Using silicone-treated

TABLE 1
DIFFERENCES BETWEEN THE TWO TYPES OF BILE PIGMENT[a]

Property	Bilirubin	Conjugated bilirubin
Van den Bergh reaction	Indirect	Direct
Solubility in aqueous solution at acid or neutral pH	—	+
Solubility in chloroform and lipid solvents	+	—
Occurrence in icteric urine	—	+
Occurrence in bile	—	+
Affinity for brain tissue	+	—
Attachment to plasma albumin	+	+
Affinity for denatured plasma proteins	—	+
Ease of oxidation	+	++
Photosensitivity	++	+
Association with hemolytic jaundice	++	(+)
Association with obstructive jaundice and hepatitis	+	+++

[a] Data from reference B10.

kieselguhr and a solvent system containing chloroform, methanol, carbon tetrachloride, and pH 6.0 phosphate buffer, they were able to separate slow-moving, indirect-reacting bilirubin from a fast-moving, direct-reacting pigment, which was water soluble. The authors were unable to make bilirubin behave as a fast-moving pigment by the addition of sodium deoxycholate, a naturally occurring substance that can replace alcohol in the indirect reaction. Mixing the two pigments with sera or with tissue extracts did not alter their subsequent movement on chromatograms, so that there was no evidence that the direct or indirect reaction might be due to the presence of inhibitors or catalysts. Since neither of the pigments contained protein, it was concluded that two separate pigments existed.

The direct-reacting pigment was later shown to consist of two components (pigment I and pigment II), both of which are water soluble and react promptly in the van den Bergh reaction in aqueous solution. They were separated from each other and from bilirubin by reverse phase partition chromatography, using a butanol/pH 6 phosphate buffer system (C6). The absorption curves of the three pigments were observed to be similar but not identical. Pigment II, the more polar pigment, moved with the solvent front on the chromatogram and consequently was obtained only in an impure state. Attempts to isolate and characterize this pigment were unsuccessful.

Separation of the direct pigment from the indirect pigment has also been achieved using absorption chromatography on alumina (P3) and silica gel (S1), and by paper chromatography (G4, K2, M3). Pure pigments have not been obtained using these procedures. Sakamoto *et al.* (S2, S3) have employed ion exchange resins to separate two forms of direct bilirubin from dog bile, namely, the ester form of bilirubin and the salt form of bilirubin. These two pigments were first described by Yamaoka and Kosaka (Y1) in 1951. They showed that when a protein-free filtrate of icteric serum or urine, or bile is saponified it no longer possesses the characteristics of direct bilirubin but behaves like indirect bilirubin. They therefore concluded that the greater part of direct bilirubin is in the form of an ester; a small part they considered existed as a salt of bilirubin, since the addition of hydrochloric acid to a solution containing chloroform resulted in some of the pigment passing into the chloroform as bilirubin. The ester form of bilirubin was obtained free from bile acids using an Amberlite IR 4B column followed by an IRC 50 column. The structure of this ester was not determined, but as it did not contain glucuronic acid, it is not identical with pigments I and II of Cole *et al.* (C6).

2.3. Azo Derivatives of Bile Pigments

The instability of pigments I and II has made their isolation a difficult task. Attempts to characterize them by indirect means were therefore undertaken and a study was made of the different pigments formed in the coupling reaction of bile pigments with diazo reagents.

Overbeek *et al.* (O2) have recently re-examined the original work of Fischer and Haberland (F1) on the formation of azo pigments from bilirubin. They showed that if the coupling reaction is carried out in a mixture of 60 % ethanol, 30 % chloroform, and 10 % water, in which all the components of the reaction are soluble, then one molecule of bilirubin will form two molecules of azobilirubin. After a preliminary splitting of the bilirubin molecule at the methene bridge, this reaction takes place in two stages.

Bilirubin + *p*-diazobenzenesulfonic acid → azobilirubin + hydroxypyrromethene carbinol (*1*)

Hydroxypyrromethene carbinol + *p*-diazobenzenesulfonic acid → azobilirubin + formaldehyde (*2*)

The constants for these second order reactions have been determined and they indicate that the first reaction proceeds at a considerably faster rate than the second reaction. It is possible that the van den Bergh reaction, as used in clinical chemistry, takes place in the same way but, as it is not taking place in a homogeneous medium, it is probable that the kinetics will be more complicated.

The formation of two azo pigments in the coupling reaction of diazobenzenesulfonic acid with alcoholic extracts of icteric serum has been demonstrated by Kawai (K2) and Billing (B6). The less polar azo pigment (pigment *A*) is identical with the azo pigment formed from bilirubin and consists of a mixture of diazonium salts of neoxanthobilirubinic acid and isoneoxanthobilirubinic acid (S7). Pigment II yields only the more polar azo pigment (pigment *B*) and pigment I forms, in alcoholic solution, a mixture containing equal quantities of both azo pigments, *A* and *B* (Fig. 1). On suitable chromatograms two isomers of azo pig-

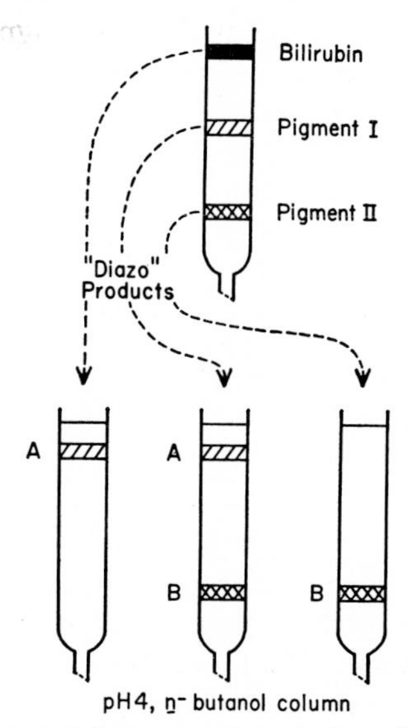

FIG. 1. Behavior of plasma bile pigments and their diazo products on partition chromatograms (L1).

ment *B* can be identified and under certain conditions of diazotization a brown product, probably related to hydroxypyrromethene carbinol, is formed. The close relationship between bilirubin and pigment II was apparent, and it was concluded that since one molecule of bilirubin forms two molecules of azo pigment *A*, one molecule of pigment II would form two molecules of azo pigment *B*. It was therefore anticipated that the difference between bilirubin and pigment II could be predicted from a knowledge of the difference in structure of azo pigments *A* and *B*.

Identification of azo pigment *B* as a glucuronide of azo pigment *A* was demonstrated independently by Billing *et al.* (B12) and Schmid

(S7). The British workers isolated and purified the pigment from diazotized bile by means of counter-current procedures, while Schmid effected his purification using ascending paper chromatography. Quantitative glucuronic acid determinations and enzyme studies, with various β-glucuronidase preparations, have established that one molecule of azo pigment *B* is equivalent to one molecule of azo pigment *A* conjugated with one molecule of glucuronic acid. The ease with which azo pigment *B* is hydrolyzed by mild alkali treatment to azo pigment *A* suggested an ester linkage between the propionic acid side chain of azo pigment *A* and the C-1 hydroxyl of glucuronic acid, rather than a glycosidic linkage. Confirmation of the nature of the glucuronide has been provided by Schachter (S4), whose observation that azo pigment *B* reacts with hydroxylamine with the liberation of glucuronic acid indicates the existence of a carboxyl glucuronide.

2.4. Structure of Bile Pigments

Gray *et al.* (G3) have identified the products of alkaline potassium permanganate oxidation of bilirubin by chromatographic comparison with authentic compounds. The only compound obtained was 4-methyl-3-carboxyethylpyrrole-2,5-dicarboxylic acid, and they have therefore concluded that bilirubin is a lactam compound and in common with other bile pigments has a IXα structure (Fig. 2).

The separation of pigment I from the other bile pigments has been achieved by partition chromatography, and the molar ratio of pigment (expressed as bilirubin) to glucuronic acid has been found to be approximately one. It has therefore been proposed that pigment I is bilirubin monoglucuronide. This structure is in agreement with the finding that pigment I forms both azo pigments *A* and *B* in the van den

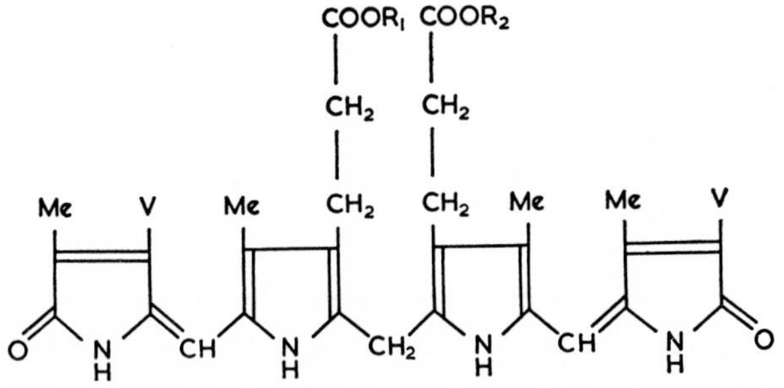

Fig. 2. Structure of bile pigments. (Bilirubin, R_1 and R_2 = H; pigment I, R_1 = glucuronyl and R_2 = H; pigment II, R_1 and R_2 = glucuronyl; in all, *Me* = methyl and *V* = vinyl).

Bergh reaction. Evidence that pigment II is the diglucuronide of bilirubin comes indirectly from the identification of azo pigment *B* as the monoglucuronide of azo pigment *A* and from the studies of Talafant (T1, T2). Using dog bile treated with acetone to remove the conjugated bile acids, Talafant isolated "direct bilirubin" by means of electrophoresis at pH 6.8. He found that this pigment could be converted to bilirubin by spleen and bacterial β-glucuronidases and liver and pancreatic enzymes. Alkaline hydrolysis with ammonia yielded glucuronic acid in amounts equivalent to two molecules of glucuronic acid for each molecule of bile pigment. Talafant (T3) has since obtained pure specimens of the sodium salt of bilirubin diglucuronide by first precipitating the pigment as the lead salt and then converting it to the sodium salt by the use of a suitable cation exchange resin. No infrared studies have yet been reported on this pigment, but it is to be expected that they will confirm the ester structure assigned by several groups of workers (B12, S4, S8).

3. Occurrence of the Three Bile Pigments in Biological Material

3.1. Bile

Pigment II is the main pigment found in bile. In gall bladder bile, up to 30 % of the pigments may occur as pigment I, but in fistula bile not more than 10 % is found in this form. Unconjugated bilirubin appears to be an artifact and can only be detected in grossly infected specimens of bile or in specimens that have been left at room temperature for several hours.

Verschure *et al.* (V1) have demonstrated by paper electrophoresis that in gall bladder bile the pigments form part of a lipoprotein complex to which cholesterol and bile acids are also bound. In hepatic bile this complex is not found in appreciable amounts, and the bile pigments are bound to two separate proteins.

3.2. Plasma

It is generally accepted that in normal subjects, most, but not necessarily all, of the bilirubin formed results from the breakdown of hemoglobin in the reticuloendothelial system. The bilirubin is then conjugated in the liver and excreted into the bile as a water-soluble pigment. The capacity of the liver to conjugate bilirubin is limited (W6), so that in cases of overproduction (e.g., hemolytic jaundice) free bilirubin will appear in the plasma. A similar result will be obtained if the ability of the liver to conjugate bilirubin is diminished (e.g., in the newborn infant). On the other hand, if the excretion of the bile is for some reason obstructed (e.g., in hepatitis and obstructive jaundice), conjugation will still take place and pigments I and II, as well as bilirubin, will accumulate in the plasma. In nonpathological plasma from humans and domesticated animals (B4) small amounts of bilirubin are found. The presence of conjugated bilirubin is questionable and in most instances is probably an artifact, produced during the estimation by the coupling of small amounts of bilirubin in aqueous solution.

Electrophoretic studies (G2, M2) have shown that both types of bilirubin are bound to albumin and sometimes α-globulin. Klatskin and Bungards (K6) found that this binding occurs between pH 6 and pH 9 and that below pH 5 separation occurs almost completely with respect to bilirubin but only partially with the conjugated pigment. This suggests that bilirubin and conjugated bilirubin are attached in different ways to plasma proteins.

3.3. Urine

In fresh specimens of icteric urine the conjugated pigments but not bilirubin can be detected. This is in keeping with the observation that bilirubinuria is absent in patients whose plasma contains large amounts of free bilirubin but no conjugated pigments. Techniques for the quantitative determination of bile pigments in urine have yet to be developed and it is therefore not known whether pigments I and II are excreted in the same or different ways. Nizet and Barac (N2) in their studies on the icteric dog have given us an over-all picture of bile pigment excretion. They have shown that it is an active process involving first the splitting of the protein complex. This takes place in the proximal tubules and in the descending loop of Henle. They did not characterize the enzymatic processes involved.

Greater amounts of bile pigments are excreted by patients with obstructive jaundice or acute hepatitis than by patients with chronic hepatitis or hepatocellular damage (D5). In the latter group of patients the jaundice persists and a decreased excretion affords a poor prognosis. Whether this failure to excrete bile pigments can be explained on the basis of renal failure or whether it is due to a diminished ability to conjugate bilirubin remains to be investigated. The presence of higher proportions of bilirubin and pigment I in plasma from patients with chronic liver disease would tend to support the latter hypothesis since pigment I is less water soluble than pigment II and therefore less easily excreted.

3.4. Feces

From experiments involving the administration of N^{15}-bilirubin to three cases of obstructive jaundice, Gilbertsen *et al.* (G1) have demonstrated that the conjugation of bilirubin is necessary for its conversion to fecal urobilinogen. In adults this conversion is almost complete so that, according to Watson (W5), only about 5–20 mg bilirubin (or mesobilirubin) appear daily in normal feces; it is not known if this bilirubin is present in a conjugated form. In infants, the formation of fecal urobilinogen does not occur until a suitable intestinal flora has developed, so that appreciable amounts of bilirubin and some biliverdin have been recovered from infant feces. The use of broad spectrum antibiotics in adults will also cause the appearance of bilirubin in feces, and Lowry and Watson (L7) have employed this technique to isolate large quantities of crystalline bilirubin from the feces of patients with hemolytic jaundice.

3.5. Tissues

Suitable techniques for the analysis of bile pigments in tissues have not yet been devised. There is some evidence that the bile pigments are bound to elastin (W9), and it has been suggested that the staining of the tissues may partly be due to the presence of pigments attached to the extravascular plasma albumin (B10). The fact that a higher level of plasma bile pigments is needed before jaundice is detected visually in infants with hemolytic disease than in adults with obstructive jaundice (W8) suggests that bilirubin and conjugated bilirubin are probably bound in different ways in the tissues. The observation of Stempfel *et al.* (S16) that exchange transfusions for hemolytic disease of the newborn are less efficient when conjugated pigments are present gives further support for this hypothesis.

The great binding capacity of the tissues for bile pigments can be appreciated if one studies the time taken for patients to lose their jaundice when an obstruction has been surgically relieved. Rundle *et al.* (R2) calculated that the serum bilirubin falls exponentially so that it reaches its half-value in 17.4 days.

Bilirubin, but not the conjugated pigments, has an affinity for brain tissue (B5, C2, W3), and in newborn infants, with high plasma bilirubin levels due to erythroblastosis or prematurity, staining of the brain nuclei may occur. This causes brain damage and subsequent neurological difficulties, a detailed account of which is given in a review by Claireaux *et al.* (C3).

3.6. Cerebrospinal, Amniotic, Pleural, and Ascitic Fluids, and Lymph

Both free and conjugated bilirubin have been detected in the cerebrospinal fluid (CSF) using diazotization techniques. The amounts are rarely greater than 1 mg% and cannot be directly correlated with levels in the plasma or the duration of the jaundice. Amatuzio *et al.* (A3) found that 8 of 10 decompensated cirrhotics had traces of bile pigment in their CSF but no increased amount of protein, while 10 of 11 patients in coma had greater amounts of bile pigment (mean, 0.35 mg/100 ml CSF, mainly as conjugated bilirubin) and also increased protein concentration. Stempfel (S14) examined the CSF from 25 patients with hemolytic disease of the newborn and, in spite of considerable variation in the CSF protein concentration, was able to detect a significant correlation between it and the CSF bilirubin concentration. In one case of "inspissated bile syndrome" Stempfel and Zetterstrom (S15) demonstrated the presence of conjugated bilirubin in the CSF. Bile pigments can also be detected in the CSF following intracranial hemorrhage and, when there is obstruction to the flow of CSF, cystic fluids and subdural effusions (B3).

Bilirubin is often present in the *liquor amnii* when the fetus is affected by hemolytic disease of the newborn (B5). Walker (W1) examined fresh specimens of amniotic fluid before the 35th week and found that the presence of bilirubin correctly indicated the disease in 94.5 % of his afflicted cases.

Bile pigments also occur in the pleural and ascitic fluids of jaundiced patients. According to Keller (K3) tumor exudates of the inflammatory type may contain more bilirubin than is present in the plasma.

The concentration of bile pigment in lymph is usually less than that in plasma. Obstruction of the bile duct in cholecystectomized animals results in a significantly greater rise in hepatic lymph bile pigment concentration compared with that in blood for the first 24 hours. Ritchie *et al.* (R1) found, however, that there was no evidence for regurgitation after the first 24 hours in cholecystectomized animals, in acute obstruction when the gall bladder was present, or during chronic obstruction in animals with or without a gall bladder; under these conditions the bile pigment concentration is greater in the plasma than the lymph.

4. Biosynthesis of Bilirubin Glucuronide

4.1. Intrahepatic

It has now been established that in common with many other glucuronides, bilirubin glucuronide can be synthesized in the adult rat, rabbit, mouse, and guinea pig liver by the enzymatic transfer of glucuronic acid from uridine diphosphate glucuronic acid to an acceptor substance, namely, bilirubin (Fig. 3) (B17, G5, L3, L5, S8). The transferring enzyme (glucuronyl transferase) has been shown by Schmid *et al.*

(S8) to be located in the microsome fraction of the liver. The neonatal development of this conjugating system has been studied by Brown *et al.* (B17) in the guinea pig. They were unable to detect the presence of bilirubin glucuronyl transferase in liver microsomes from fetal and newborn animals and, using *o*-aminophenol as a glucuronyl acceptor, were

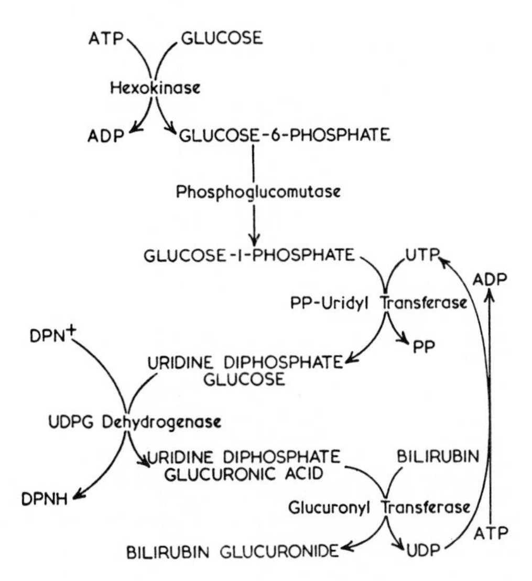

FIG. 3. A possible mechanism for the conjugation of bilirubin with glucuronic acid (B10).

able to demonstrate a deficiency of UDPGA dehydrogenase, and thus of UDPGA, during neonatal life. A similar deficiency has been described by Dutton (D8) in the mouse as well as the guinea pig and by Lathe and Walker (L5) in the rabbit.

Lathe and Walker (L5) have raised the problem as to whether the conjugation of bilirubin and such substrates as *o*-aminophenol as glucuronides involves the same or different enzymes. They studied the rates of conjugation of these two substances in liver tissue slices and suspensions from a large number of animals and found that they did not parallel one another. They did not consider that the inhibition caused by the addition of *o*-aminophenol to a bilirubin conjugating system could be explained entirely in terms of competitive inhibition, which is in agreement with the observations of Grodsky and Carbone (G5) using borneol. *In vivo,* both animals and humans with a congenital defect in bilirubin glucuronide formation also have a decreased ability to excrete other aglycones as glucuronides. Some workers therefore consider that a single enzyme system is involved in all types of glucuronide formation, so that the jaundice is due to an unsuccessful competition of the bilirubin with other substrates for the deficient glucuronic acid transferase system (S10).

Preliminary studies by Arias *et al.* (A5) using 4-methylumbelliferone as a substrate indicate that glucuronides can also be synthesized from α-glucuronic acid-1-phosphate by an enzyme system present in the soluble fraction of rat liver homogenates. It is not yet known whether such a system is capable of conjugating bilirubin.

4.2. Extrahepatic

The classic work of Bollman and Mann (B14) on the effect of hepatectomy on dogs gave the first indication that the liver was not the only organ capable of converting indirect bilirubin to direct bilirubin. They observed that, initially, bilirubin accumulated in the blood but that as the amount of pigment increased, the reaction of the plasma to the van den Bergh test became a direct one, and bile pigments began to appear in the urine. Bollman (B13) has recently reinvestigated this problem and found that the direct pigment behaves like pigment I on partition chromatograms. Whether this pigment is bilirubin monoglucuronide or some other conjugate of bilirubin has not been established. He has also observed that the injection of hemoglobin into hepatectomized dogs increases the rate of accumulation in the plasma of bilirubin and pigment I but not of pigment II. Similar observations have been made in hepatectomized rats. In these animals, Bollman found that bilirubin injections did result in a very small accumulation of pigment II as well as the other pigments. He was not able to account for this difference.

In complementary experiments Billing and Weinbren (cited in B10) infused bilirubin in amounts sufficient to saturate the capacity of the liver to excrete the pigment and noted that, in addition to bilirubin, pigment I but no pigment II accumulated in the plasma. The liver poisons ethionine and icterogenin also cause an accumulation of pigment I but not of pigment II in the blood of experimental animals (B13).

It is not known where this extrahepatic conjugation of bilirubin takes place. Grodsky and Carbone (G5) have demonstrated that rat kidney suspensions can conjugate bilirubin but at only half the rate of liver tissue. They also showed that brain tissue has very slight activity while muscle, spleen, and blood have none. Glucuronide conjugation of several substrates is known to occur in the intestinal tract (D8, H2), but as Bollman has shown that pigment I accumulates in hepatectomized dogs which have also been nephrectomized, there is at present no proof that the kidney or gastrointestinal tract are involved in bilirubin conjugation in the absence of a functioning liver. These organs may, of course, play a part in normal bilirubin metabolism, and further work is needed to determine whether pigment I is formed intra- or extrahepatically in patients with hepatitis or obstructive jaundice.

5. Nonglucuronide Conjugation of Bilirubin

Bilirubin is a lipid-soluble substance and for it to give a direct van den Bergh reaction it must be converted into a water-soluble state. This can be achieved in alkaline solution or by the use of solvents, such as ethanol, which are soluble in both water and chloroform (C1). Conjugation with glucuronic acid also causes the solubilization of bilirubin and in view of the well-known versatility of the liver's detoxifying systems, the possibility that bilirubin may be conjugated with other substances has been considered.

Jirsa *et al.* (J2) have esterified bilirubin with taurine and obtained both mono- and ditaurobilirubin, which behave like the corresponding glucuronide derivatives in the van den Bergh reaction, chromatography, electrophoresis, and ease of oxidation. These synthetic compounds have not yet been demonstrated in nature. The synthesis of a water-soluble bilirubin sulfate which gives a direct van den Bergh reaction has also been reported (W5), and recently a preliminary communication from Isselbacher (I1) provides evidence that such a substance occurs in bile. It is unlikely that bilirubin sulfate normally accounts for more than 10 % of the total bilirubin excreted. This finding is in keeping with the observation that 10–15 % of the conjugated bilirubin in human fistula bile is not converted to bilirubin by mild alkali treatment (B12). The ester form of bilirubin isolated by Sakamoto *et al.* (S2, S3) from both dog bile and icteric urine appears to be another form of conjugated bilirubin since it is alkali labile but does not contain glucuronic acid.

6. Toxicity of Bilirubin

The toxic action of bilirubin on tissue metabolism has been demonstrated both by *in vitro* and *in vivo* experiments. Lathe (L1) added bilirubin to brain brei, in concentrations such as occur in hemolytic disease of the newborn and prematurity, and observed a decrease in oxygen consumption. Day (D3) was able to reverse this depressed respiration by the addition of cytochrome c or methylene blue. Bowen and Waters (B16) then showed that this reversal could also be effected by diphosphopyridine nucleotide. Zetterström and Ernster (Z1), using isolated rat liver mitochondria, showed that bilirubin in concentrations of about 20 mg % causes uncoupling of oxidative phosphorylation. A similar effect can be demonstrated in brain mitochondria (E1). Unlike the respiratory depression, which occurs simultaneously, it cannot be reversed by the addition of DPN or cytochrome c. Conjugated bilirubin has no appreciable effect on these enzyme systems.

The large reserve capacity of the adult liver for the excretion of bilirubin has made the maintenance of plasma bilirubin levels sufficiently high to cause brain pigmentation and neurological symptoms in experimental animals a difficult procedure (L1). A further problem has been that adult brains are less susceptible to bilirubin than the brains of newborn animals (D4, F3, W2). These difficulties have largely been overcome by Johnson *et al.* (J5) who used a strain of homozygous rats with familial, nonhemolytic jaundice for their experiments. Injections of small amounts of bilirubin into newborn Gunn rats produced canary staining of areas of gray matter in the brain and spinal cord; intracellular yellow pigmentation could be seen microscopically. Johnson and Day (J4) were able to reduce the plasma bilirubin concentration necessary to cause death in these animals from kernicterus by the administration of Gantrisin or sulfadiazine. It is not known whether the drugs have a direct neurotoxic effect or whether they cause a change in the permeability of the blood brain barrier for bilirubin.

The part played by the blood brain barrier in determining the toxic effect of bilirubin has been examined by Ernster *et al.* (E1). They showed that, in the adult rabbit, the penetration of bilirubin into the central nervous system can be greatly enhanced by intracisternal treatment with *p*-chloromercuribenzoate. The injection of 24 mg bilirubin per kilogram resulted in the test animal's brain becoming yellow, especially in the cerebellum, the pons, and the basal ganglia, even though the bilirubin level in the plasma was only 7.8 mg % compared with 12.2 mg % in the control animal, whose brain was unstained. The increase in the concentration of the bilirubin in the CSF paralleled an increase in total protein concentration, which suggested to the authors the possibility that the bilirubin had entered the CSF as an albumin-bilirubin complex. These experiments give support to the suggestion of Stempfel (S14) that the functional maturity of the blood brain barrier, as well as the concentration of bilirubin in the plasma, may be significant in determining the development of kernicterus in the newborn infant.

7. Disorders of Bile Pigment Metabolism Causing Jaundice

A summary of the distribution of bile pigments in the different types of jaundice is given in Table 2.

TABLE 2
DISTRIBUTION OF BILE PIGMENTS IN JAUNDICE

Type of jaundice	Plasma			Urine
	Bilirubin	Pigment I	Pigment II	Conjugated bilirubin
Cirrhosis, hepatitis, and chronic obstruction	+	+++	++	++
Acute biliary obstruction	+	++	+++	+++
Hemolytic jaundice in adult	+	(+)	(+)	—
Neonatal jaundice	++	—	—	—
Hemolytic disease of newborn	+++	—	—	—
Inspissated bile syndrome	+++	++	+	+
Familial hyperbilirubinemia				
Gilbert's disease	++	—	—	—
Crigler and Najjar's disease	+++	—	—	—
Gunn rats	++	—	—	—
Chronic idiopathic jaundice	++	+	+	+

7.1. Obstructive Jaundice and Hepatocellular Disease

These forms of jaundice are all characterized by raised amounts of bilirubin, pigment I, and pigment II in the plasma and the presence of conjugated pigments in the urine. Reports by Billing (B8), Bollman (B13, B15), and Baikie (B1), indicate that the relative proportions of the three pigments are very variable and bear no relation to the total amount of bile pigments present. The observations of Billing and Baikie in small series of patients did not show any characteristic pattern in the distribution of the pigments for the different diseases, since pigment I was always the dominant pigment.

Bollman *et al.* (B15) made a detailed study of 56 cases of hepatocellular disease (26 with acute hepatitis, 3 with homologous serum jaundice, 22 with Laennec's cirrhosis, and 5 with postnecrotic cirrhosis) and 91 cases of biliary obstruction (37 with carcinoma of the pancreas, 13 with carcinoma of the bile duct, 18 with stones in the bile duct, 22 with benign structures and 1 with intrahepatic biliary atresia). In their cases 90 % of the patients with obstructive jaundice but only 16 % of the patients with hepatocellular disease had more pigment II than pigment I in the plasma. They therefore concluded that the presence of more than 50 % of the conjugated pigments in the form of pigment II supports a diagnosis of biliary obstruction rather than hepatocellular disease (see Fig. 4).

The efficiency of steroid therapy in causing a dramatic fall in plasma bile pigment concentration in hepatitis is well known; the mode of action of these drugs has not, however, been established. It appears unlikely that they have a chloretic effect (C4, S11), and the report of Katz *et al.* (K1) that treatment results in a decreased excretion of fecal urobilinogen suggests that there is either a decrease in the rate of hemoglobin breakdown or that another pathway of catabolism is employed which does not involve bilirubin.

7.2. Hemolytic Jaundice in the Adult

Bilirubin is the main bile pigment in the plasma of patients with hemolytic disease. In spite of the breakdown of hemoglobin being increased 5–15 times, bilirubin concentrations rarely rise to more than 2–5 mg/100 ml plasma, due to the enormous reserve capacity of the adult liver to conjugate bilirubin in excess of normal requirements. Small amounts of pigment I, with or without pigment II, are often detected but inadequate analytical techniques have so far prevented their quantitative determination. Their presence indicates some failure in pigment conjugation and excretion, the nature of which is at present unknown.

7.3. Neonatal Jaundice

So-called "physiological jaundice" in infants is attributable to a rise in plasma bilirubin, the conjugated bile pigments being absent. Billing *et al.* (B11) found that the maximum increase in bilirubin concentration

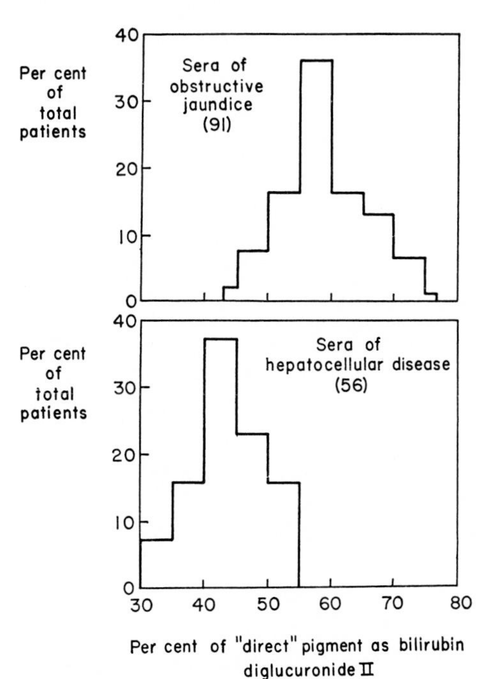

Fig. 4. Distribution of direct-reacting pigment II in the serum of patients suffering from liver disease (B15).

was inversely related to the birth weight and that in very small infants the jaundice persisted for longer periods of time than in the more mature babies (Fig. 5). These observations caused the authors to suggest that the accumulation of bilirubin in the plasma might be due to a defect in the bilirubin conjugating mechanism in the immature liver. This hypothesis has proved to be correct, for in 1957 Lathe and Walker (L3) examined the livers of 3 newborn infants and found that their capacity to conjugate bilirubin was very low even in the presence of added UDPGA. Dutton (D9) has also found almost complete absence of glucuronyl transferase and UDPG dehydrogenase in two human fetuses (3–4 months). A deficiency of glucuronyl transferase activity is therefore a probable explanation for the jaundice that appears in newborn infants, particularly if they are premature. During fetal life the bilirubin formed is presumably removed via the placenta so that it is not until birth that the infant's own liver takes over this function and the enzyme deficiency becomes apparent. Bilirubin then tends to accumulate until adequate enzyme levels have developed to permit conjugation of the bilirubin and its excretion into the bile.

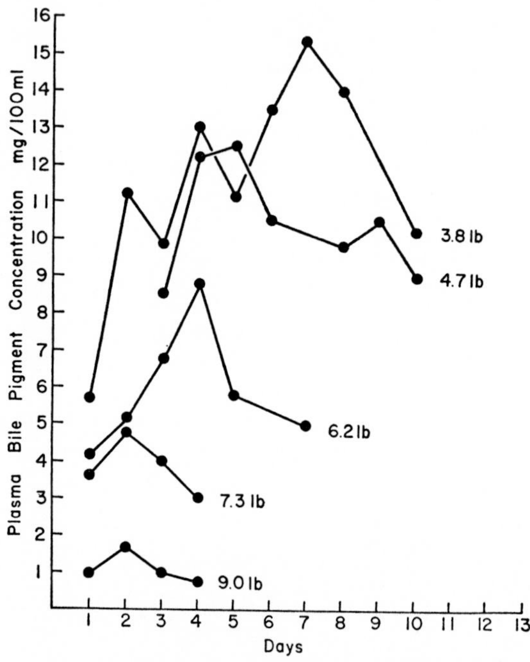

FIG. 5. The course of plasma bile pigment concentration during the neonatal period of five "typical" babies of varying birth weights (B11).

7.4. Hemolytic Disease of the Newborn

The concentrations of bilirubin in the plasma of infants with hemolytic disease tend to be considerably higher than those found in the premature infant since there is an increased pigment formation superimposed upon the neonatal enzyme deficiencies preventing bilirubin conjugation. Jaundice appears sooner and, if present on the first day of life, is indicative of a serological incompatibility rather than functional hepatic immaturity. The experience of many groups of investigators has resulted in the acceptance of the critical level of 20 mg bilirubin/100 ml plasma or over as likely to cause kernicterus; repeated exchange transfusions are, therefore, used to prevent this level being attained. Exposure to UV light (C7) and treatment with glucuronic acid (D2) have been proposed as possible means of reducing plasma bilirubin levels, but their usefulness has yet to be confirmed. Since the glucuronyl transferase system requires UDPGA and not glucuronic acid as a glucuronyl donor, it is difficult with our present knowledge to understand the rationale and mode of action of the latter treatment (D9, S9).

In a small proportion of cases of hemolytic disease of the newborn (inspissated bile syndrome), large amounts of conjugated bilirubin as well as free bilirubin accumulate in the plasma due to obstruction (J3, L2, S16). Conjugated bilirubin is apparently nontoxic to the brain so that exchange transfusions in such infants are indicated only if the free bilirubin and not the total bile pigment concentration is rising above 20 mg/100 ml plasma.

7.5. Familial Hyperbilirubinemia

Recurrent jaundice from childhood has been reported in several pathological conditions and has been shown, with one exception, to be due to abnormal amounts of unconjugated bilirubin in the plasma. Arias and London (A4) have examined liver tissue from two patients with constitutional hepatic dysfunction (Gilbert's disease) and found that they had a deficiency in bilirubin glucuronyl activity but that their UDPGA content appeared to be normal (Fig. 6). Their ability to form *o*-aminophenol glucuronide was also impaired. Neither patient had conjugated pigments in the plasma but the gall bladder bile of one of the patients was found

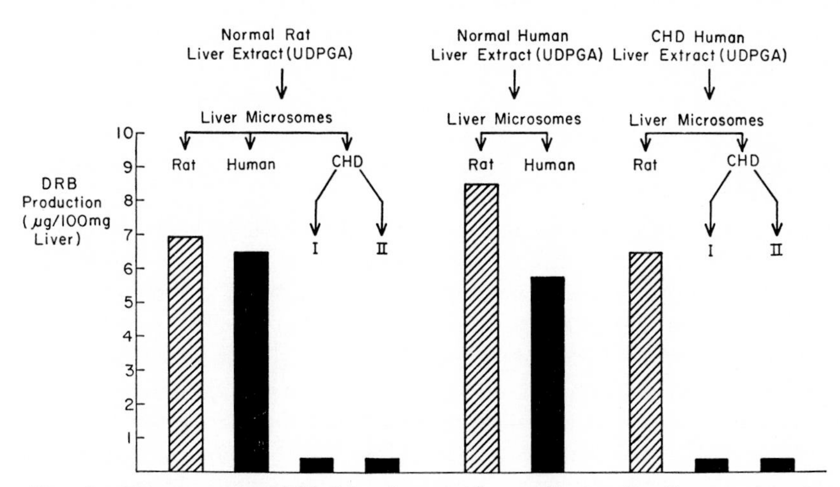

Fig. 6. Direct-reacting bilirubin glucuronide production by liver preparations from normal rats and human beings and from patients with constitutional hepatic dysfunction (Gilbert's disease) (A4).

to contain conjugated bilirubin. The authors considered that this might be due to a slight degree of transferase activity or to the presence of an alternative pathway for glucuronide synthesis.

Schmid (S6) has examined the bile of 3 children with Crigler and Najjar (C8) type of jaundice, whose plasma bilirubin levels were over 20 mg %, and found that in contrast to the case of Arias and London it was colorless and contained only traces of direct bilirubin. He observed that the ability of these children to conjugate menthol, salicylic acid, and metabolites of hydrocortisone as glucuronides was markedly depressed but not abolished. It seems likely that there is a deficiency of glucuronyl transferase activity in these patients. However, this cannot be regarded as the whole explanation of familial hyperbilirubinemia for it has now been reported that in mild cases of Gilbert's disease the excretion of such substrates as salicylamide (B2a), *n*-acetyl-*p*-aminophenol (S7a), and menthol (I2) as glucuronides occurs in a normal way. Further, in these patients, analysis of duodenal drainage showed that bile pigment excretion appeared to behave normally and to involve conjugation to the glucuronide. It is probable that there is a defect in the mechanism whereby bilirubin is transported from the plasma to the site of glucuronide formation in the liver cell, though this has still to be proved.

Familial hyperbilirubinemia is not limited to the human species and has been found in a homozygous strain of Wistar rats (G6). These Gunn rats have plasma bilirubin levels of 5–11 mg %, with no appreciable amount of conjugated pigment in the plasma, bile, or urine. Their impairment in glucuronide formation has been shown *in vitro* by Carbone and Grodsky (C1), Schmid *et al.* (S10), and Lathe and Walker (L3) to be due to the absence of glucuronyl transferase activity. This defect has also been demonstrated *in vivo* by Schmid *et al.* (S10), who injected bilirubin and conjugated bilirubin (in the form of sterile bile or jaundiced serum) into these Gunn rats and studied their excretion in the bile (Fig. 7). Bilirubin did not alter the bile pigment excretion significantly while the bilirubin glucuronide was removed almost completely in the bile. This experiment demonstrated that in these animals the hepatic excretory apparatus is normal and that the impaired bilirubin excretion can only be explained by its lack of conjugation. In spite of the nonexcretion of the free bilirubin in the bile, examination of the plasma levels showed that it was removed from the blood. This observation and the fact that the jaundice does not normally increase was considered by these investigators to indicate the existence of alternative pathways of heme or bilirubin metabolism than those already described.

Chronic idiopathic jaundice, which is characterized by heavy black pigmentation of the parenchymal cells in the centrilobular zones of the liver, differs from other forms of familial jaundice so far described in that conjugated bilirubin as well as unconjugated bilirubin is present in the plasma and bilirubinuria occurs. The type of conjugated pigment in the plasma has not been characterized, and up to the present no adequate explanation for the jaundice has been put forward (D6).

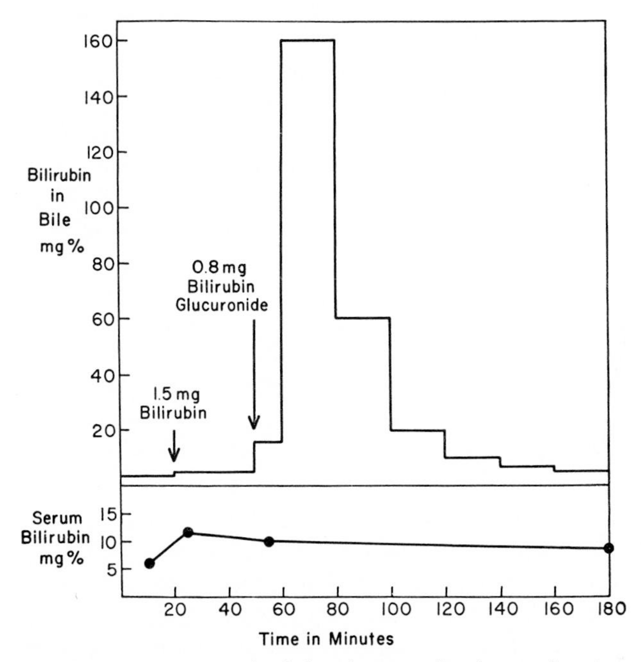

FIG. 7. Bilirubin excretion in the bile of a jaundiced rat after intravenous injections of bilirubin and bilirubin glucuronide (S10).

8. Methods

8.1. DIAZOTIZATION

The original method of Hijmans van den Bergh and Snapper (H6) for the estimation of bilirubin in serum, involving the coupling of the pigment with diazotized sulfanilic acid, still forms the basis of most modern methods. Numerous modifications have been introduced in attempts to speed up the rate of the coupling reaction and to differentiate between the two types of pigment. Details of these up to 1954 are given in With's monograph (W9). Knowledge of the association of kernicterus with a high concentration of free bilirubin in the plasma has emphasized the need for a satisfactory method, which will give a quick, accurate answer on small quantities of plasma, and recent methods have been designed to help solve this problem.

The method of Malloy and Evelyn (M1) has been used by many investigators to give an estimate of the amounts of bilirubin and conjugated bilirubin in plasma. For infants, this has been modified for 0.1–0.2 ml plasma by the use of proportionately smaller amounts of diazo reagent and volumes of solution than originally prescribed (A2, H3, L6). In this procedure the final dilution of plasma is 1 in 40 or more, so that the protein remains in solution. In spite of the use of a blank, interference due to turbidity and hemolysis may be encountered, and with low pigment concentrations the high dilution employed makes the method somewhat inaccurate. Thirty-minute color development in aqueous solution was selected by Malloy and Evelyn as a measurement of conjugated bilirubin. As mentioned by these authors, color development is not always complete in this time, and Ducci and Watson (D7) drew attention to the fact that it occurred in two phases. The rapid phase they considered represented the amount of "direct" pigment present and they somewhat arbitrarily chose a "1-minute" prompt direct reading to represent this. It has not been possible to associate this value with either pigment I or II (B7). The method has been severely criticized by Klatskin and Drill (K5), but it has been widely used in clinical practice. Watson (W4) has reviewed his findings in a large series of cases and has discussed the importance of the "1-minute bilirubin" in diagnosis.

In a critical evaluation of the Malloy and Evelyn method, Lathe and Ruthven (L4) have pointed out that a sharp distinction between the rate of reaction of bilirubin and conjugated bilirubin in coupling with diazotized sulfanilic acid in aqueous solution, can be drawn only at a low pH and at low concentrations of diazotized sulfanilic acid. The coupling of serum bilirubin may be almost complete at pH 3.5 and a concentration of 0.52 % diazotized sulfanilic acid. It is therefore probable that the small amount of the color that develops with normal serum in aqueous solution at pH 1.5 and a concentration of 0.0037 % diazotized sulfanilic acid in the Malloy and Evelyn method is due to bilirubin and the apparent presence of small amounts of "conjugated" bilirubin in plasma from infants with physiological jaundice or erythroblastosis, when assessed by a "direct" reading, is probably an artifact. These investigators found that if they used the usual diazotized sulfanilic acid concentration and increased the concentration of undiazotized sulfanilic acid from 0.0063 % to 0.097 %, then the rate of coupling of both conjugated bilirubin and bilirubin in aqueous solution was not significantly altered.

The addition of methanol to give a 50 % solution did, however, result in a very considerable increase in the rate of coupling of bilirubin. Under these conditions they considered that a 5-minute reading for conjugated bilirubin and a further 5-minute reading for total bile pigments, after the addition of methanol, gave satisfactory values.

Experiments with pigment I, isolated chromatographically (B12), suggested that the coupling of the pigment with diazotized sulfanilic acid was not complete in aqueous solution, since the addition of ethanol increased the color development sometimes as much as 40 %. This matter needs to be investigated further if a full understanding of the substances measured in the "direct reaction" is to be made.

Methods of determining bile pigment concentration which involve protein precipitation have been criticized because of the losses of conjugated pigment onto the precipitate. They have, however, the advantage that dilution of the plasma need not be greater than 1:10 (K4) and that less trouble is incurred with turbidity. Stoner and Weisberg (S17) have avoided the precipitation of the conjugated pigment by adding the serum to the diazo reagent and then adding concentrated hydrochloric acid, before precipitating the protein with ammonium sulfate and ethanol in the usual way. Although the authors did not make any observations on this point it is probable that the hydrochloric acid hydrolyzed the azo pigment *B* formed to the less polar azo pigment *A* and so prevented its coprecipitation with the protein.

Perryman *et al.* (P2) found that the pigment losses on protein precipitates could also be prevented by diluting the plasma 1:100 instead of 1:10 and then estimating the color by means of a transistor amplifier coupled to a simple photoelectric absorptiometer. In order to get speedy color development and final stability of color they used the technique of Patterson *et al.* (P1) and added sodium sulfamate to the diazo reagent to remove nitrous acid and sodium azide to remove excess diazo reagent at the end of the reaction. Absorptiometer readings were made at 425 and 530 mμ so that a correction could be made for the heme pigments likely to be present in plasma from infants with erythroblastosis.

The method of Jendrassik and Gróf (J1), with its many modifications, is in popular use in many countries. The use of sodium benzoate as a solubilizing agent in an alkaline pH gives a speedy result for total bile pigment concentrations, and high dilutions of plasma are unnecessary. This procedure cannot, however, be used to give values for conjugated bilirubin since in alkaline solution free bilirubin will give a direct reaction. Dangerfield and Finlayson (D1) have added caffeine to the sodium benzoate in phosphate buffer solution in order to facilitate the coupling and to stabilize the azobilirubin, since this pigment rapidly decomposes in alkaline solution. Hsia *et al.* (H4) have also used caffeine with the sodium benzoate to speed up the coupling reaction and, in addition, have increased the strength of the sulfanilic acid to 0.3 %. After the addition of methanol, readings are made in 5 minutes. They considered that a 50-second reading, before the addition of the methanol, gave a reading equivalent to the "1-minute prompt bilirubin" but did not report what effect their reagents had on the coupling of bilirubin in aqueous solution, and it is doubtful if this reading is of any significance.

A modification of Powell's method (P4), with sodium benzoate and urea solution to speed up the coupling reaction, has been used by O'Hagan (O1) to investigate the bilirubin content of serum from a large number of blood donors. He stated that color development was maximal in half a minute and that errors due to hemoglobin and ferrihemalbumin could be eliminated by the use of suitable blanks. The mean values for serum bilirubin in 200 donors were 0.45 ± 0.19 mg/100 ml serum for males and 0.36 ± 0.18 mg/100 ml serum for females. Of 2500 people screened, only 108 had unaccountably high levels of over 1.5 mg %, and this level was arbitrarily taken as the upper limit of normal.

Sims and Horn (S13) did not consider that the addition of urea to the sodium benzoate solution was of any advantage and observed speedy coupling at pH 6.0–7.0 without it. The addition of nitrite to the diazo blank was found to eliminate errors due to hemolysis, and colorimetric readings were made at 525 mμ since this is the isobestic wavelength for oxyhemoglobin and methemoglobin. Urea-ethanol solution has also been used to facilitate rapid coupling of bilirubin with diazotized sulfanilic acid. Patterson *et al.* (P1) have been able to eliminate the cloudiness that tends to develop with this method by the addition of concentrated phenol.

8.2. Direct Spectrophotometry

The use of the icterus index, as described by Meulengracht, for the assessment of jaundice has fallen into disrepute because of the errors caused by the presence of lipochromes, carotenoids, and other yellow pigments. Josephson (J6) in his survey found that the correlation coefficient between icterus index and serum bilirubin concentration was 0.69 in 360 healthy subjects and 0.84 in 40 jaundiced subjects. In newborn infants however, bilirubin is the only yellow pigment likely to be present and the possibility of determining serum bilirubin concentrations by direct measurement has again been re-examined. Abelson and Boggs (A1) diluted serum from infants with erythroblastosis 1 in 50 and studied the absorption curves. They found that in addition to the bilirubin peak at 460 mμ there was one at 405–415 mμ due to hematin, which disappeared with clinical improvement, and another at 622–625 mμ due to metalbumin. Of the 85 cases examined, only 7 showed this latter peak and of these only 2 survived. For bilirubin determinations readings were taken at 415 and 460 mμ and a correction factor applied.

White *et al.* (W7) have criticized this last method and others using the Sorêt band of hemoglobin as one of the wavelengths for their 2 readings, since this may vary considerably. These investigators take their readings at 455 mμ and 575 mμ since the absorbancy of heme pigments at 455 mμ (the maximum for bilirubin) is almost identical with that at 575 mμ and a correction factor can easily be applied (S12). A precision spectrophotometer capable of isolating a narrow band of light is essential for this method, which can be used for as little as 20 μl serum. No differentiation between conjugated and free bilirubin is afforded so that the method has limitations which need to be appreciated by the pediatrician. Mertz and West (M4) have overcome this difficulty by precipitating any conjugated pigment that might be present with 80 % acetone and then determining the free bilirubin spectrophotometrically; the presence of any conjugated pigment can be detected by the visual examination of the protein precipitate.

8.3. Pigments I and II

So far, only one method for the quantitative determination of the two types of conjugated pigment has been described (B7). After a preliminary protein precipitation, which involves some loss of conjugated pigment, the alcoholic supernatant is taken to dryness and pigments are separated by partition chromatography on silicone-treated kieselguhr, using a butanol/pH 6 buffer system. The column is separated into 3 portions, each containing one of the 3 pigments; these are eluted with an alcoholic solution of diazotized sulfanilic acid and the pigments are estimated colorimetrically. Recoveries with this procedure are about 90 % and, as some tailing of the pigments on the columns is experienced, separation may not be complete. A more exact method is needed for these determinations.

8.4. Bilirubin Standards

The variation in the extinction coefficients of different bilirubin preparations is well known, and O'Hagan (O1) has suggested that the source of the bilirubin should always be stated. The relative extinction of the azobilirubin formed on coupling with diazotized sulfanilic acid depends on the solvent and the pH of the solution and may be altered by the presence of protein (L4). It is therefore essential for accurate plasma determinations that the standard is prepared using pooled human plasma, in normal amounts, and that exactly the same procedure is carried out on the standard as the unknown. Bilirubin glucuronide appears to behave in a similar manner to bilirubin as regards its spectral characteristics in human plasma and upon diazotization, and until stable preparations of the pigment have been obtained it will be necessary to give all results in terms of bilirubin.

8.5. Collection of Samples

The importance of keeping plasma samples away from the light prior to carrying out bile pigment determinations has rightly been re-emphasized in several recent publications (C7, O1, S13), for losses of up to 30 % in one hour may occur, especially in bright sunlight. Bilirubin appears to be 2–3 times as photosensitive as bilirubin glucuronide so that particular care is needed with infant plasma; the nature of the compounds formed under the influence of light has not been satisfactorily examined. In whole blood, the losses are smaller and Nauman (N1) has stated that the bilirubin concentration does not appear to be immediately affected by post-mortem changes although difficulties in analysis may be encountered due to hemolysis. Harris *et al.* (H1), on the other hand, have observed decreases in serum bilirubin in infants, just before death, which have continued in the immediate post-mortem period. Slight diurnal variations in bilirubin levels in normal subjects, which are not greatly affected by food have been reported by Balzer (B2) but it does not appear that the time at which blood is drawn for bilirubin determinations is of great importance.

8.6. Urine

No satisfactory method for the quantitative determination of bile pigments in urine has yet been evolved (W9). Direct spectrophotometry is impossible on account of the presence of yellow pigments other than bile pigments. Diazotization procedures necessitate the application of correction factors, which are not entirely adequate. Some purification can be effected by absorption of the bile pigments onto barium sulfate but their subsequent elution is by no means quantitative.

For the qualitative examination of bile pigments in urine the use of proprietory tablets (Ictotest) has become increasingly popular (F2). The procedure is simple, involving only the putting of 5 drops of urine onto an asbestos mat and then placing on it a moist tablet containing a stable diazonium compound, *p*-nitrobenzenediazonium *p*-toluenesulfonate. Color development occurs in 30 seconds if the urine contains at least 0.05 mg bile pigments per 100 ml. The sensitivity of the test is equal to that of Fouchet's test and few false positives are given.

References

A1. Abelson, N. M., and Boggs, T. R., Plasma pigments in erythroblastosis fetalis. I. Spectrophotometric absorption patterns. *Pediatrics* **17**, 452-460 (1956).

A2. Altmann, V., Nelson, M., and Pernova, M., Determination of plasma bilirubin by a simple micromethod. *Am. J. Clin. Pathol.* **26**, 956-959 (1956).

A3. Amatuzio, D. S., Weber, L. J., and Nesbitt, S., Bilirubin and protein in the cerebrospinal fluid of jaundiced patients with severe liver disease with and without hepatic coma. *J. Lab. Clin. Med.* **41**, 615-618 (1953).

A4. Arias, I. M., and London, I. M., Bilirubin glucuronide formation in vitro: demonstration of a defect in Gilbert's disease. *Science* **126**, 563-564 (1957).

A5. Arias, I. M., Levy, B. A., and London, I. M., Studies of glucuronide synthesis and of glucuronyl transferase in liver and serum. *J. Clin. Invest.* **37**, 875-876 (1958).

B1. Baikie, A. G., The chromatographic separation of the serum bile pigments in jaundice. *Scot. Med. J.* **2**, 359-362 (1957).

B2. Balzer, E., 24 Stunden-Rythmus des Serumbilirubins. *Acta. Med. Scand. Suppl.* **278**, 67-70 (1953).

B2a. Barniville H. T. F., and Misk, R., Urinary glucuronic acid excretion in liver disease and the effect of a salicylamide load. *Brit. Med. J.* **I**, 337-340 (1959).

B3. Barrows, L., Hunter, F. T., and Barker, B. Q., The nature and clinical significance of pigments in the cerebrospinal fluid. *Brain* **78**, 59-80 (1955).

B4. Berger, H. J., Quantitative analysis of direct and indirect bilirubin in the sera of domesticated animals. *Zentr. Veterinärmed.* **3**, 273-280 (1956).

B5. Bevis, D. C. A., Icterus of the brain in the newborn. *Lancet* **265**, 1357-1358 (1953).

B6. Billing, B. H., Quantitative determination of bile pigments in serum using reverse phase partition chromatography. *Biochem. J.* **56**, Proc. xxx (1954).

B7. Billing, B. H., A chromatographic method for the determination of the three bile pigments in serum. *J. Clin. Pathol.* **8**, 126-129 (1955).

B8. Billing, B. H., The three serum bile pigments in obstructive jaundice and hepatitis. *J. Clin. Pathol.* **8**, 130-131 (1955).

B9. Billing, B. H., and Lathe, G. H., Excretion of bilirubin as an ester glucuronide giving the direct van den Bergh reaction. *Biochem. J.* **63**, 6P (1956).

B10. Billing, B. H., and Lathe, G. H., Bilirubin metabolism in jaundice. *Am. J. Med.* **24**, 111-121 (1958).

B11. Billing, B. H., Cole, P. G., and Lathe, G. H., Increased plasma bilirubin in newborn infants in relation to birth weight. *Brit. Med. J.* **ii**, 1263-1265 (1954).

B12. Billing, B. H., Cole, P. G., and Lathe, G. H., The excretion of bilirubin as a diglucuronide giving the direct van den Bergh reaction. *Biochem. J.* **65**, 774-784 (1957).

B13. Bollman, J. L., Bile pigments in serum in disease of liver. *In* "Hepatitis Frontiers." International Symposium (sponsored by the Henry Ford Hospital, Detroit, Michigan), (F. W. Hartman, G. A. LoGrippo, J. G. Mateer, and J. Barron, eds.), Chapter 37, pp. 467-474. Little, Brown, Boston, 1957.

B14. Bollman, J. L., and Mann, F. C., Studies on the physiology of the liver. XXII. The van den Bergh reaction in the jaundice following complete removal of the liver. *A.M.A. Arch. Surg.* **24**, 675-680 (1932).

B15. Bollman, J. L., Whitcomb, F. F., and Hoffman, H. N., Exhibition at Meeting of American Medical Association, San Francisco, 1958.

B16. Bowen, W. R., and Waters, W. J., Bilirubin encephalopathy: Studies related to the site of inhibitory action of bilirubin on brain metabolism. *A.M.A. J. Diseases Children* **93**, 21 (1956).

B17. Brown, A. K., and Zuelzer, W. W., Studies on the neonatal development of the glucuronide conjugating system. *J. Clin. Invest.* **37**, 332-340 (1958).

C1. Carbone, J. V., and Grodsky, G. M., Constitutional nonhemolytic hyperbilirubinemia in the rat; defect of bilirubin conjugation. *Proc. Soc. Exptl. Biol. Med.* **94**, 461-463 (1957).

C2. Claireaux, A. E., Cole, P. G., and Lathe, G. H., Icterus of the brain in the newborn. *Lancet* **265**, 1226-1230 (1953).

C3. Claireaux, A. E., Lathe, G. H., and Norman, A. P., *In* "Recent Advances in Pediatrics" (D. Gairdner, ed.), 2nd ed., Chapter II. Churchill, London. 1958.

C4. Clifton, J. A., Ingelfinger, F. J., and Burrows, B. A., Effect of cortisone and hydrocortisone on hepatic excretory function. *J. Lab. Clin. Med.* **51**, 701-708 (1958).

C5. Cole, P. G., and Lathe, G. H., The separation of serum pigments giving the direct and indirect van den Bergh reaction. *J. Clin. Pathol.* **6**, 99-104 (1953).

C6. Cole, P. G., Lathe, G. H., and Billing, B. H., Separation of the bile pigments of serum, bile and urine. *Biochem. J.* **57**, 514-518 (1954).

C7. Cremer, R. J., Perryman, P. W., and Richards, D. H., Influence of light on the hyperbilirubinemia of infants. *Lancet* **i**, 1094-1097 (1958).

C8. Crigler, J. F., and Najjar, V. A., Congenital familial nonhemolytic jaundice with kernicterus. *Pediatrics* **10**, 169-179 (1952).

D1. Dangerfield, W. G., and Finlayson, R., Estimation of bilirubin in serum. *J. Clin. Pathol.* **6**, 173-177 (1953).

D2. Danoff, S., Grantz, C., Boyer, A., and Holt, L. E., Jr., Reduction of indirect bilirubinaemia in vivo. *Lancet* **i**, 316 (1958).

D3. Day, R., Inhibition of brain respiration in vitro by bilirubin. Reversal by various means. *Proc. Soc. Exptl. Biol. Med.* **85**, 261-264 (1954).

D4. Day, R., Kernicterus: Further observations on the toxicity of heme pigments. *Pediatrics* **17**, 925-928 (1956).

D5. Deenstra, H., L'excrétion des pigments biliaires par les reins. *Ann. méd. (Paris)* **51**, 685-700 (1950).

D6. Dubin, I. N., Chronic idiopathic jaundice: a review of 50 cases. *Am. J. Med.* **24**, 268-292 (1958).

D7. Ducci, H., and Watson, C. J., Determination of serum bilirubin with special reference to the prompt reacting and chloroform soluble types. *J. Lab. Clin. Med.* **30**, 293-300 (1945).

D8. Dutton, G. J., Foetal and gastro-intestinal glucuronide synthesis. *Biochem. J.* **69**, 39P (1958).

D9. Dutton, G. J., Glucuronide synthesis in foetal liver and kidney. *Lancet* **i**, 49 (1958).

E1. Ernster, L., Herlin, L., and Zetterström, R., Experimental studies on the pathogenesis of kernicterus. *Pediatrics* **20**, 647-652 (1958).

F1. Fischer, H., and Haberland, H. W., Ueber die Konstitution des Bilirubins sowie die seiner Azofarbstoffe und die Gmelinsche Reaktion. *Z. physiol. Chem.* **232**, 236-258 (1935).

F2. Free, A. H., and Free, H. M., A simple test for urine bilirubin. *Gastroenterology* **24**, 414-421(1953).

F3. Fröhlich, A., and Mirsky, I. A., Susceptibility to convulsions in relation to age. II. Influence of bile in rats. *Proc. Soc. Exptl. Biol. Med.* **50**, 25-28 (1942).

G1. Gilbertsen, S., Campbell, M., and Watson, C. J., Some observations on the variable fate of bilirubin depending on conjugation and other factors. *J. Lab. Clin. Med.* **50**, 818 (1957).

G2. Gray, C. H., and Kekwick, R. A., Bilirubin serum protein complexes and the van den Bergh reaction. *Nature* **161**, 274 (1948).

G3. Gray, C. H., Nicholson, D. C., and Nicolaus, R. A., The IX-α structure of the common bile pigments. *Nature* **181**, 183-185 (1958).

G4. Gries, G., Gedik, P., and Georgi, J., Zur Trennung des direkten und indirekten Bilirubins durch Papierchromatographie. *Z. physiol. Chem.* **298**, 132-139 (1954).

G5. Grodsky, G. M., and Carbone, J. V., The synthesis of bilirubin glucuronide by tissue homogenates. *J. Biol. Chem.* **226**, 449-458 (1957).

G6. Gunn, C. K., Hereditary acholuric jaundice in a new mutant strain of rats. *J. Heredity* **29**, 137-139 (1938).

H1. Harris, R. C., Lucey, J. F., and Maclean, J. R., Kernicterus in premature infants associated with low concentrations of bilirubin in the plasma. *Pediatrics* **21**, 875-883 (1958).

H2. Hartiala, K. J. V., Leikkola, K., and Savola, P., Further studies on intestinal glucuronide synthesis. *Acta Phys. Scand.* **42**, 36-40 (1958).

H3. Hsia, D. Y. Y., Allen, F. H., Jr., Diamond, L. K., and Gellis, S. S., Serum bilirubin levels in the newborn infant. *J. Pediat.* **42**, 277-285 (1953).

H4. Hsia, D. Y. Y., Hsia, H. H., Gofstein, R. M., Winter, A., and Gellis, S. S., Determination of concentration of bilirubin in serum. I. Rapid micromethod employing photoelectric colorimeter. *Pediatrics* **18**, 433-435 (1956).

H5. Hijmans van den Bergh, A. A., and Müller, P., Ueber eine indirecte Diazoreaktion auf Bilirubin. *Biochem. Z.* **77**, 90-103 (1916).

H6. Hijmans van den Bergh, A. A., and Snapper, I., Die Farbstoffe des Blutserums. *Deut. Arch. klin. Med.* **110**, 540-561 (1913).

I1. Isselbacher, K. J., Demonstration of bilirubin sulphate in bile. *J. Clin. Invest.* **37**, 904 (1958).

I2. Isselbacher, K. J., Unpublished evidence advanced at the 1958 meeting of the American Association for the study of Liver Diseases in Chicago.

J1. Jendrassik, L., and Gróf, P., Vereinfachte photometrische Methoden zur Bestimmung des Blutbilirubins. *Biochem. Z.* **297**, 81-89 (1938).

J2. Jirsa, M., Večerek, B., and Ledvina, M., Di- and monotaurobilirubin similar to a directly reacting form of bilirubin in serum. *Nature* **177**, 895 (1956).

J3. Jirsová, V., Jirsa, M., and Janovský, M., Importance of the quantitative determination of direct and indirect bilirubin in hemolytic disease of newborn. *Acta Pediat.* **47**, 179-186 (1958).

J4. Johnson, L., and Day, R., Personal communication to Harris, R. C., Lucey, J. F., and Maclean, J. R., Kernicterus in premature infants associated with low concentrations of bilirubin in the plasma. *Pediatrics* **21**, 875-883 (1958).

J5. Johnson, L., Blanc, W. A., Lucey, J. F., and Day, R., Kernicterus in rats with familial jaundice. *A.M.A. J. Diseases Children* **94**, 548 (1957); also see Johnson, L., Sarmiento, F., Blanc, W. A., and Day, R., Kernicterus in rats with an inherited deficiency of glucuronyl transferase. *A.M.A. J. Diseases Children* **97**, 591-608 (1959).

J6. Josephson, B., The icterus index as a measure of serum bilirubin concentration. *Acta Genet. Statist. Med.* **4**, 231-235 (1953).

K1. Katz, R., Ducci, H., and Alessandri, H., Influence of cortisone and prednisolone on hyperbilirubinemia. *J. Clin. Invest.* **36**, 1370-1374 (1957).

K2. Kawai, K., On the paper chromatography of bilirubin. *Igaku Kenkyu* **23**, 572-581 (1953).

K3. Keller, C., Clinical chemistry of the pleural cavity. II. Electrolytes, glucose and bilirubin. *Deut. Arch. klin. Med.* **201**, 539-552 (1954).

K4. King, E. J., and Coxon, R. V., Determination of bilirubin with precipitation of the plasma proteins. *J. Clin. Pathol.* **3**, 248-259 (1950).

K5. Klatskin, G., and Drill, V. A., The significance of the "one minute" (prompt direct reacting) bilirubin in serum. *J. Clin. Invest.* **29**, 660-670 (1950).

K6. Klatskin, G., and Bungards, L., Bilirubin-protein linkages in serum and their relationship to the van den Bergh reaction. *J. Clin. Invest.* **35**, 537-551 (1956).

L1. Lathe, G. H., The chemical pathology of bile pigments. Part I. The plasma bile pigments. *Biochem. Soc. Symposia* **12**, 34-45 (1954).

L2. Lathe, G. H., Exchange transfusion as a means of removing bilirubin in haemolytic disease of the newborn. *Brit. Med. J.* **i**, 192-196 (1955).

L3. Lathe, G. H., and Walker, M., An enzyme defect in human neonatal jaundice and in Gunn's strain of jaundiced rats. *Biochem. J.* **66**, 9 P (1957).

L4. Lathe, G. H., and Ruthven, C. R., Factors affecting the rate of coupling of bilirubin and conjugated bilirubin in the van den Bergh reaction. *J. Clin. Pathol.* **11**, 155-161 (1958).

L5. Lathe, G. H., and Walker, M., The synthesis of bilirubin glucuronide in animal and human liver. *Biochem. J.* **70**, 705-712 (1958).

L6. Lawrence, K. M., and Abbot, A. L., Micromethod for the estimation of serum bilirubin. *J. Clin. Pathol.* **9**, 270-273 (1956).

L7. Lowry, P. T., Bossenmaier, I., and Watson, C. J., A method for the isolation of bilirubin from feces. *J. Biol. Chem.* **202**, 305-309 (1953).

M1. Malloy, H. T., and Evelyn, K. A., The determination of bilirubin with the photoelectric colorimeter. *J. Biol. Chem.* **119**, 481-490 (1937).

M2. Martin, N. H., Bilirubin-serum protein complexes. *Biochem. J.* **42**, Proc. xv (1948).

M3. Mendioroz, B. A., Charbonnier, A., and Bernard, R., Étude chromatographique de la bilirubine en solution pure. *Compt. rend. soc. biol.* **145**, 1483-1485 (1951).

M4. Mertz, J. E., and West, C. D. A rapid method for the determination of indirect bilirubin. *A.M.A. J. Diseases Children* **91**, 19-22 (1956).

N1. Nauman, H. N., Post mortem liver function tests. *Am. J. Clin. Pathol.* **26**, 495-505 (1956).

N2. Nizet, E., and Barac, G., Localization intrarénale de la bilirubine chez le chien ictérique. *Compt. rend. soc. biol.* **146**, 1282-1284 (1952).

O1. O'Hagan, J. E., Hamilton, T., Le Breton, E. G., and Shaw, A. E., Human serum bilirubin. An immediate method of determination and its application to the establishment of normal values. *Clin. Chem.* **3**, 609-623 (1957).

O2. Overbeek, J. T. G., Vink, C. L. J., and Deenstra, H., Kinetics of the formation of azobilirubin. *Rec. trav. chim.* **74**, 85-97 (1955).

P1. Patterson, J., Swale, J., and Maggs, C., The estimation of serum bilirubin. *Biochem. J.* **52**, 100-105 (1952).

P2. Perryman, P. W., Richards, D. H., and Holbrook, B., Simultaneous microdetermination of serum bilirubin and serum "haem-pigments." *Biochem. J.* **66**, 61 P (1957)

P3. Polonovski, M., and Bourrillon, R., Metabolism de la bilirubine. II. État physico-chemique de la bilirubine dans la bile. *Bull. soc. chim. biol.* **34**, 963-972 (1952).

P4. Powell, W. N., A method for the determination of serum bilirubin with the photoelectric colorimeter. *Am. J. Clin. Pathol.* **8**, 55-58 (1944).

R1. Ritchie, H. D., Grindlay, J. H., and Bollman, J. L., Surgical jaundice: Experimental evidence against the "regurgitation theory." *Surg. Forum* **7**, 415-418 (1957).

R2. Rundle, F. F., Perry, D., Cass, M., and Oddie, T. H., Rise in serum bilirubin with biliary obstruction and its decline curve after operative relief. *Surgery* **43**, 555-562 (1958).

S1. Sakamoto, T., Studies on bile pigments. II. Separation of natural direct bilirubins. *Acta Med. Okayama* **10**, 30-46 (1956).

S2. Sakamoto, T., Yamamoto, S., Yahata, K., and Kondo, T., Studies on bile pigments. VI. Separation of natural bilirubins on ion exchange columns. *Igaku Kenkyu* **27**, 121-123 (1957).

S3. Sakamoto, T., Komuta, K., Kondo, T., Hirano, H., Monobe, T., and Kaneda, K., A form of direct reacting bilirubins appearing in jaundiced urine. *Acta Med. Okayama* **11**, 81-87 (1957).

S4. Schachter, D., Nature of the glucuronide in direct-reacting bilirubin. *Science* **126**, 507-508 (1957).

S5. Schmid, R., Direct-reacting bilirubin, bilirubin glucuronide in serum, bile, and urine. *Science* **124**, 76-77 (1956).

S6. Schmid, R., Congenital defects in bilirubin metabolism. *J. Clin. Invest.* **36**, 927 (1957).

S7. Schmid, R., The identification of direct-reacting bilirubin as bilirubin glucuronide. *J. Biol. Chem.* **229**, 881-888 (1957).

S7a. Schmid, R., and Hammaker, L., Glucuronide formation in patients with constitutional hepatic dysfunction (Gilbert's disease). *New Engl. J. Med.* **260**, 1310-1314 (1959).

S8. Schmid, R., Hammaker, L., and Axelrod, J., The enzymic formation of bilirubin glucuronide. *Arch. Biochem. Biophys.* **70**, 285-287 (1957).

S9. Schmid, R., Jeliu, G., and Gellis, S. S., Indirect bilirubinemia. *Lancet* **i**, 855 (1958).

S10. Schmid, R., Axelrod, J., Hammaker, L., and Swarm, R. L., Congenital jaundice in rats due to a defect in glucuronide formation. *J. Clin. Invest.* **37**, 1123-1130 (1958).

S11. Shay, H., and Sun, D. C. H., Possible effect of hydrocortisone on bilirubin excretion by the liver. *New Engl. J. Med.* **257**, 62-65 (1957).

S12. Shinowara, G. Y., Johnston, D. M., Herritt, M., and Lorey, E. P., Spectrophotometric studies on blood, serum and plasma. The physical determination of hemoglobin and bilirubin. *Am. J. Clin. Pathol.* **24**, 696-707 (1954).

S13. Sims, F. H., and Horn, C., Some observations on Powell's method for determination of serum bilirubin. *Am. J. Clin. Pathol.* **29**, 412-417 (1958).

S14. Stempfel, R., Serum and cerebrospinal fluid bilirubin in hemolytic disease of newborn. *Acta Pediat.* **44**, 502 (1955).

S15. Stempfel, R., and Zetterström, R., Cerebrospinal fluid bilirubin in the neonatal period with special reference to the development of kernicterus. *Acta Paediat.* **43**, 582-586 (1954).

S16. Stempfel, R., Broman, B., Escardo, F., and Zetterström, R., Obstructive jaundice complicating hemolytic disease of the newborn. *Pediatrics* **17**, 471-481 (1956).

S17. Stoner, R. A., and Weisberg, H. F., Ultramicromethod for serum bilirubin by diazo blue reaction. *Clin. Chem.* **3**, 22-36 (1957).

T1. Talafant, E., Properties and composition of the bile pigment giving a direct diazo reaction. *Nature* **178**, 312 (1956).

T2. Talafant, E., The nature of direct and indirect bilirubin. V. The presence of glucuronic acid in the directly reacting bile pigment. *Collection Czechoslov. Chem. Communs.* **22**, 661-663 (1957).

T3. Talafant, E., Sodium salt of the directly reacting bile pigment. *Nature* **180**, 1050 (1957).

V1. Verschure, J. C. M., Electro-chromatograms of human bile. *Clin. Chim. Acta* **1**, 38-48 (1956).

W1. Walker, A. H. C., Liquor amnii studies in prediction of hemolytic disease of newborn. *Brit. Med. J.* **ii**, 376-378 (1957).

W2. Waters, W. J., and Briton, H. A., Bilirubin encephalopathy: Preliminary studies related to production. *Pediatrics* **15**, 45-48 (1955).

W3. Waters, W. J., Richert, D. A., and Rawson, H. H., Bilirubin encepalopathy. *Pediatrics* **13**, 319-325 (1954).

W4. Watson, C. J., The importance of the fractional serum bilirubin determination in clinical medicine. *Ann. Internal Med.* **45**, 351-368 (1956).

W5. Watson, C. J., Color reaction of bilirubin with sulfuric acid: a direct diazo-reacting bilirubin sulfate. *Science* **128**, 142 (1958).

W6. Weinbren, K., and Billing, B. H., Hepatic clearance of bilirubin as an index of cellular function in the regenerating rat liver. *Brit. J. Exptl. Pathol.* **37**, 199-204 (1956).

W7. White, D., Haidar, G. A., and Reinhold, J. G., Spectrophotometric measurement of bilirubin concentrations in the serum of the newborn by the use of a microcapillary method. *Clin. Chem.* **4**, 211-222 (1958).

W8. With, T. K., Über den Wert des Serumbilirubins, bei welchem Ikterus direkt sichtbar wird. *Acta Med. Scand.* **114**, 379-382 (1943).

W9. With, T. K., "The Biology of the Bile Pigments." Arne Frost-Hansen, Copenhagen, 1954.

Y1. Yamaoka, K., and Kosaka, K., Studies on bile pigments. I. On the nature of indirect and direct bilirubin. *Proc. Japan. Acad.* **27**, 715-720 (1951).

Z1. Zetterström, R., and Ernster, L., Bilirubin, an uncoupler of oxidative phosphorylation in isolated mitochondria. *Nature* **178**, 1335-1337 (1956).

AUTOMATION*

Walton H. Marsh

Department of Pathology, State University of New York,
Downstate Medical Center, Brooklyn, New York

Page

1. Introduction

1.1. Growth of Clinical Chemistry Laboratories

The development of the chemical laboratory as an essential functional unit in hospital procedure has been a slow and obscure process. Most hospitals, with the exception of those directly connected with universities, did not have clinical chemistry laboratories until after 1900 (J1). Thus, the development of clinical chemistry as a separate branch of hospital service has mainly occurred within the last 60 years. Even today in smaller institutions, the technical and professional staff in the chemical laboratory must be responsible for running certain tests which in larger institutions have been subdivided and classed under the headings of serology, hematology, bacteriology, virology, parasitology, etc. The definition of "clinical chemistry laboratory" is therefore still a function of local circumstance. However, the entity of clinical chemistry has developed to such a degree that a chapter on the development of auto-

* Illustrations are given of instruments and equipment readily accessible to the author. It must be appreciated that similar apparatus will be found in the catalogues of other leading commercial firms throughout the world. Certain microequipment (M. Sanz, *Clin. Chem.*, **3**, 406, 1957) will be reviewed in a subsequent volume. *The Editors.*

matic techniques and devices appears warranted even though automation in clinical chemistry is still young.

As the clinical chemistry laboratory evolved there was also a corresponding increase in the number of tests requested per patient. These represent the increases which have come about by the more prolific use of "admission or screening" tests. Some hospitals are now running chemical tests on almost all admissions, i.e., more patients are having some form of chemical examination. The more prolific use of the diagnostic tests usually ordered as "profiles" or "patterns" results in more tests being requested per patient. The net result is that chemistry laboratories are being faced with ever increasing workloads. Our own statistics show an average 10 % increase per year for the last 10 years. This increasing load may tend to level off, but it is likely that this will not come about for some time. The information available to the clinician comes from three sources: patient history, physical examination, and laboratory examination. Of the three sources of data, the latter at present is the most feasible one to provide additional information on any one case. It is for this reason that the leveling off in the laboratory workload may be slow in coming.

The complexity of tests is generally increasing. Although many established test methods are undergoing modifications which tend to make them simpler, in general, the advent of newer tests, such as measurement of the various enzyme activities and the trend toward analysis of substances occurring in traces only, has more than counteracted the simplifying modifications of more established methods.

Also, as is probably only too painfully certain, the nonuniformity of test results (W2), as established by various international surveys, points to the need for a greater accuracy with less variability and more standardization in chemical testing methods.

One result of the increasing number of tests being requested for each patient is that pressure is exerted on the chemist to scale down his test methods to microproportions. To obtain the same degree of accuracy when working with volumes of 100 μl or less usually requires a more meticulous and time-consuming procedure.

1.2. Scope of Chapter

The clinical chemist, thus faced with increasing workloads, greater complexity of tests, trends toward microanalysis, and the need, in some cases, for greater accuracy, might well contemplate efficiency measures. Any new approach would be welcome which would help increase his efficiency without loss of accuracy and thus allow his technical staff to devote more of its time to the tests requiring greater personal care for their performance. Automation may provide a partial answer to some of these problems.

This chapter presents eight major activities which need to be performed in conducting tests more generally found in clinical chemistry laboratories. In each of these activities a certain amount of automation is found. Types of automated equipment for each of the following categories are presented: glassware washing; filtration and centrifugation; pipetting and transferring; titration and buretting; mixing, agitating, shaking; weighing; evaporation and distillation; and colorimetric reading and recording of results. Certain other activities which already are, or may become, important in clinical chemistry laboratories are not covered. The activities not included are: chromatography, electrophoresis, ion exchange, and radioactivity. The final material presented includes two separate automated systems, both of which automatically complete chemical analyses by combining and coordinating some of the eight activities presented previously. A fair amount of space has been devoted to describing the two systems, as they represent the highest level of automation which has been or might be applied to clinical chemistry tests. Additional presentation is made of some of the tests which have already been adapted for automatic analysis along with some general comments concerning utility and current restrictions. Speculations have been included about some of the future possibilities of automation in this field.

2. Automation as a Laboratory Aid

2.1. Definition of Automation

The very definition of the word automation is in a state of flux, or at least it can be used in a number of connotations by any writer on the subject. Quoting from Cunningham (C2): "To a control engineer, automation is the application of feedback principles in devices for doing work or making decisions. To a machine-tool manufacturer, automation is the use of machines in automatic production. To a producer of materials handling equipment, automation is the use of special conveyors and similar apparatus for moving materials about a factory . . . . To a small business man, automation is, according to at least one advertisement, something he can acquire by purchasing a single desk calculating machine for his office."

Since automation in clinical chemistry has not yet come up with a device in which feedback principles have been applied and since the adjustments of the device are made by the operator rather than by the machine, in the present discussion the word automation (from the adjective automatic) will be used in its broadest dictionary sense: "having a self-acting or self-regulating mechanism that performs a required act at a predetermined point in operation, especially machinery or devices which perform work formerly or usually done by hand." (Webster's *New International Unabridged Dictionary*, 2nd Ed.).

2.2. Major Laboratory Activities

There are eight major steps which are normally carried out in doing the various standard clinical chemistry tests: glassware washing; filtration and centrifugation; pipetting and transferring; titration and buretting; mixing, agitating, shaking; weighing; evaporation and distillation; and colorimetric reading and recording.

2.3. Activities Not Discussed

Chromatography has already entered the clinical chemistry laboratory in some places. A number of methods for the separation of normal and abnormal trace components from body fluids have been proposed. With the advent of chromatography have also come various forms of densitometers and even centrifuges for the more rapid development of chromatograms. Although these devices represent some element of automation and are extremely useful for those working with chromatography, they have not been included in the general discussion.

Gas phase chromatography has shown a very rapid development. Small units with more sensitive effluent gas analyzers are being marketed. Although its potentialities with respect to utility in clinical laboratories have not been fully assessed, it may well make possible the rapid and quantitative analysis of blood oxygen, carbon dioxide, carbon monoxide, methanol, ethanol, and fatty acids.

Paper electrophoresis has also made some entry into the clinical laboratory. Advances in instrumentation have provided for development of electropherograms in a continuous manner under the combined influence of electrical potential and gravitational force as well as under the influence of electrical potential and centrifugal force. These developments have not been included because as yet they are in the periphery of clinical chemistry.

The advances in ion exchange analytical methods have also had limited application in some laboratories. Chromatographic and ion exchange methods of purification and identification of trace components in body fluids have occasionally appeared in clinical chemical literature. The development of fraction collectors and even the fully automatic amino acid analyzer are outside the scope of this presentation.

Finally, radioactivity with its counters, scalers, sample changers, and print-out devices has not been covered in spite of intriguing possibilities in the clinical use of radioactive enzyme substrates.

2.4. Discussion of Individual Activities

2.4.1. *Glassware Washing*

This activity, while not directly concerned with the running of clinical chemical tests, is nevertheless of sufficient importance in obtaining accurate chemical analyses to merit some discussion in an article on automation in clinical chemistry. In fact, it is likely that many laboratories which have made only incipient moves in the direction of automating the chemical procedures have already gone a long way toward automating the glassware washing section.

2.4.1.1. *Pipet Washers.* One of the most irksome problems in washing glassware formerly was the handling of pipets. The manual operation of washing and rinsing was generally accompanied by considerable breakage. Most of these problems have been minimized with the use of the automatic pipet washer (Fig. 1) which is available in most laboratory

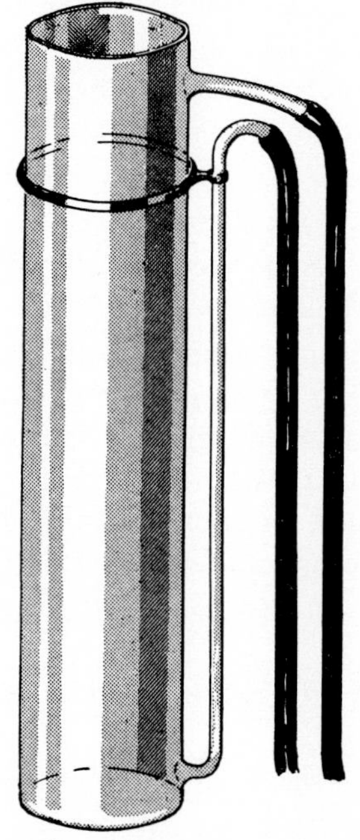

Fig. 1. Automatic pipet washer. The siphon principle is utilized to empty the chamber after it has been filled to the preset point. This is a schematic diagram. Many varieties are commercially available.

supply houses. The principle is quite simple. The incoming water stream from the tap or reservoir fills the main chamber and, concurrently, an outlet tube connected to the chamber bottom. The height of the outlet at its highest point determines the highest level in the chamber. The discharge point of the outlet is below the level of the bottom of the chamber. When the outlet tube is filled, a siphon action is initiated in it. The diameter of the outlet is sufficiently great to cause a greater flow through the outlet than is coming in through the inlet. After the main chamber has emptied, the flow of air through the outlet breaks the siphon and the filling cycle is initiated.

This type of automatic siphon pipet washer is available in a range from relatively low-cost, simple designs to more expensive units in metal or plastic (polyethylene). The plastic construction helps reduce breakage in the handling of the pipets. At least one washer for test tubes[1] uses the siphon principle. In conjunction with the filling and emptying cycle of the tank the test tube carrier also tilts up and down, automatically filling and emptying the test tubes. Another type of pipet washer[2] supplies its own pump and heater element. The heater supplies hot water which is then forced into and around the pipets by the pump action. After drainage of the wash solution the pipets are automatically rinsed by the same device in a separate chamber. The entire wash and rinse action requires no more than five minutes.

2.4.1.2. *General Glassware Washers.* Many of the currently available laboratory washing machines have followed the styles and improvements found in the automatic dishwashers which are so popular with the modern housewife, in restaurants, and in hotels.

In one of the least automated types of washing device[3] (Fig. 2) which can be mounted either over or in a sink, nylon brushes rotating at less than 1000 rpm supply inside and outside scrubbing action to the glass item which is manually held. More than twenty brushes of assorted size and shape can be supplied to allow cleaning of the variously shaped laboratory pieces. These brushes are interchangeable and easily mounted in the chucks provided. An accessory spray rinsing attachment is also available.

More fully automated glassware washers are available in cabinet

[1] Amherst-Boekel, Scientific Glass Apparatus Co., Bloomfield, New Jersey.

[2] VirTis, laboratory supply houses.

[3] Southern Cross, Will Corp., Rochester, New York; Washer Junior, Scientific Products, Evanston, Illinois.

design.[4a,b,c] Some of these are movable on casters while others are designed for more permanent attachment to water and to steam lines. The glassware in all of these types is mounted in or on removable baskets or trays. There is provision for heating the water to about 160°F with either electrical heat or by means of steam.

FIG. 2. Simple glassware washer. The diagram shows a simple spinning-brush washing attachment (Southern Cross), which facilitates routine washing procedures. Other brushes are also supplied for variously shaped glassware.

There is usually provision for a rinse prior to the washing cycle. The washing period is followed by two rinses of varying duration, and in some models the final rinse is followed by a period of electrical or steam heating which removes a great deal of the residual moisture.

Apart from the flexibility of some models in adjusting the sequence and timing of the various cycles, the major differences among the models lie in the manner in which the scrubbing action of water is utilized. In one of the models,[4a] the water jet is propelled against the glassware in a spiraling motion, while in another,[4b] there is an impeller (propeller) in the bottom of the cabinet which throws a high-speed spray against the glassware. The third type[4c] contains the glassware in metal baskets attached to a rotating drum (Fig. 3). The baskets and glassware are rotated several hundred times through the water in a tank. While passing one end of the tank the glass is further subjected to water or steam jet action. The filling and emptying with water plus the jet action is de-

[4a] James, Inc., Independence, Missouri.

[4b] Chemical Rubber Co., Cleveland, Ohio.

[4c] Fisher Scientific Co., New York, New York.

pended on for cleaning. In all of these types of washers a special non-foaming detergent is employed to aid the scrubbing action of the water.

The larger and more expensive glassware washers[5] generally have employed pumping and heating systems which build up the water pressure and temperature and then force the water out of nozzles at high velocity.

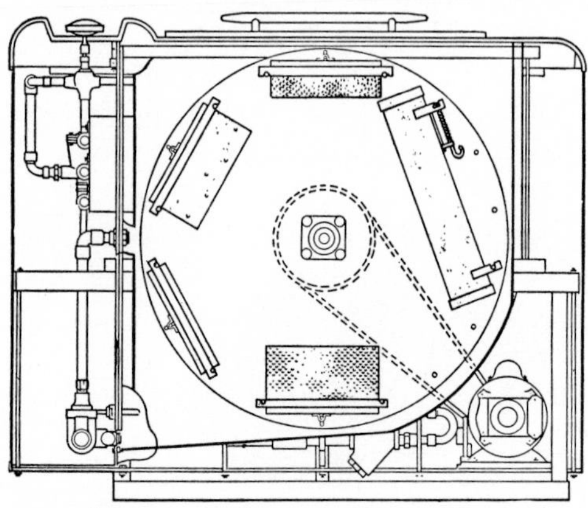

FIG. 3. Laboratory glassware washing machine. The washing machine (Fisher Scientific Co.), which is automatic in operation, supplies timed cycles of washing and rinsing, using hot water under thermostatic temperature control.

The high-velocity jet streams striking the glassware, combined with the surface active properties of detergents, supply ample force and cleaning action for most laboratory glassware. The timing of the washing and rinsing cycles can be preset, depending on local convenience. The glassware is placed in baskets or on removable racks and is usually hot enough at the end of the rinsing cycle to complete the drying as it cools. Some laboratories still have occasion to use an oven for the final drying operation.

A comparatively new device[6] which has recently become available, utilizes the scrubbing action set up in water by cavitation. The cavitation is produced in the cleaning tank by ultrasonic vibration of about 40 kc. The device is simple to operate. Glassware to be cleaned is placed in a small tank containing detergent. This tank containing a transducer is connected to the power unit. When the switch is turned on, the proper cavitation for sonic cleaning is generated. Rinsing may be accomplished

[5] Heinicke Instruments, Hollywood, Florida; Rochester Scientific Co., Rochester, New York.

[6] Ultrasonic cleaner, Scientific Products, Evanston, Illinois; Glennite ultrasonic cleaner, Standard Scientific Supply, New York, New York.

either in the same tank after replacing the wash water or in a supplementary one. The tank capacities listed (maximum approx. 860 cubic inches) are so small that the use of this device would be restricted to micro glassware, syringes, etc.[6a]

2.4.1.3. *Summary.* For washing pipets the desk-top pipet washer using the siphon principle has gained wide acceptance. For small, difficultly cleaned glassware, the newer sonic washers are available.

Although simple and relatively inexpensive spinning brush attachments can be used for general glassware washing, each piece of glassware must be individually handled. This tends to maintain the breakage problem.

The more intricate and more fully automated washers have both advantages and disadvantages. The advantages of the cabinet type of washer are: the moderate purchase and installation cost, the reduction in handling of glassware, and the fact that higher water temperatures can be employed than are considered tolerable by human hands. One of the difficulties with these devices is in the removal of adhesive materials (e.g., dried blood). Another appears in the problem of getting water exchange through narrow-bore orifices, such as tubes, cylinders, or volumetric flasks. Although this appears to have been taken care of with the rotating drum washer, during rotation the movement of poorly placed glassware against itself or the walls of the basket might cause etching after a period of time.

Finally, the most expensive of the washers which can adequately wash most of the varieties of glassware with the high pressure water jets, have certain space, installation, and purchase cost requirements which might be difficult to achieve in many laboratories.

2.4.2. *Filtration and Centrifugation*

Most of the problems which the clinical chemist has in separating solids from liquids in routine analysis are relatively easily solved by use of non-automated methods of filtration or centrifugation. Except for certain reagents which require filtering or centrifuging in bulk, the main problems from the standpoint of automation are that multiple, relatively small samples must be filtered or centrifuged in a manner which prevents one sample from contaminating another.

In general, the material in blood to be tested must be obtained in a cell-free condition. Serum must be obtained free of cells and clots. Likewise, plasma either must be obtained without cells or possibly the whole blood may be deproteinized and the test substance may be determined

[6a] Since the submission of this article larger sized tanks have become available.

from the protein-free filtrate. In any event, almost all specimens must undergo some form of separation procedure *before* the testing can be started. Before the test can be *completed* certain routine methods require precipitation and separation of the products as, for example, some enzyme procedures, traditional calcium methods, etc. A quite considerable time is used up in these separations, but in the absence of adequate automated procedures, the best means of increasing efficiency has been to devise suitable alternatives. A need for precipitation has been removed by flame photometry, complexiometric calcium methods, some electroanalytical or indirect chloride methods, alkaline and acid phosphatase (K1,P1), along with numerous others proposed. In one current device (Autoanalyzer), the need for precipitation has been circumvented by the use of dialysis (see below).

2.4.2.1. *Filtration.* There have been many developments in the field of filtration, but the use of filter paper, gauze, cotton or glass wool along with a suitable funnel still fills the requirements for most clinical laboratory filtrations.

2.4.2.2. *Centrifugation.* There has been very little automation of value to the clinical chemist in the development of centrifuges. Design and accessories have improved since the days of the first step toward automation when the motor or turbine drive was added to centrifuges. A subsequent step in automation might be described as the incorporation of the timing devices with automatic cut-off switch which are featured on virtually all modern centrifuges.[6b] Although centrifugal force has been applied to cause more rapid development of paper chromatographic and electrophoretic patterns, this is more a refinement in technique than a step toward automation.

For the present, the best way to decrease the time needed for precipitation and removal of solids is to devise methods not requiring filtration or centrifugation.

2.4.3. *Pipetting and Transferring*

Automation in the pipetting and transferring of solutions has been suggested for many years (see, for instance, Rowe (R1), 1916). No attempt will be made here to cover the historical developments in this field. Any automatic pipet must supply the positive or negative pressure (with respect to atmospheric pressure) needed to move a liquid from the reservoir into the pipet. This has variously been supplied by such means as hydrostatic pressure, external vacuum or pressure pumps, in

[6b] Since the submission of this article, a more fully automated centrifuge has become available.

microwork by capillary forces, or by manual procedures such as are found in squeezing rubber bulbs, pulling syringe plungers, pistons, etc. One critical unit found in such automated systems is the valve action traditionally performed by a finger. This action has been supplied by such valves as clamps, stopcocks, and various forms of check and cut-off valves. In turn, these valves may be self-seating or actuated by electromagnetic force. In filling, the zero adjustment of volume may be accomplished by the overflow method or by stopping the movement of the piston at a predetermined point. The metering or discharge mechanism may be supplied by hydrostatic pressure or positive pressure exerted by reversal of filling pressure.

Three types of pipet or pipet attachment in which rubber bulbs are employed for filling are shown in Fig. 4. In the first type shown,[7a] after

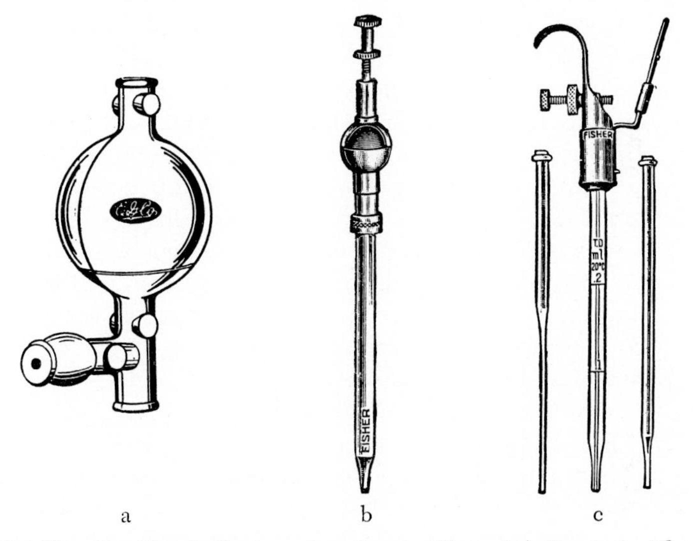

FIG. 4. Use of rubber bulbs in automatic pipetting. (*a*) Propipet. The rubber bulb is supplied with three manually operated ball valves which control the filling and emptying. (*b*) and (*c*) Examples of rubber bulbs which are compressed respectively by plunger and lever action (lever, Fisher Scientific Co.).

collapsing the rubber bulb, control of emptying and filling is by means of two agate-ball valves which are opened by squeezing the valve between the fingers. The squeezing of the rubber around the valve allows movement of air past the ball valve, which movement is stopped as soon as finger pressure is released. This is a convenient extension of the

[7a] Propipet, laboratory supply houses.

principle of the glass bead in the rubber tube design which has been used by many as a simple valve system. A second type[7b] is one where the amount of compression of the rubber bulb is controlled by a plunger. The plunger can be set to travel between two fixed distances. The distance the plunger travels determines the amount of compression of the rubber bulb which in turn determines the volume intake and discharge. This process may be repeated any number of times and has been found convenient for certain laboratory purposes. The third type shown in the figure[7c] employs the same principle except that in this case the bulb is compressed by means of a lever. The length of travel of the lever is controlled by a setscrew and locknut. This adaptation has been scaled to microvolume delivery. There are many other modifications of this type of pipet available which might have some degree of utility in clinical laboratories.

There is a greater popularity of the syringe or piston type of pipet. This is probably due to the fact that the volume metered from a cylinder of constant bore is proportional to the length of travel of the piston. Mechanically this is a simpler movement to control and in micro work it is found more accurate to measure volumes as changes in units of length. There are a good many designs of the syringe pipet which are available from most laboratory supply houses.[8] Figure 5 is a simple diagram of such a pipet, which is designed for repetitive metering by hand. There is no question that the use of such pipets in routine chemical analysis can produce greater speed and probably accuracy (D1) than was formerly accomplished by traditional pipetting methods.

The syringe pipet has been automated even further by the addition of a motor to drive the piston. A number of types (Fig. 6) are currently available.[9] No attempt will be made to discuss all of these types. They are usually supplied with syringes of different capacities which can be used for addition of almost all convenient volumes. The syringe piston is connected to a rotating cam. The degree of eccentricity of the cam may be adjusted to allow for volume adjustments. The syringes and valves are available in various materials which allow a much larger range of potentially corrosive or solvent reagents to be conveniently dispensed.

[7b] Automatic 2-ml pipet, laboratory supply houses.

[7c] Microautomatic pipet, Fisher Scientific Co., New York, New York.

[7d] Volupettor, Standard Scientific Supply, New York, New York.

[8] Continuous syringe pipet, Aloe, St. Louis, Missouri; Aupette, Clay Adams, New York.

[9] Brewer automatic pipetting machine, laboratory supply houses; Filamatic, automatic pipetting machine, laboratory supply houses; Volustat, Fisher Scientific Co., New York, New York.

The machines are usually designed for continuous operation to deliver unit samples at a specific rate or for "single shot" operation where the operator presses a foot switch to obtain one sample.

Just as there are many needs around the laboratory for repetitive accurate addition of reagents for which the manually operated syringe

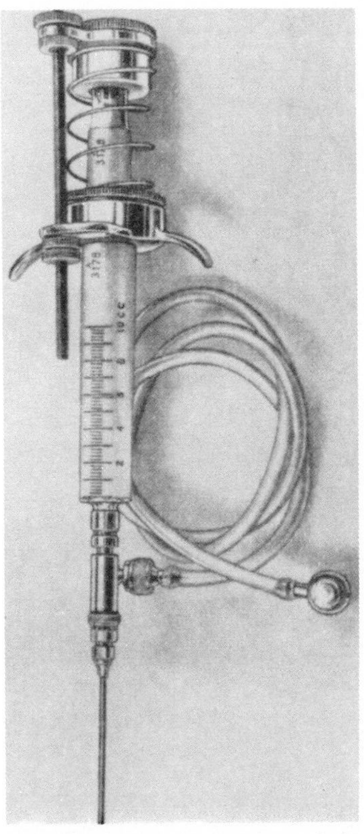

FIG. 5. Syringe pipet. There are many variants of the syringe pipet. The one shown (Aupette, Clay-Adams) has a spring action return of the plunger to a preset position. The backflow of reagent is prevented by a check valve in the side arm. The accuracy of metering is of the order of plus or minus 1.5%. This type of pipet is successfully employed to speed routine reagent additions.

pipet is suited, there are many for which the mechanically driven pipettor is suitable. One problem occurring in larger institutions is the need for large quantities of containers to which some anticoagulant has been added. Non-technical personnel with virtually no training in the use of such a device have very successfully fulfilled this need.

From what has already been presented, it seems that automation has supplied part of the answer to multiple sampling of constant aliquots of any one reagent. The problem of removing constant volumes from reagents of differing composition is only partially answered by these techniques. At least two rinsings of the syringe pipet are needed (D1) before it is sufficiently free of contamination to be applied to another sample.

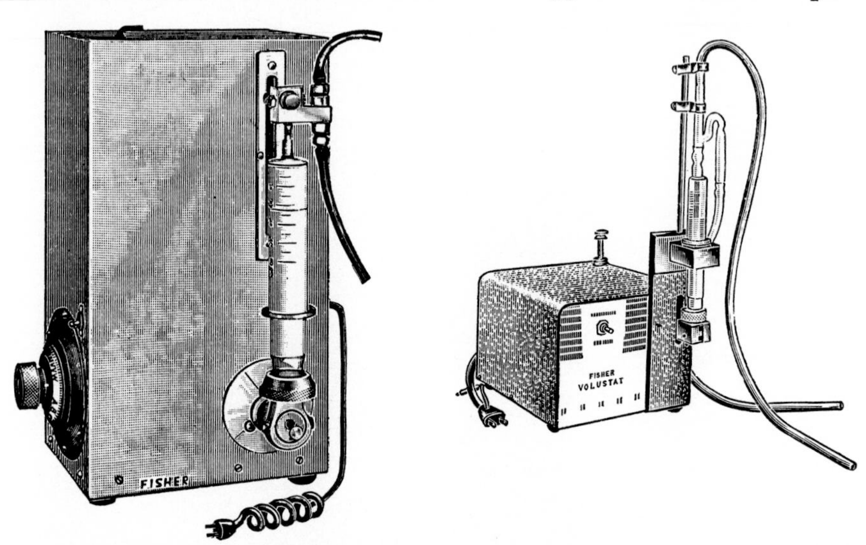

FIG. 6. Automatic pipetting machines. Two types of automatic pipetting machines are shown (*a*) Brewer and (*b*) Volustat (Fisher Scientific Co.). These machines serve the same function as the hand-operated syringe pipets but eliminate the tedium of multiple manual reagent additions. They may be operated by foot switch to deliver a single sample or may be operated in a continuous fashion to deliver samples at a rate of slightly more than forty five per minute. They may be successfully run by unskilled operators.

As one proceeds into the area of microchemistry, the problems of cross contamination, drainage error, and speed of operation of automated pipets become increasingly important. A well-trained technician can sample various specimens using lambda pipets, filling by capillarity with or without mouth suction, wipe off the tip, deliver and rinse out, in about 30 seconds per specimen. However, since multiple pipets are used, the cleaning, washing, and drying operations must be included in the overall assessment of the speed of operation. If these latter procedures are done by the technicians, then the use of automatic micropipets or pipet-buret combinations might readily be considered. Seligson (S3) has recently described a new adaptation of the overflow method for automatic zeroing of pipets which can be used in the microrange. A schematic representation of this is shown in Fig. 7. The apparatus is available commercially,[10] but it may be constructed readily from available glassware. With this arrangement it is possible to remove constant aliquots serially from samples having varying composition. The operation is relatively simple. Gentle vacuum applied to the overflow arm aspirates a sample

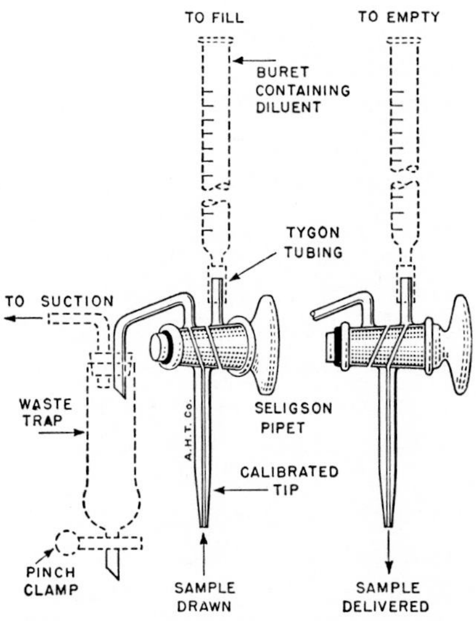

FIG. 7. Combination of pipet and buret. This combination (Seligson, A. H. Thomas Co.) is usable for multiple microdeterminations in which the sample is first picked up and then diluted. The pipet tip is calibrated to contain samples as small as 50 μl. The sample is quantitatively rinsed out and diluted by a reagent added from the buret.

through the capillary tip and stopcock. The stopcock is rotated 90° when the sample column has reached a bend in the overflow arm. The contained volume in the bore of the stopcock and in the capillary tip is the calibrated volume. The leading portion of the sample column rinses out the residual reagent from the previous sampling and is rejected in the overflow. The stopcock is turned to the next position to allow a measured amount of reagent to flow from the buret through the stopcock and capillary tip of the pipet and into a suitable receptacle. In this manner the reagent washes out and at the same time dilutes the contained sample with a constant volume of reagent. The residual reagent in the tip is then sucked out through the overflow side arm, and the next

[10] Seligson automatic pipet, A. H. Thomas Co., Philadelphia, Pennsylvania; Clinalysis apparatus, Kopp Laboratory Supplies, New York, New York.

sample is admitted through the capillary tip to begin the next sampling operation. This pipet-buret combination can be used for volumes as small as 0.05 ml and has been satisfactorily used for a good many different test methods.

A combination of the pipet-buret just described with the use of syringe pipets has been reported by Saifer *et al.* (S1). The various plasma specimens are all sampled and diluted by use of the pipet-buret. All other reagents are added by syringe pipets. Supernatant solutions are transferred by decantation rather than by volumetric transfer, which speeds the process of transferring without causing significant reduction in over-all accuracy. The net effect of these time-saving principles and devices is reported to have been a rapid, accurate scheme for microanalysis within the scope of persons of limited technical skill, with considerable monetary savings from reduced glassware breakage.

When one considers the sample volumes needed to perform microchemical analyses using the pipet-buret device compared to the use of individual pipets, it is apparent that there is a variable amount of sample which is used for washing out contamination in the pipet and then rejected. This rejected volume of sample should be added to that actually needed for the determination to obtain the total volume of sample required for the performance of any particular test. This volume lost due to rinsing of the pipet tip of the pipet-buret would assume importance only in cases where the original sample volumes were marginal in quantity. This might not be a prime consideration in laboratories dealing exclusively with adult blood, but might have to be considered when dealing with infant blood.

From what has been presented on automatic pipets, there is no question that a considerable variety is available. The range covers micro- to macroscale, with complexity varying from simple adaptations of hand-squeezed rubber bulbs to completely mechanical dispensing units in which the operator merely employs a switch. There also is no doubt that the employment of these automatic aids to correct metering of solutions can be and is of tremendous aid in the operation of a clinical laboratory engaged in routine analysis. The range of cost covers possibilities for almost every budget. As has been mentioned, some laboratories have converted many routine tests so that automatic pipetting devices of relatively low cost are used for all reagent additions as well as for serial sampling of blood. Re-use of a single calibrated volume for all specimens can obviously reduce variability of test results. There is a reduction in breakage of glassware and a reduction in the need for washing glassware which can more than make up for the initial and replacement expense of these automatic pipets. In using these devices, almost all of which require some form of valve or stopcock, one is not free from the occasional problems of leaking or sticking, the latter also being applicable to reagent-lubricated pistons. However, the fact that this type of automated metering is not universally employed is probably more a function of human resistance to change than to any other single factor.

2.4.4. *Buretting and Titration*

Much of what has been said about automatic pipets can be extended to burets. However, since the buret is conventionally used in titrations, it is felt necessary to include additional discussion of automatic titrators which have shown considerable advance in the production of automated designs.

Burets in the automatic category range through many types. Most chemists are thoroughly familiar with the Squibb[11] and Schellbach[12] automatic burets. In these, liquid is forced from a glass reservoir into the glass buret with pressure supplied by a rubber bulb. Pressure control may be provided by finger action on an air escape tube or may be provided by a glass check and escape valve. The burets are zeroed either by a backflow siphon action of the excess titrant or by overflow of excess titrant. Discharge of the titrating solution occurs under the influence of gravity through a stopcock or manually controlled pinchclamp. This type of automatic buret has been well established for some time and further discussion will be brief. Lately, polyethylene adaptors have become available which allow the buret to fit directly into the concentrated acid or alkali bottles supplied by the manufacturers of these reagents. In many cases this makes the dispensing of concentrated acid or alkali a much simpler process than formerly. Among the many variants of this type of automatic zeroing buret is another one in which the reservoir is of polyethylene.[13] The need of a rubber bulb is eliminated since compression of the reservoir itself forces the liquid into a glass buret with an overflow leveling device whence it can be buretted by stopcock control.

For microchemical work there is quite a variety of automatic or semiautomatic buret available. The Schellbach and Squibb types which have already been described have been adapted to microdispensing. These

11 Automatic buret, laboratory supply houses.

12 Squibb automatic buret, laboratory supply houses; Schellbach automatic buret, laboratory supply houses.

13 Squeezomatic, Emil Greiner, New York, New York.

burets, which discharge by gravity, show increases in drainage error as the bore of the buret decreases in size.

Another type of buret which has been popular for a number of years is the capillary buret which is equipped with a reservoir for simple filling. Examples of this type are the Grunbaum-Kirk and Sanz[14] microburets. A recent modification allows for reservoirs containing ion exchange resins which are capable of generating either standard acids or bases by the addition of a standard potassium chloride solution. These ion exchange filled reservoirs are usable for many determinations before exhaustion of the exchange characteristics occur. In this type of buret the reading of a meniscus is still required, the meniscus being provided by the introduction of a small air bubble in the capillary which moves at the same rate as does the column of titrating solution.

Most of the available types of microburets are of the displacement types. In the Rehberg type, the turning of a screw moves a piston which in turn displaces mercury into one end of the graduated glass capillary. The readings are all made from the changes in position of the mercury/titrant interface in the graduated tube which is calibrated in 0.01-ml divisions. Many modifications of this type of buret have been proposed. The two general aims have been to remove the need for reading an interface or meniscus and to get away from having a direct contact between mercury and the solution to be titrated. Most of the more recent types which are commercially available have achieved the first aim. By converting the volume displaced from the buret into distance moved by the piston, the readings can be made by micrometer,[15a] by dial reading,[15b] or by a mechanical counter[15c] which reads directly in volume units. Outside of the techniques of placing air or another inert fluid between the mercury meniscus and the titrant meniscus, the most popular means of eliminating the direct contact between mercury and titrant is by use of the glass syringe piston[15b,d] or, in the case of the Gilmont pipet-buret[15d] by use of a glass plunger (Fig. 8). In the latter case the reservoir volume is kept to a minimum to reduce temperature effects and the over-all accuracy is about 0.1 % with titrating volumes of about 0.1 ml.

The burets mentioned in the previous paragraphs have all used

[14] Grunbaum-Kirk and Sanz capillary buret, Microchemical Specialties, Berkeley, California.

[15a] Micrometer screw buret, Microchemical Specialties, Berkeley, California.

[15b] Microburet, Will Corp., Rochester, New York.

[15c] Gilmont ultramicroburet, E. Greiner, New York, New York.

[15d] Gilmont micropipet-buret, E. Greiner, New York, New York.

manually operated stopcocks or manually operated plungers for controlling the liquid flow from the buret. More elegant ones for macrotitrations have an automatic cut-off valve which operates in conjunction with a sensing device to determine the endpoint of a titration.[16] This type of titrating device is employed in potentiometric titration methods. In these methods a reaction occurs which involves the addition or removal of some ion which in turn causes a change of potential which is measured by an indicator electrode. The potential from the indicator

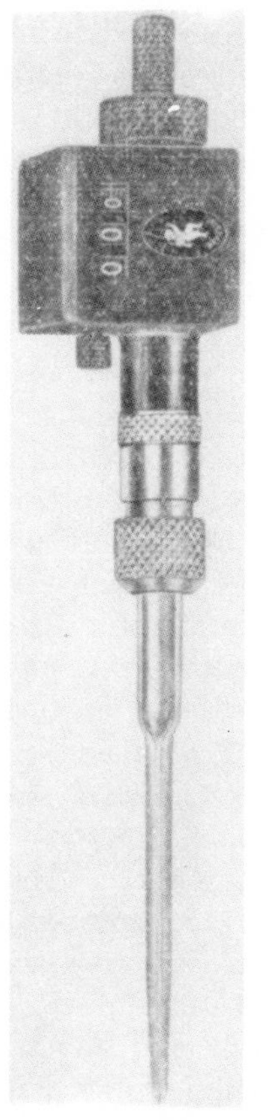

FIG. 8. Micropipet-buret. This design of the moving piston type of metering of reagent is shown as adapted to microscale (Gilmont, E. Greiner). This buret is filled and discharged by movement of a piston, the volume being registered on a mechanical scale. It may be obtained with a total capacity of 100 μl, read to 0.01 μl with a stated accuracy of 0.1%.

[16a] Automatic titrimeter, Fisher Scientific Co., New York, New York.

system, after amplification, activates the electromagnetic valve which controls the flow of titrant. The lag inherent in the system requires that an anticipation unit be incorporated to prevent overshooting of the endpoint. This anticipation unit can be preset to a potential change occurring near the endpoint. When this preset potential is reached, the rate of addition of free-flowing or piston-controlled titrant is reduced until the equivalence point is reached. This type of titration is useful in the absence of a good colorimetric endpoint or in the presence of turbidity or fluorescence, which would interfere in the proper determination of a colorimetric endpoint. It can be used in carrying out titrations such as acid-base, oxidation-reduction, precipitation, and complex formation, for which the potential at the endpoint is known. A reasonable time required for titration is about 3 minutes, with some reduction in time possible when accuracy permits. These automatic titrators also feature some form of automatically controlled stirring which starts when the titration vessel is positioned. In the least automated of the titrimeters,[16a,b] the flow of liquid from the buret is by gravity. At the endpoint the final reading of the meniscus is by eye. In a slightly more automated system,[16c] the filling of the buret is automatically stopped by a liquid level sensing unit, and after discharge the final buret reading is by means of a photoelectric meniscus finder which allows the result to be read and recorded. In another system,[16d] no anticipation unit is employed, and the titration is carried past the endpoint. In this latter device, the addition of titrant is controlled by a motor-driven micrometer screw which meters the outflow at a constant rate. The changes in potential in the titrating vessel can be recorded as a function of time and therefore as a function of the volume of titrant added.

The types of automatic titrator which have just been described are not usually found in a clinical laboratory. There have, however, been some recent reports on the use of such devices for the determination of chloride in body fluids. For comparative purposes, the following discussion will deal with some of the newer modifications in the determination of chloride by electroanalysis.

A short report (K2) has appeared in which an automatic titrator has been used in a modified version of the Schales and Schales procedure (S2). In the modified method which employs 3 ml of a 1:10 blood filtrate, the standard mercuric nitrate in nitric acid is used as a titrant. The recovery values reported are sufficiently accurate for clinical use;

[16b] Automatic titrator, Beckman Instruments, Fullerton, California.

[16c] Analmatic titrator, Baird-Tatlock, Chadwell Heath, Essex, England.

[16d] Recordomatic titrator, Precision Scientific Co., Chicago, Illinois.

however, the relatively large volume of initial sample plus the length of time necessary to complete the titration would seem to prevent its universal acceptance in its present form.

Another electrometric method has been recently described (S4) which has been successfully employed in three laboratories for 1 to 4 years. In this method the titrant, a standard solution of silver nitrate, is manually admitted to the titration medium by means of a micrometer screw microburet. The indicator electrodes are attached to a sensitive galvanometer and measure the rapid change in potential at the endpoint as a result of excess silver ions. The use of a motor driven titration table for stirring reduces to 2–3 seconds the equilibration time after each addition of silver nitrate. A graph is constructed showing galvanometer units compared to buret units. The technique, which requires about 2 minutes for a titration, can be used on serum samples of 0.10 ml or less with an accuracy of about 0.5 %. Further details of construction and use are presented in the original article. This method appears to be a little fussy with respect to the preparation of the electrodes as well as galvanometer settings. A further drawback might be found in the number of galvanometer readings required to establish the endpoint precisely.

A much simpler and more rapid chloride titration unit has been described by Cotlove *et al.* (C1, see also L1). This or similar design has been commercially produced by a number of companies.[17] The titration principle is that of a coulometric titration. In coulometric methods the titrant is prepared by electrolysis rather than added as a solution from a buret.

A simplified diagram of the design is shown in Fig. 9. Basically the unit consists of a clock, a stirrer, a silver ion generating circuit, and a silver ion indicating circuit containing a meter relay for stopping the clock, stirrer, and generator current on completion of the titration. With the stirrer on, the clock and the silver ion *generating* current are started together. Silver ions from a silver electrode (silver wire is used as the anode) are released into the solution to be titrated at a constant rate under the influence of a constant direct current. As long as chloride ion is present, silver ion is removed from solution as AgCl. After exhaustion of chloride from the solution, the increasing silver ion concentration increases the current flow between the *indicator* electrodes. When this current reaches a preset value, the meter relay is activated to stop the clock and ion generating system. Since the rate of generation of silver

[17] Chloride titrator, American Instrument Co., Silver Springs, Maryland; Coulomatic titrimeter, Fisher Scientific Co., New York, New York.

ion is constant, the amount of chloride precipitated is proportional to the elapsed time.

The actual steps performed by the operator are simple and comprise starting the stirrer and raising the sample into position. With the timer reset to zero, the adjustable pointer of the meter relay is set to 10 μa

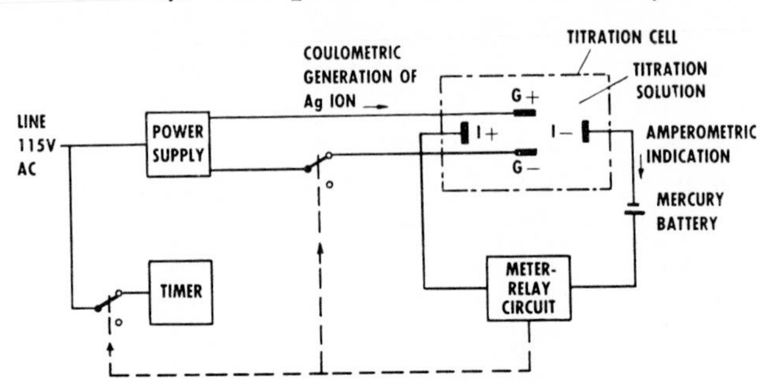

Diagram of Functional Operation

FIG. 9. Automatic coulometric titrator. This relatively simple design (Cotlove, Aminco) has been successfully employed for the rapid and accurate coulometric titration of chloride in serum samples of 0.1 ml or less.

above the reading of the indicating pointer (initial current indicator). When the titration switch is turned on, the timer is started and generation of silver ion is begun. At the endpoint the indicating pointer reaches the setting of the adjustable pointer which closes the switches they carry and actuates the relays to shut off the timer, generator current, and stirrer. The elapsed time is then read to the nearest 0.05 seconds. In general, titrations of chloride in sera require less than 60 seconds when employing serum volumes of 0.1 ml. Smaller volumes of serum can be easily accommodated.

The titration medium is also simple to prepare. It contains gelatin to give a smooth indicator current and prevent reduction of silver chloride. With a preservative added, the reagent may be made up in quantity and stored for some time when refrigerated.

There are some interferences with this titrating method which should be mentioned but which are negligibly small or which can be circumvented. In normal plasma the effect of sulfhydryl, sulfide, or other silver reactive material is quite small, yielding a falsely high result of about 0.60 meq/liter. In urine, the errors while more variable are still considered negligible. It is only in tissue extracts that large interferences are encountered, and these may be largely removed by alkaline oxidation with H_2O_2 as well as by other methods.

The stated variability is certainly within acceptable limits, as are the recovery values reported. It would seem likely that there is room for this rapid, accurate automatic device for chloride determination in many laboratories. The cost of the silver wire "reagent" is low. There is some need for cleanliness and proper care of both the generator and indicator electrodes.

Still another electroanalytical method for the determination of chloride has recently been reported (B1). This technique is a polarographic method employing the dropping mercury electrode (general reference M4). By means of an external potentiometer, a potential difference is applied between the large anode (mercury pool) and the small cathode (mercury drop). The current strength is indicated by a galvanometer. The current density across each of the electrodes is dependent on the exposed area. This means that the cathode (mercury drop) is the determining electrode in the current-voltage relation. As the applied potential difference increases, various ions in the solution begin to plate out on the mercury drops. When this occurs, there is a rapid increase in current flow between the electrodes. The magnitude of the applied potential at which deposition occurs is specific for any one ion. In determining chloride polarographically, soluble iodate ion is displaced from silver iodate by the chloride present in 0.1 ml of serum. The amount of soluble iodate is determined on a polarograph in the presence of lead ion as an internal standard and protein as a maximal suppressor. The intimate details can be obtained from the original article. The need for centrifugation at 5000 rpm for 15 minutes as well as the 5-minute development time for the polarograms would tend to cause some impediment in the universal acceptance of this method, particularly in view of the more rapid coulometric method.

Another method for determination of chloride by automatic means is discussed in the section on the Autoanalyzer.

Automatic titrators; colorimetric endpoint. Automatic titrators[18] sensitive to a colorimetric endpoint have not been on the market long enough to have made any impression in clinical chemistry.

Summary. Most of the automatic burets which have been discussed have been primarily refinements of models that were introduced many years ago. In the adaptations of automatic burets to microchemistry, some burets have eliminated major drainage errors and contact of titrant with mercury and have provided readily readable volumes on dials or mechanical counters. These improvements are reflected in increased

[18] Color-matic endpoint detector, Central Scientific Co.; Analmatic, Baird-Tatlock, Chadwell Heath, Essex, England.

cost; however, the increased ease and accuracy of operation should very well offset the higher cost in those laboratories doing titrations in the region of 100 μl or less.

One of the developments in electrometric titrations has produced at least one instrument for the determination of chloride which is simple, rapid, and accurate. For those institutions large enough to run many chloride determinations and with chloride allowed as an emergency test, the acquisition of such an instrument would be readily justified. Future development in electrometric methods might increase the importance of any of these instruments, particularly if it were adapted to more than one clinically important test.

2.4.5. *Mixing, Agitating, Shaking, Homogenizing*

In clinical chemical practice, most of the mixing activities in making up reagents are done either manually or by the use of some small auxiliary device, such as a stirring motor. Not much space will be devoted to description of various electrical mixing or stirring devices. Any electrical motor with convenient stirring attachment is usually sufficient to replace manpower for this operation. Air agitation, direct or by way of air-driven motors, is a well-known aid in titrations in test tubes, centrifuge tubes, or other containers in which manual swirling is difficult. The air-*driven* motors are preferred in the presence of solvents where fire hazard exists. They also function well when water vapor, as in steam evaporation setups, would soon cause corrosion of the simple unprotected air-cooled motors.

Possibly the most significant advance in agitation devices in recent years has been the use of motor-driven magnets. The magnetic field of an externally driven rotating magnet is used to rotate an iron rod or secondary magnet within the mixing vessel. This iron rod or magnet is usually enclosed in a glass or plastic envelope to protect it from reacting with and contaminating whatever is being agitated. The size and design of the internal rotating piece can be almost whatever the operator wishes. This type of agitator has gained wide acceptance not only by itself but as an accessory mixing unit for such devices as the automatic titration units previously described. A more recent modification of magnetic stirring has incorporated it with a hot plate to allow the double process of heating and stirring to be conveniently carried out.[19]

Another design places the driving magnet[20] at the top of the flask. The stirring magnet within the flask is connected to a suitable stirrer. The

[19] Magnetic stirring hot plates, laboratory supply houses.

[20] Scientific Glass Apparatus Co., Bloomfield, New Jersey.

location of the magnetic field above the flask allows for stirring in vacuum or pressure operations concurrently with heating or cooling the main body of the flask from the bottom. The elimination of the older types of vacuum seal represents considerable advance.

Another ingenious use of the magnetic stirrer is of more interest to clinical chemists. The two forms of the Van Slyke blood gas apparatus have been constructed with a plastic-coated magnetic bar within the extraction chamber[21] (Fig. 10). The extraction of gases is facilitated by

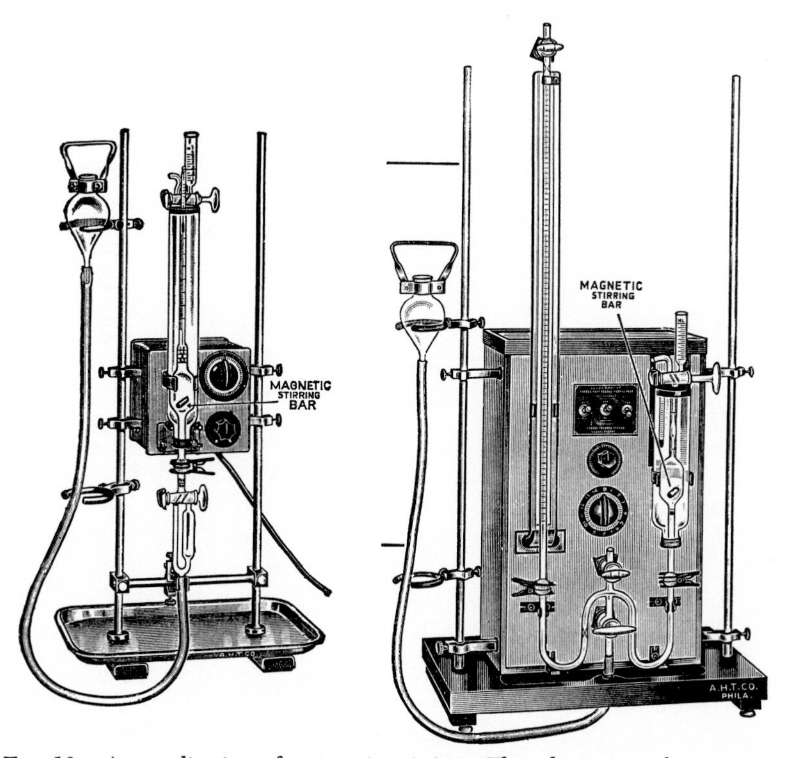

FIG. 10. An application of magnetic stirring. The placement of magnetic, externally driven stirring bar in the gas chamber of the Van Slyke volumetric and manometric blood gas apparatus (Magne-Matic, A. H. Thomas) has eliminated some of the instrumental and breakage problems found in using the more conventional designs.

the magnetic agitation which can be controlled by controlling the rate of rotation of the driving magnet. A timer has also been incorporated which allows for a uniform length of agitation. The fact that the extrac-

[21] Magne-matic Van Slyke apparatus, A. H. Thomas, Philadelphia, Pennsylvania.

tion chamber remains stationary and in a vertical position allows for the use of ball joints at various locations throughout the apparatus. This in turn reduces the breakage cost which in the past has been quite significant, especially with the volumetric design. The introduction of the timer and motor allows one sample to be agitated in one apparatus while another sample is being admitted to a second apparatus. Thus, with two such devices it is possible to dovetail multiple determinations under the care of only one operator.

Finally, one other application of the magnetic principle of agitation has been in the design of water baths.[22] Placing a magnetic agitator in the bottom of the bath has allowed for the movement of the bath water needed for uniformity of temperature without sacrificing bath space at the top of the bath formerly needed for a motor stirring device. In the two designs of bath which are available, the movement of water in the baths is accomplished by a perforated circular plate which is pulled downward at intervals by a powerful electromagnet in the base. The pulse intervals are controlled by a timing relay preset at the factory.

More space will not be devoted to applications of the magnetic principle of agitation. The advances already made have greatly simplified many mixing, stirring, and agitation procedures, and it is likely that many further applications will be made.

The nonmagnetic, motorized shaking machine has come into somewhat wider use in clinical chemistry laboratories. Many types and designs are available, depending on the type of agitation desired. Many come with variable speed, variable amplitude, and even movement in three planes. A thorough discussion will not be attempted here. In larger hospital laboratories, there is some utility for such devices. Many of the extraction procedures, particularly where multiple two-phase extractions are required, are handled more conveniently and even more accurately by mechanical shaking. In another application, where multiple samples of plasma are simultaneously being equilibrated with CO_2, it has been found extremely convenient to gas the samples in a train on a shaker using a tank source of 5.5 % CO_2. The gentle shaking in this case is to provide a more rapid change of plasma surface to give greater speed in equilibration.

One other device allows for the simple combination of swirling and heating.[23] This design might have some limited uses in clinical laboratories since it combines two activities normally used simultaneously.

[22] Magne-whirl Blue M water bath, E. Greiner, New York, New York.
[23] Oscillating hot plate, Fisher Scientific Co., New York, New York.

Another type which is the variant of an oscillating hotplate is the water bath with incorporation of a shaking machine.[24] The combination of shaking during incubation in enzyme determinations is not a usual feature of clinical chemistry laboratories. It is considered likely that greater reproducibility of such determinations might be obtained if this were so utilized.

Before concluding the section on mixing, a short coverage is warranted of the type of mixer which employs high-speed rotating blades for mixing or homogenizing action. A number of designs are available covering virtually all useful laboratory volumes[25] from fractions of a milliliter to over three liters. Many come with variable speeds up to 45,000 rpm. Although there is occasionally demand for the clinical chemistry laboratory to run analyses on tissue, the homogenization of such tissue is best accomplished by the low-speed, rotating glass pestle in a test tube homogenizer. The use of the high-speed homogenizer in clinical chemistry is usually restricted to mixing such reagents as the olive oil emulsion used as substrate in some lipase determinations.

In summary, it may be said the greatest change in the complexion of mixing and stirring devices has been due to the application of the magnetic stirring devices. These have simplified many such activities and the trend may be expected to continue. The design of nonmagnetic agitators has mainly undergone improvement, and in some cases it has been combined with methods of simultaneously applying or controlling heat input.

2.4.6. *Weighing*

2.4.6.1. *Analytical.* In the course of the years, there has been considerable improvement in the design of the various analytical balances, which has given some of them semiautomatic features. The free-swinging oscillations of the equal arm beam have been damped by means of magnetic dampers or by use of air or oil dashpots. This has facilitated analytical weighing by decreasing the amount of time necessary for the pans to come to rest. Further improvements have come as a result of removing the need for manual handling of weights. The chainomatic and keyboard balances have been quite popular for a number of years. One of the latest innovations, the Swiss-designed Mettler balance (Fig. 11), has incorporated a number of simplifying innovations which make analytical

[24] Dubnoff shaker, Aloe, St. Louis, Missouri; Water bath shaker, G. S. Co., E. Greiner, New York, New York.

[25] Waring blendor, Osterizer, VirTis Omnimixer, laboratory supply houses.

weighing a relatively simple procedure. In this design[26] only one pan is employed. The beam is loaded with a constant weight at all times. The weighing of an unknown is accomplished by removal of an equivalent weight from the preloaded beam. This substitution weighing gives a constant sensitivity for all weighings up to the capacity of the balance.

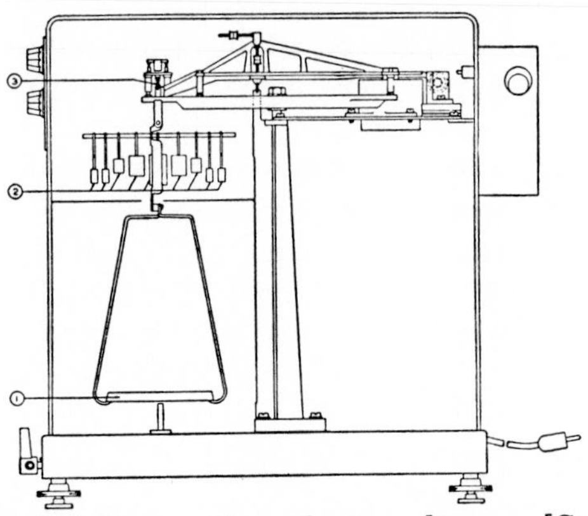

FIG. 11. Analytical balance. The side-view diagram [Gram-atic (Mettler), Fisher Scientific Co.] illustrates a method of manipulation of analytical weights by knob controls as well as the technique of substitution weighing. Other details are supplied in the text. The use of such a balance, which requires little training, considerably shortens the time needed for analytical weighings.

The removal of the weights is accomplished by turning appropriate knobs on the front of the balance case. Turning of the knobs not only removes the appropriate weight but also indicates the value of the weight on a mechanical scale which is mounted on the front panel. The weight values at the extreme low end of the weighing range are projected on a frosted glass plate adjacent to the mechanical scale. These values can be read from an optical scale, a vernier scale, or a micrometer scale, depending on the desired readability. Air damping of the pan oscillations brings the pan rapidly to rest, and the unknown weight can easily be read from the counter and projected scale. These balances are offered in a number of weight ranges from micro to large capacity. In the microrange the balances have a standard deviation for differential weighings of not more than 2 μg. In the range more used in clinical laboratories, i.e. up to 200 g, a standard deviation of 50 μg is obtained. There is no question that this balance design allows semiskilled persons to perform rapid and accurate analytical weighings.

[26] Ainsworth Right-A-Weigh; Quik-Chex, American Balance Co., New York; Gram-atic (Mettler), Fisher-Scientific Co., New York, New York.

Two other manufacturers have produced balances with the single pan design.[27] In these balances, as in the Mettler, the counter weights are loaded in such a fashion that the knife edges are protected during the coarse weighing process. These balances also feature semicircular, clear front panels for better visibility and numerous additional features which provide for simpler operation.

In summary it may be said that for an additional price increment a great deal of additional convenience may be obtained in what was formerly a formidable process of analytical weighing.

Although microweighings (i.e. 100 mg or less) are not common in routine laboratories, there is available a moderately priced microbalance[28] employing a relatively new principle in weighing (Fig. 12). This balance has a capacity of 100 mg. It is portable and battery (dry cell)

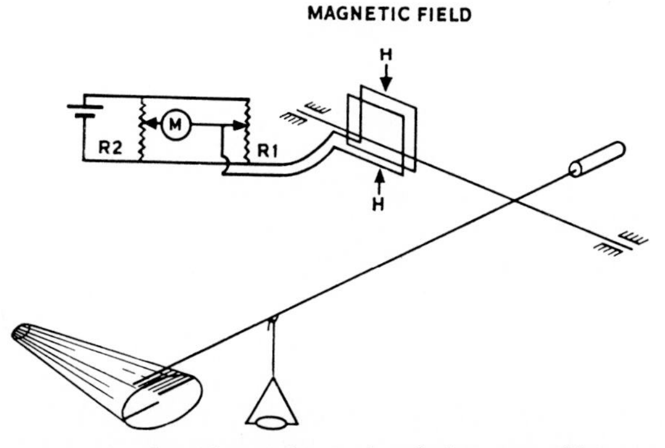

Fig. 12. For microanalytical weighing this balance, usable up to 100 mg, dispenses with the use of weights except for calibration purposes. The cone shown in the lower left portion of the diagram represents the conical projection of light which illuminates two fine fiduciary lines on a glass screen. The beam is balanced when the end of the beam is lined up with the two fine lines on the screen. The beam is shown here with a pan stirrup hooked through a small loop on the beam. The opposite end of the beam is counterbalanced by a weight, indicated in the drawing as a cylinder. The beam is counected at right angles to the arm going through the magnetic field, H. The combination of the magnetic field and a coil attached to the beam is called a torque motor. When the beam is balanced with the empty pan, the meter M is nulled by the zero control dial R1. When it is balanced with the calibrating weight in the pan the meter is nulled by a calibrate control. Unknown weights are read directly in micrograms from R2 which is a precision potentiometer dial.

[27] Sartorius-Selecta, laboratory supply houses; Becker-NA-1, laboratory supply houses.

[28] Cahn Electrobalance, Cahn Instrument Co., Downey, California.

operated. It operates on the null principle with the magnetic torque from an electric coil offsetting the torque from the unknown weight. The current is proportional to the torque, and the setting of the balance potentiometer is proportional to the weight. With this null method of weighing, leveling of the balance, vibration, and temperature have negligible effect (standardization or calibration of the balance accompanies each weighing). In its most sensitive range the accuracy of this balance is about $\pm$ 2.5 μg. Weighing usually requires three steps: zeroing with an empty pan, calibration against a standard weight, and reading the unknown weight. The zero and calibration adjustments need be made only occasionally.

2.4.6.2. *Utility*. In the utility range of weighing, the single-pan Mettler is the primary balance in which manual weight handling is eliminated by use of weight-lifting devices. The balance uses the same principles as found in the analytical balances.

2.4.7. *Distillation and Evaporation*

Very little automation has taken place in the clinical laboratory in the areas of distillation and evaporation. Some improvements in distillation have been offered in the use of electrically heated tapes, rings, and mantles. The nitrogen distillation apparatus designs have improved somewhat. The use of infrared light bulbs as sources of gentle heating in evaporations has increased in popularity. Oscillating hot plates have been mentioned. These could be of use in hastening evaporation procedures. In the realm of vacuum evaporation procedures the employment of the rotating evaporators has proved very successful. In general, the types of distillation and evaporation activities usually found in clinical laboratories is of such a nature that the presence or absence of automation makes relatively little difference. The use of the paper chromatographic and electrophoretic techniques has introduced an additional need for ovens for solvent evaporation but this has merely required the extension of the use of existing ovens or possible purchase of new ones designed primarily for that purpose.

The major change in distillation procedures has possibly been due to the decreased need in supplying distilled, ion-free water. This has been brought about by the ever-increasing use of ion exchange resins. These now come packaged in convenient cartridges for almost any output desired and can be used in conjunction with polyethylene squeeze bottles, with distilled water reservoirs, or hooked directly into tap water supplies. In laboratories already supplied with metal stills, the use of ion exchange resins can remove the need of redistillation of water which was formerly required for determinations involving trace metals.

2.4.8. *Colorimetric Reading and Recording*

It has proven feasible to take the electrical output from photocells or phototubes and either with or without amplification record the magnitude and duration of the output. The recording may be made either in the form of a line tracing on a moving chart or may be converted to numerical values and printed by a read-out device. Further refinements can be supplied in which the instrument converts optical densities (or transmittance units) to concentration values. More intricate recording colorimeters or spectrophotometers are also available for continuous scanning and recording of complete spectra from ultraviolet to infrared. No further discussion of these will be attempted since they do not serve a normal function in routine clinical chemistry laboratories.

The simpler recording colorimeters or spectrophotometers have usually been employed as an adjunct to other automated procedures. Analysis and integration of color values from developed paper chromatographic or electrophoretic strips is possible by use of recording densitometers. Continuous recording of ninhydrin colors in amino acid analyses from ion exchange eluates has been devised. In one instance, a very simple recording colorimeter has been used in the determination of prothrombin time (H1). The recording in this case is employed to establish the clotting rate. The time interval between mixing of reagents and final completion of clotting is obtained from the graphical record. The instrument has apparently been successfully and accurately used in a clinical chemistry laboratory for more than 4 years.

It can be said that to date there has been relatively little use for automatic colorimetric reading and recording devices in clinical laboratories. Further use of these will undoubtedly come about as laboratories begin to accept the more fully automated processes of analysis of the type described in the following sections.

2.5. Automatic Analysis; Sequential Summation of Unit Steps

With the current methods of chemical analysis in clinical laboratories, it is obvious that for a completely automatic analysis to be performed some of the unit steps which have been discussed in the previous sections must be prearranged in proper sequence, so that the sum of the individual steps constitutes a complete analysis. The ultimate in automatic analysis would be where each unit process would contain its own monitoring system to make its own necessary corrections for alterations in physical or chemical variables and where all known tests would be programed to the extent that the operator might merely push the proper button and wait for the answer. Although the technological know-how is probably available to construct such a device, the cost and complexity would be prohibitive.

However, certain of these processes have been partially automated. In some other instances (e.g. Cotlove chloride titrator) a simple alternative to a conventional procedure has been supplied which allows a simpler device to accomplish the same result as an otherwise more complicated one. In this case, the titrating "solution" is the outflow of silver ions which is kept constant by the electrical current with the amount of "solution" added becoming a function of time. It is simpler to control and automate such an electrical "buret" than a corresponding volume buret. Thus, otherwise technologically difficult manipulations (from the standpoint of adaptation to automation) are likely to have simpler counterparts by using a suitable alternative process. Some of the steps leading to full automation have already been accomplished. Thus, automatic colorimetric and amperometric titrations are already available. Recording potentiometers have already been used successfully in such applications as recording colorimeters, titrimeters, etc.

It is only in few cases that the complete sequence of steps involved in a chemical analysis has been arranged to work in an automatic fashion. Two of these systems will be covered in the following discussion.

Analmatic[29] *and Autoanalyzer.*[30] The Analmatic system appears to have been initially designed for industrial laboratory operation, yet its flexibility of design if not the cost could allow for adaptation to clinical chemical procedures. The Autoanalyzer was initially designed for clinical chemical procedures with sufficient flexibility for industrial laboratory adaptation. Since more information concerning the use of the Autoanalyzer in clinical chemistry is presently available, relatively more discussion will be devoted here to its application.

2.5.1. *Analmatic*

The Analmatic comprises a number of units, each unit designed to accomplish at least one of the laboratory steps described in the previous sections. Each unit may be operated separately or in combination and sequence under the programing supplied by a sequence controller unit which works by means of auxiliary devices, such as power supply and timing pulse units. The major activities which are supplied by separate units are: sampling, dispensing pipetting, transfer pipetting, volumetric metering, mixing and reaction units, titration (amperometric or colorimetric), extraction, and recording.

29 Analmatic, Baird and Tatlock (London) Ltd., Chadwell Heath, Essex, England.
30 Autoanalyzer, Technicon Instruments Corp., Chauncey, New York.

Sampling Unit. Sampling may be obtained on a continuous or intermittent basis with sample inlets and outlets controlled by electromagnetically operated valves. These valves are of the diaphragm type. The sample fluid comes into contact with the diaphragm, the glass vessel, valve block, and the intake and outlet tubes. The intake and outlet tubes are of plastic but could also be of rubber, or possibly glass. The composition of the diaphragm is usually acid-resisting butyl rubber but may be changed, depending on the chemical nature of the fluid to be handled. The valve block is made of Perspex but may be made of other substances if desired.

The sampling units have also been designed to receive samples which are already in containers, such as test tubes, and to carry out some form of automatic treatment on these samples.

Dispensing Pipet. This unit (Fig. 13) is composed of a glass syringe mounted on a block which contains 2 of the diaphragm valves. The syringe piston is weighted and is forced up by the pressure of the incoming liquid until it reaches a preadjusted stop. The valve controlling the entering liquid closes, and the valve controlling the delivery opens,

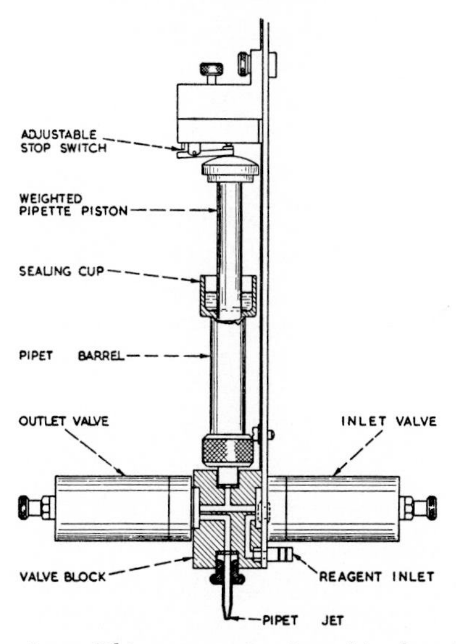

FIG. 13. Automatic pipet. This automatic pipet (Analmatic; Baird and Tatlock) is another application of the syringe pipet. This pipet acts by means of electrical controls as compared to the mechanical or motor-driven ones. The filling is accomplished by hydrostatic pressure and the discharge by means of the weighted pipet piston. The inlet and outlet valves are electromagnetically activated. The pipet has been employed in the dispensing unit shown in Fig. 14.

which allows the weighted piston to fall and discharge the liquid. The satisfactory completion of filling and delivery may be signaled by the piston at either end of its travel, the signal being used to actuate switches, if necessary.

A combination of three of the single pipet units and valves is available (Fig. 14). This combined unit will simultaneously add as many as three

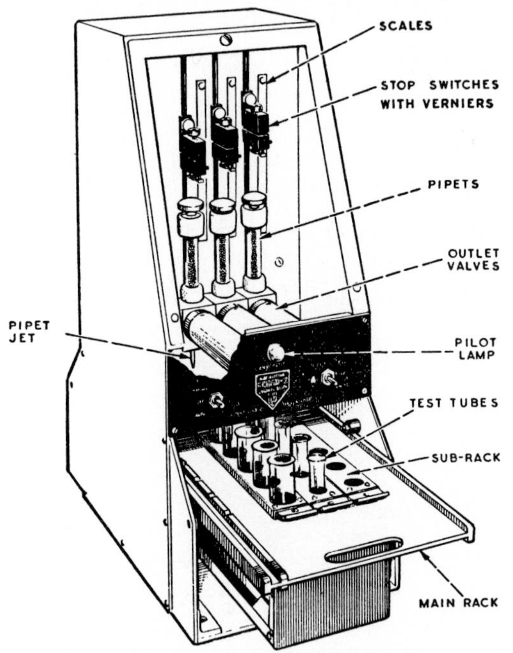

FIG. 14. Automatic dispensing unit. This multiple dispensing unit (Analmatic; Baird and Tatlock) may be employed for the simultaneous addition of three reagents to three separate test tubes. The test tube rack is moved manually along the glides and dispensing is accomplished at each position. The unit will meter samples at the rate of about 24 per minute.

reagents. Test tubes are placed in rows of three in a guided test tube rack. The rack is pushed manually in steps and as each set of three tubes positions under the three corresponding orifices, a preset volume of reagent is added to each tube. The cycle is initiated by the operator moving the rack forward one step. An interlocking device prevents the pipet from dispensing unless the tubes are properly positioned. A small light indicates when the cycle is complete and the rack should be moved again. With nonviscous liquids a rack containing three rows of 10 test tubes (i.e., 30 tubes) can have all additions completed in 75 seconds.

The test tube rack can be obtained for any convenient-sized test tube and interchangeable pipets with maximum volumes of from 1 to 20 ml are available. The actual volume metered is preset with the adjustable stop.

Transfer Pipet. This is a modification of the three-unit dispensing pipet. The unit automatically takes predetermined aliquots from three test tubes and transfers them into three others. The donor and receiver test tubes are in racks locked together and moved manually in stepwise fashion along the guides. The cycle automatically commences when the tubes are properly positioned. The piston-type pipets powered by electrically controlled air-driven cylinders dip into the donor tubes, fill to volume, lift and move sideways, and finally dip into the recipient tubes and discharge. After discharge, the pipets return to the starting position above the donor tubes. Adequate safety devices are incorporated to prevent improper moving of test tubes or pipets. Thirty aliquots of 2 ml each can be transferred in the above manner in about 3 minutes.

Volumetric Metering. This unit is employed where volumes exceeding the capacity of the automatic pipets are desired. Liquid flowing into the unit rises until it contacts a prepositioned probe. The contact probes are for electrically conducting liquids. Capacitance-sensitive probes are available for nonconducting or inflammable liquids. A number of liquids may be admitted in sequence by employing multiple probes and valves.

Mixing and Reaction Unit. The volumetric metering unit (see above) has been combined with a mixing and reaction unit, so that after multiple reagents have been added to the sample by the metering device, gas is passed in to provide for proper mixing and reaction is allowed to occur. The contents are then passed on to the next unit, and the reaction chamber is rinsed before another sample is admitted (Fig. 15).

Extraction. A system has been devised which allows for performing successive liquid-liquid extractions followed by reagent addition and color formation. In one application the test substance in water, after being extracted into another solvent, is backwashed with color-forming reagent, and the solution is then passed into the colorimeter for reading and recording.

Titration. An automatic titrating device with an "anticipation" unit is available for titrations having an electrometric endpoint. The flow of titrant into the reaction vessel proceeds continuously until the amplified potential produced by the sensing electrodes in the solution reaches the first preset value in the anticipation unit. This actuates the valve controlling the inflow, causing it to reduce the continuous flow to one which is dropwise. The titrant is added dropwise to prevent overshooting until the potential reaches the second preset value (endpoint) at which time the titrating solution is cut off completely by the valve. A delay system prevents the unit from responding to false endpoints prior to the true one.

A unique feature of this unit is the method used for reading and recording the results. The reading is accomplished by a photoelectric

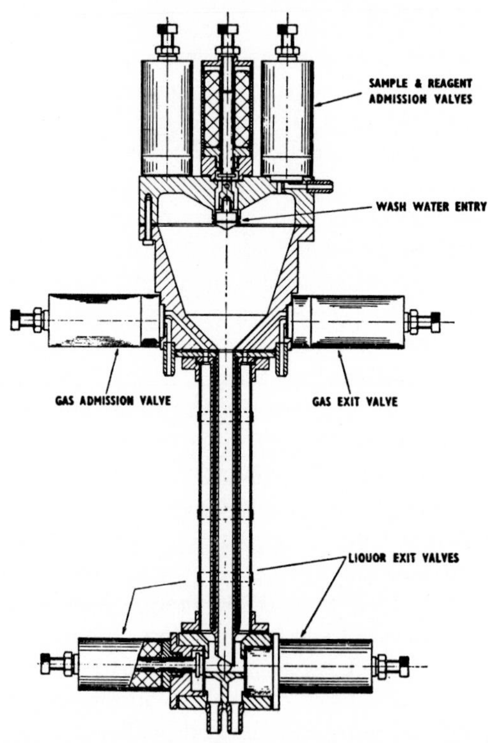

FIG. 15. Metering, mixing, and reaction unit. This unit has been designed for the automatic admission of sample and reagents into a reaction chamber. Agitation is supplied by air or gas. After a preset time for completion of the reaction, the mixture is passed on to a recording colorimeter, not shown (Baird and Tatlock).

meniscus follower system which can be calibrated to read in volume units of titrant used or any unit value which is related to volume. The result is recorded on a print-out tape. The meniscus follower is also set up to control the refilling of the buret.

Further refinements provide for automatic draining and washing of the reaction vessel.

Similar units adapted for colorimetric endpoints are currently in the development stage.

Recording Absorptiometers. At least two types are available. In the first unit there are two test sample cells and one standard reference cell. Successive samples are metered alternately into the two test cells. While the cell containing one sample is being read and recorded, the other cell is being automatically rinsed in preparation for arrival of the succeeding sample.

In the second type of recording absorptiometer, a test tube containing the sample for colorimetric determination is placed manually into a holder on the instrument. The holder automatically grips the test tube and draws up a portion of the contents into the cell. After the proper reading and recording is made, the sample is returned to the test tube and a green panel light indicates the end of the sequence. The instrument now releases its grip on the test tube, which can be replaced by another. The complete sequence can give recorded results at a rate of slightly more than four samples per minute. To attain this rate, rinsing of the cell and pick-up unit have been dispensed with; however, the design is such that residual volume adhering to the walls is minimal and does not introduce significant errors. Recording with these absorptiometers can be either graphical, dot diagram, or digital.

Sequence Controller. Full automation of chemical analysis by the Analmatic system involves combining the individual unit processes into an orderly sequence and providing for repetition of that sequence.

Thus, the mechanical and sensory (or analytical) procedures which are necessary for completion of a chemical analysis have been set up as a series of electrical procedures. This sequence of electrical events is handled in stepwise fashion in a uniselector. All physical manipulations involved in sampling, handling of samples, measurement and recording of results are related to or controlled by corresponding electrical events. These unit processes correspond to one position on a uniselector, and by means of relays the actual opening and closing of appropriate valves, starting and stopping of motors, or sensitizing liquid-level probes are accomplished. Timing devices working in conjunction with the relays provide the needed time sequence. After completion of each unit operation the uniselector steps the circuits into the correct condition for the next one. In this way the whole sequence is completed and repeated automatically.

Adequate safeguards are installed to prevent mishaps from occurring. The details will be found in the literature supplied by the manufacturer.

Summary. The Analmatic system as a method for chemical analysis comprises a series of mechanical unit processes which are controlled by and put out electrical "information." The sequence of this electrical information is controlled by an electrical device called a sequence controller. The electrical "information" produced as the end result of the analysis is converted to the mechanical process involved in graphing or writing the result.

As can readily be seen, there is a considerable technological and electrical complexity involved in this automated scheme of chemical analysis. The cost of complete automation of a clinical laboratory would be outside the reach of most clinical laboratories. However, there are certain units described which might be applicable in large laboratories as individual aids, e.g., automatic dispensing pipets.

2.5.2. *Autoanalyzer.*

This instrument is the first design to become available commercially which gives a completely automated chemical analysis and the cost of which is within the reach of institutional laboratories. The balance of this section will be devoted to features of its construction, applicability to clinical testing methods, limitations, and potentialities.

Some of the basic features of the instrument have been previously described (S5). Since that time quite a few modifications have been introduced which give the device greater flexibility. As in the Analmatic, a number of unit processes are combined to give the proper over-all analytic scheme. The five essential components are as follows.

Sampling Unit (Fig. 16). The current sample plate contains forty

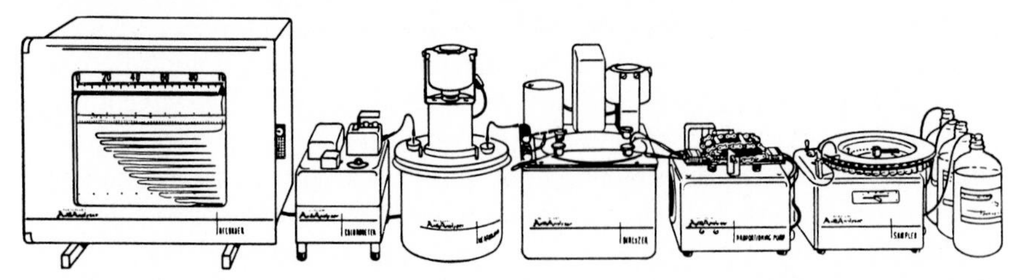

FIG. 16. Autoanalyzer. The diagram indicates the six component parts of the Autoanalyzer (Technicon Instruments Corp.) which are from right to left: sample turntable, pump, dialyzer in constant temperature water bath, heating bath, dual-beam flow colorimeter, and recorder. The reagent bottles would normally be placed adjacent to the pump.

holes in the periphery of a horizontally held plastic disc. Small plastic cups which contain the individual samples or standards fit into the holes in the plastic disc. A sample to be analyzed is merely poured into the cup in sufficient volume to complete the analysis. The little plastic cups can be washed and re-used or are sufficiently inexpensive to be considered disposable. The plastic disc is locked in any convenient position to the drive shaft of an indexing motor by a simple compression lock. The drive shaft can be set to rotate at three speeds, corresponding to analysis rates of twenty, forty, or sixty samples per hour. The most universally satisfactory rate to date appears to be the forty sample per hour rate.

Mounted to the same box is a sleeve through which the plastic sample pick-up tube is inserted. The movement of this sleeve controls the dipping in and withdrawal of the sample tube from each sample cup as it is properly positioned. Thus, sampling is accomplished by automatic rotation of a sample cup into position, dipping in of the sample pick-up tube, holding until adequate volume has been removed, withdrawal of the pick-up tube followed by rotation of the sample plate to bring the subsequent sample into position to repeat the process. A preset warning device is included to provide a buzzing sound to signal the completion of all the analyses.

The removable cups in the disc allow for more than forty samples to be run in a sequence, providing the operator manually reloads the disc during the run. A possibly simpler technique is to have more than one disc available should more than forty specimens need processing and to replace the disc when the buzzer sounds. This introduces a negligibly small time lag.

It has been shown that evaporation of moisture from the samples in the plastic cups can cause significant changes (e.g., greater than 5 %) in the recorded values. After samples are poured into the plastic cups, it may be as long as one hour before they are analyzed. If the room temperature is sufficiently high and air currents are strong, losses of moisture by evaporation may be appreciable. A plastic plate is supplied which fits over the samples to reduce this evaporation to a small value and to prevent contamination of the otherwise exposed samples with dirt or spillage of chemicals. In moderate climates, evaporation effects after one hour are usually within the range of clinical acceptability. However, as often occurs, the sample cup is filled with a large enough volume for more than one test. After it is analyzed on the machine for one test, it is reutilized either by transferring the cup to another plate or by allowing it to remain in place. The combined period of time for two or more tests during which moisture loss may occur might require the use of the plastic cover at all times.

Usual procedure would include having standards in one to three cups of the forty available positions, although more or less would be possible depending on the test which is being run and the desired accuracy.

Pump. The pump is of the peristaltic action type. A constant-speed motor drives two continuous sprocket chains. Between the chains, equidistant from each other, are spaced five metal rollers. The roller assembly is hinged to allow it to be swung up. This disconnects the rollers from the driving motor and gets them out of the way. When the roller assembly is swung down and locked, the rollers roll across a laminated plastic plate which is supported from beneath by springs. On the plate are located up to eight elastic tubes in parallel, which are mounted in end blocks. These end blocks space the tubes, hold them under slight tension, and keep them from slipping. The tubes are then compressed between a roller and the plate, the point of compression moving lengthwise along the tubes as the roller moves along. As each roller is lifted off the plate on completion of its compression stroke, the succeeding roller has compressed the tubes at a point further back, and the entrapped volume between the rollers is forced into the system. The tubes are supplied in polyvinyl plastic, which has sufficient elasticity to rebound rapidly after compression. Rubber tubes have also been used. At a constant pumping *rate,* the internal diameter of the tubes determines the volume flow of fluid through the tubes. The interruption of *sample* flow is provided between samples by admission of air through the sample line, i.e., for a sampling rate of one analysis per minute (sixty samples per hour) the pump aspirates sample for 2/3 minute and air for 1/3 minute. The air admitted between samples causes depressions in the recorded curves as the color developed in the analysis returns toward base line values. Air is also admitted into the system at one or more other points to break up the flowing liquid into small aliquots. This air bubble pattern reduces intermixing between aliquots and supplies a scrubbing action which results in a more discrete transport of the various fluids. Disruption of this pattern is usually productive of "noisy" recordings.

Dialyzer. After the sample has been pumped into the system, generally some diluent (or reagent) is added simultaneously with air. The admission of three substances simultaneously is accomplished in a glass "cactus" having approximately the same internal diameter as the transfer tubes used throughout the process. Other additions where two streams flow together have more conventional *h* or *y* joins. After each addition there is usually a mixing coil which is merely a glass coil placed horizontally to prevent layering of solutions of differing densities.

The pump moves the diluted and now segmented sample into a dialyzer. The dialyzer consists of a cellophane sheet compressed between two plastic plates. The plastic plates have superimposed grooved channels through which fluid is pumped. The two fluids in the channels then both come in contact with opposite sides of the cellophane membrane which allows dialysis to occur. The upper solution is generally the sample stream and the lower is either water or a reagent to act as acceptor of all materials dialyzed. In this continuous flow dialyzer, protein and other high molecular-weight colloidal and particulate materials are separated from substances which are diffusible. For a more complete discussion of dialysis, see Stauffer (S6). Although dialysis rate may be much affected by many physical and chemical factors, sufficient control of these factors has been achieved in this flow stream dialyzer to maintain constant proportionality between the concentration of test substance in the sample stream and in the recipient stream. In some instances ambient temperature fluctuations have caused variable dialysis rates, requiring the dialyzer unit to be kept immersed in a constant temperature (38°C) bath. Since, with adequate precaution, the cellophane membrane is continuously usable for at least a month, changing membranes does not constitute a major problem.

Baths

(*a*) *Heating bath.* After the diffusible materials from the sample stream have transferred to the acceptor stream, the sample stream is rejected. Where removal of protein is not essential, the sample stream is retained until the final result is recorded. Where heating is necessary for color development, the test fluid passes through 460 inches of glass coil in a motor-stirred glycol bath at 95°C. The bath temperature is thermostatically controlled at 95°C to keep expansion of the air bubbles to a minimum and vaporization of water down to a point where surging of the solution leaving the bath does not destroy the segmentation phenomenon established throughout the remainder of the system.

(*b*) *Incubation bath.* For enzyme activity determinations in the Autoanalyzer, an alternate bath for incubation at 37.5°C is available. This water-filled bath is identical to the 95°C bath described above with the thermoregulator set at the lower temperature. The incubation bath is placed in the flow stream immediately after mixing of the sample, air, and buffered substrate.

Recording Colorimeter. The flowing stream which now has attained its appropriate color development is passed into a recording colorimeter. The air is removed in a small precell attachment feeding the flow cell. The colorimeter employs the dual beam principle and compares the electrical output of the photocell from the reference side with that from the flow cell. This difference is then traced as per cent transmittance on a conventional chart recorder. The solution which has been read and recorded is passed into a drain or suitable waste container. The use of a dual beam gives the colorimeter the greater stability usually associated with this principle. The filters supplied with the colorimeter have been narrow band-pass interference types. The balancing of circuits between reference and flow sides is accomplished by a selection of manually inserted orifices of different sizes in combination with a potentiometer.

2.5.3. *Tests Currently Adapted*

General Discussion. Since a number of chemical methods have been adapted to this automated device, the method of changing from one test to another is important. This is accomplished by maintaining intact as much of the tubing, and as many of the joints and mixing units as possible, and disconnecting only at the dialyzer, heater, or colorimeter. The tubing under the pump, still connected to the rest of the flow lines, is freed by lifting the end blocks from positioning pins. The entire flow line circuit for each test complete with end blocks but not including the dialyzer, heater, or colorimeter is then laid aside until needed again. The connect-disconnect procedures are simplified by small tapered connectors. If any of the connections leak due to fatigue of the polyvinyl tubing, it is usually sufficient to snip off a short length of tube to get a tight connection again. The manifold and associated tubes for the next test are placed in the system by making the proper connections and placing the new manifold in position. If all the tubes are marked by tape to show where they connect or into what reagent they are immersed, the change-over between tests can be accomplished readily in five minutes or less and pumping of reagents for the next test can be commenced.

The end blocks containing the tubes in the pump each have three positioning holes. This is to allow the tubes to be initially stretched before pumping is started. If the tubes are not put under adequate initial tension, one or more of them begin to "S" or "snake" under the influence of the rollers and are thus useless as far as pumping is concerned. The amount of initial tension must therefore be adjusted by the operator before each run. The degree of stretching to obtain adequate tension alters as a result of permanent elongation of the tubes due to wear. The stretching of the tubes affects the wall thickness and internal diameters but the resulting changes in the proportions of the various reagents pumped do not cause large variations in the over-all color produced, and with proper use of standards the system shows sufficient constancy to give accurate results. For this reason the volume flow per minute of the different diameter tubes as indicated by the manufacturer of this instrument are to be used as approximate guides only and not absolute quantities. The pumping rate is not only influenced by the internal diameter of the tubes but also to some extent by the position of the liquid levels in the various reagents and the back pressure exerted by the system.

The wearing out of the plastic tubes under the rollers does not affect tubes of different diameter to the same extent. The tubes with smaller internal diameter wear faster for reasons which are not completely clear. Thus one is faced with replacement problems. Although it would be possible to wait until each tube in a manifold either fails to pump or else breaks in use, the consequences of this happening in an unattended machine are quite unpleasant. If a tube should break during pumping and is immediately spotted, the number of samples in the process of analysis—anywhere from three to five—would have to be repeated after the worn-out tube is replaced. With this prospect in mind, it is advisable to change all the tubes in any test manifold once every 2 or 3 weeks as a prophylactic measure. Longer periods are feasible, depending on the amount of use. By having duplicate circuits available for each test, the replacement problem need not cause delays in running analyses.

In the pouring of samples into the cups, the presence of clotting or clot tendency must be observed. With a firm or extensive clot in a sample cup, the pick-up tube becomes clogged immediately. This is not too serious since the analyses previous to the clotted specimen will continue in a normal fashion through the balance of their analytical procedures. However, in samples having relatively small clots or having "soft" clots, the sample including the clot will be drawn into the flow stream. The resistance of the clot to flow changes the flow stream rate and probably the proper proportioning. The clot may cause complete clogging at any point if its progress stops. The clot may also progress slowly until it is finally rejected. Although partial or complete clogging will generally be easily spotted on the record, subsequent sampling action will continue but the results will be erroneous.

It is obvious that close examination of each specimen poured by the operator is essential to prevent such an occurrence. Should a sample show a clotting tendency, enough usable sample may be obtained by pouring the sample into the cup through cotton gauze or other open mesh filter device.

Methods Adapted. At the present time, six methods of interest to clinical chemists are available for analysis with the Autoanalyzer. They are methods for glucose, urea nitrogen, chloride, calcium, inorganic phosphate, and alkaline phosphatase. Micromodifications are available for glucose and urea. These methods are undoubtedly undergoing further modification and change and will be presented, therefore, only in sufficient detail to indicate what has already been done.

It will also be noted that, to date, the methods employed by the automatic device are adaptations (therefore modifications) of previously described methods. Certain test methods which have been too difficult to control by usual routine manual procedures have been adequately adapted to the device due to its generally greater precision in controlling the variables of timing, heating, and volume addition. The machine thus gives no greater specificity than the original methods unless specificity is a function of the variables (or dependent ones) quoted above, or is provided by the process of dialysis or the narrow band pass filters employed.

2.5.3.1. *Glucose.* The flow diagram of this method is shown in Fig. 17.

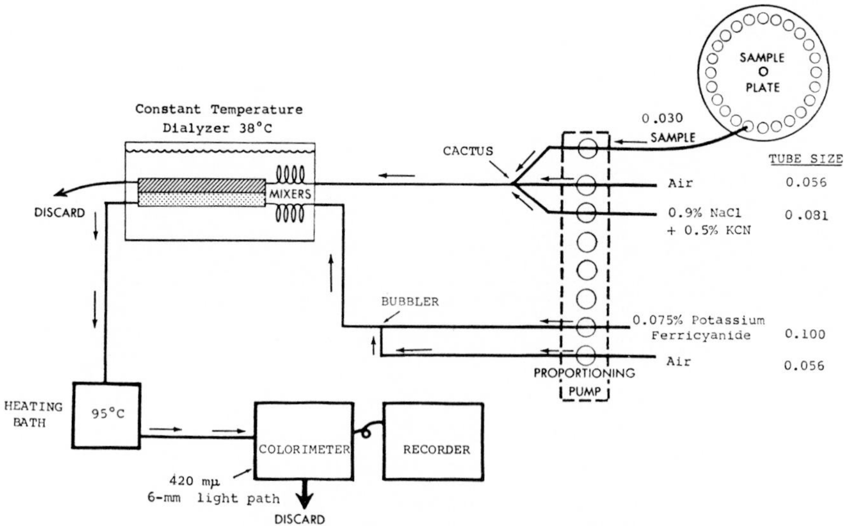

Fig. 17. Flow diagram of glucose method

The method employed is a modification of that proposed by Hoffman (H2). In this modification the disappearance of color is measured when yellow alkaline ferricyanide is reduced to nearly colorless ferrocyanide in the presence of glucose and heat. The reaction is catalyzed by the presence of cyanide ion, although this is not essential (M3).

As can be seen, the sample is diluted with one of the reagents, air is also admitted, and the material, after mixing, is pumped through the dialyzer above the cellophane membrane and then discarded. The potassium ferricyanide, after the addition of air, is pumped in contact with the under surface of the membrane and acts as acceptor for the glucose. The test stream then enters the heating bath and ferricyanide reduction occurs. The extent of color disappearance is measured at 420 mμ in a cell with a 6-mm light path, and a record is made of per cent transmittance change. The calibration curve is obtained by adjusting the instrument to 100 % transmittance with water, recording the per cent transmittance with reagents and then with standards.

The determinations can be run using whole blood, serum, plasma, urine, or cerebrospinal fluid. The volume of sample used with controlled temperature dialysis at the recommended rate of forty determinations per hour is approximately 0.32 ml.

The method is fairly specific for glucose in that samples which have had the glucose destroyed by fermentation yield color values corresponding to 1 or 2 mg % glucose. Unpublished data[31] concerning precision of the method and comparison with results of the Somogyi-Nelson procedure indicate that the automated results are generally as good as those obtained by conventional manual procedures.

The macromethod just described has been modified for microanalysis. Manually prediluted samples are used. Fifty microliters is diluted to 1 ml (1:20 dilution). The preferred rate of analysis is twenty samples per hour or forty per hour with a water specimen between each test specimen. The slower rate of doing the determinations is to allow greater time between samples, which allows for the system to be flushed out with reagents for a longer time. This reduces the effects of contamination. A larger sample tube is employed to smooth out the small pumping fluctuations which are negligibly important in the macromethod. Another difference is the need for two dialyzers in series. It was not pointed out previously that with glucose only a small portion of that initially present in the sample was transferred to the recipient stream. With the micromodification it is necessary to obtain a larger percentage of the glucose of the initial sample in the recipient stream. This is accomplished by doubling the dialyzer path length which doubles the effective dialyzing surface.

2.5.3.2. *Urea Nitrogen: Macro and Micro.* The flow diagram of the urea nitrogen method is shown in Fig. 18. The method originally used for citrulline (F1) has been adapted to the determination of urea (O1). Although the method has had limited success in routine clinical chem-

[31] Personal Communication: Dr. J. J. Carr, Bird S. Coler Hospital, Welfare Island, New York.

istry laboratories, it is readily adapted to use by the automatic analyzer (M1, S5). The method is based on the condensation of diacetyl and urea in the presence of acid and an oxidizing agent to give a colored product the structure of which is possibly that of a triazine derivative. The di-

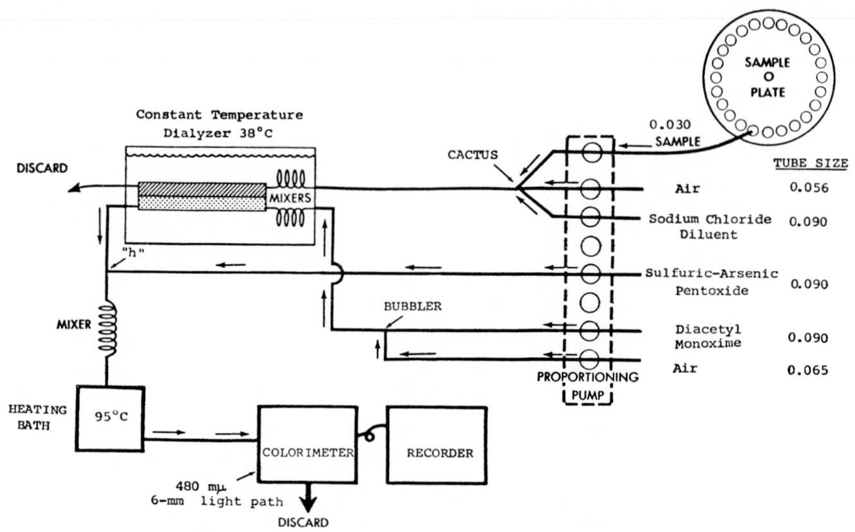

FIG. 18. Flow diagram of urea nitrogen method.

acetyl is formed by acid hydrolysis of the less odorous diacetyl monoxime under the same conditions needed for the condensation reaction.

The sample, air, and saline are mixed and pumped through the dialyzer and then discarded. The recipient fluid contains the diacetyl monoxime reagent properly segmented with air. After exiting from the dialyzer, concentrated sulfuric acid containing arsenic pentoxide is added to the recipient stream, mixed, and passed through the heating bath. The color developed is then read at 480 mμ in a 6-mm cell and recorded in the usual manner.

In this method, the addition of the strong acid-oxidant mixture shortens the life of the polyvinyl plastic tubes. This is particularly evident at the point where the flow stream exits from the heating bath. This is a case where chemically more inert tubes give better service. The disposal of this mixture also has to be given some consideration, especially in institutions which would run a relatively large number of urea samples.

The method as outlined does not give a linear standard curve throughout the entire usable range of the scale. This presents no problem with the use of adequate standards. It has been mentioned (S5) that urea recoveries with the automatic analyzer were not accurate in transmission ranges below 20 %. Urease-treated whole blood gives no detectable color with the automatic analyzer (S5) even with blood from uremic patients (M1). However, the published reports from several laboratories have indicated a precision of the automated scheme which is greater than is usually needed for clinical work. The use of a method which is not sensitive to ammonia levels allows direct urea determinations on urine specimens without prior removal of ammonia. The method can be used for determinations of urea in plasma, serum, and spinal fluid, as well as in whole blood. At the recommended rate of forty specimens per hour, about 0.3 ml of specimen is required for each analysis. At this rate of analysis, there is some contamination found when a sample concentration greater than 70 mg urea nitrogen per 100 ml is followed by one of about 10 mg per 100 ml.

In the microadaptation of this method, 50 μl of the sample is prediluted to 1.0 ml. This total volume is added to the sample cup. The analysis rate is reduced to twenty determinations per hour, which, along with the larger sample pick-up tube, results in the complete utilization of the diluted sample. Apparently the membrane area in a single dialyzer unit allows a sufficiently adequate amount of urea to transfer to the recipient stream to conduct the analysis. This compares with the need for additional dialyzing surface required in the microadaptation of the glucose method.

2.5.3.3. *Chloride.* The pattern of flow for this method is depicted in Fig. 19. The method is based on a modification proposed by Zall *et al.* (Z1). In this method, chloride ion in acid solution containing ferric ions displaces thiocyanate ion from mercuric thiocyanate. The thiocyanate liberated reacts with excess ferric ions to form the red ferric thiocyanate $(FeSCN)^{++}$ complex. The reaction is sensitive to the nature and amount of acid present. The presence of bromide would interfere with the results.

In the automated procedure, the sample is segmented with air and mixed with diluent in the usual fashion. The diluent in this case contains 0.05 % of the wetting agent,[31a] Hyamine 2389 (alkyltolylmethylammonium chloride, where alkyl equals C_9–C_{15}). Although this agent contains chloride, it adds at most a constant small value to the material dialyzing from the sample and gives smoother recordings. The acceptor stream in this case is composed solely of distilled water segmented with air. The color reagent is composed of a saturated solution of mercuric thiocyanate

[31a] Rohm and Haas Co., Washington Square, Philadelphia, Pennsylvania.

containing ferric nitrate (0.05 M Fe^{+++}) and of mercuric nitrate (0.0005 M Hg^{++}), made to a final acid concentration of 0.05 N with nitric acid. Due to the low solubility of mercuric thiocyanate, the addition of the color reagent is made through two lines to give adequate reagent for the test.

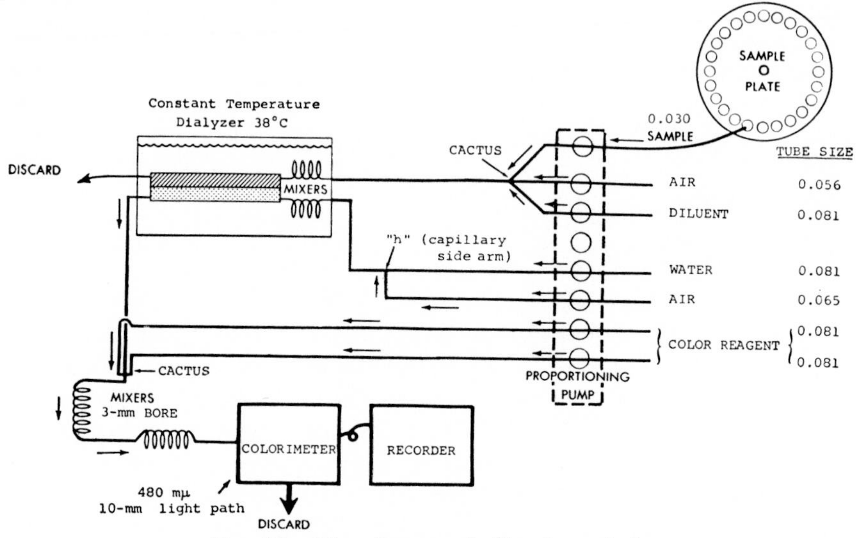

FIG. 19. Flow diagram of chloride method.

The reaction occurs rapidly at room temperature and is complete by the time the stream exits from the second of the two mixing coils and enters the colorimeter.

The method can be used for the range of about 40–150 meq/liter, but for determination of chloride in blood the usable range would be from 75 to 125 meq/liter. With the cell having a 10-mm light path this falls in the most accurate range of the colorimeter, although the curve is relatively flat (1 % transmittance is equivalent to 2.4 meq chloride per liter). In this range, the device should replicate itself to within 0.5 % transmittance, and correlation with a conventional method[32] tends to confirm this. The test may be run with serum, urine, and spinal fluid. At the recommended rate of forty determinations per hour the volume of test material is approximately 0.3 ml.

2.5.3.4. *Calcium.* The calcium procedure adapted for automation (Fig. 20) was proposed by Williams and Moser (W1). Although this method

[32] Personal Communication: Dr. P. D. Rosahn, New Britain General Hospital, New Britain, Connecticut.

has been difficult to control in many clinical chemistry laboratories due to pH effects on color formation, instability of dye, and other factors, it has been run satisfactorily by the automated process.

The basis of this method is the color change produced by the formation of a calcium purpurate complex. Although the complex will form

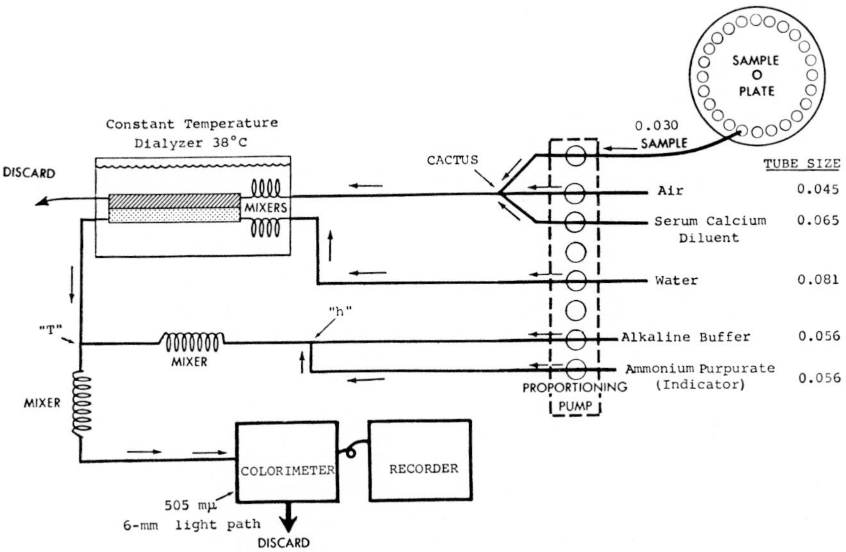

Fig. 20. Flow diagram of calcium method.

over a relatively wide pH range, the alkaline pH's have usually been chosen due to the increased sensitivity obtained.

In adapting the method for automation a problem was encountered in the diffusion of calcium from serum. In serum, a significant amount of the calcium is present in a form bound to protein. In order to liberate the bound calcium it is necessary to reduce the pH to four. The sample diluent in this case not only contains the wetting agent, Hyamine, but also includes an acetate buffer of sufficient strength to attain this pH for all samples. Apparently small pH changes in this region cause changes in the rate of dialysis of calcium.

The acceptor stream in this instance is limited to distilled water. The actual color development occurs after the alkaline buffer is premixed with the ammonium purpurate indicator and the resultant stream is admitted to the test solution. Since the pH at which color development is recorded is quite critical, the alkaline buffer used is the water soluble organic base, diethylamine, which has been adjusted to a pH of 11.7.

No heating is required for color development, so after mixing the final solution is passed into the recording colorimeter, which records the result in the usual fashion.

The recommended rate of analysis is forty samples per hour, requiring about 0.3 ml of sample. There are two points in the analysis scheme where establishment of the correct pH is quite critical. The usable range of calcium values for blood serum is quite restricted; however, with adequate care in reading the relatively flat curve and with a sufficient number of standards, results which compare favorably with the more conventional procedures have been obtained.[33]

2.5.3.5. *Inorganic Phosphate.* This method, shown in Fig. 21 is adapted

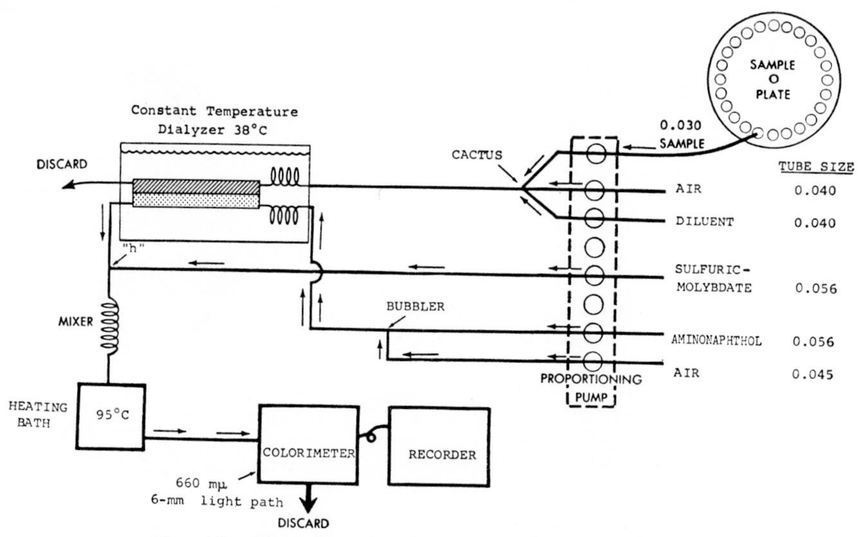

Fig. 21. Flow diagram of inorganic phosphate method.

from the Fiske-Subbarow procedure (F2). In the presence of phosphate and molybdic acid, phosphomolybdic acid is formed. On reaction with 1-amino-2-naphthol-4-sulfonic acid plus heat the characteristic blue color is formed.

As with some of the other methods, one of the reagents, i.e., aminonaphtholsulfonic acid, acts as recipient for the material dialyzed from the sample. The molybdate reagent is added to the acceptor stream after the dialyzer, mixed, and heated, and the developed color is read and recorded in the usual manner.

Due to the relatively small amount of phosphate which is dialyzed,

[33] Personal Communication: Dr. H. P. Smith, College of Physicians and Surgeons, New York, New York.

the recommended rate of analysis is twenty samples per hour, which requires slightly over 0.6 ml of sample volume. The rate of forty samples per hour is satisfactory for most clinical work except that a specimen with an abnormally high inorganic phosphate level will contaminate the succeeding specimen. The degree of contamination will, of course, be aggravated should the following sample be low in inorganic phosphate content.

The method can be used with plasma, serum, or urine, following a dilution of the latter of 1:5 or greater if necessary.

It has been stated[34] that the method, as outlined, gives results as precise as the corresponding manual application of this procedure.

2.5.3.6. *Alkaline Phosphatase*. The first reported use of the automatic analyzer for determination of enzyme activity (M2) indicates that it is feasible for some determinations of this type. The schematic flow diagram for alkaline phosphatase is shown in Fig. 22. The method which has

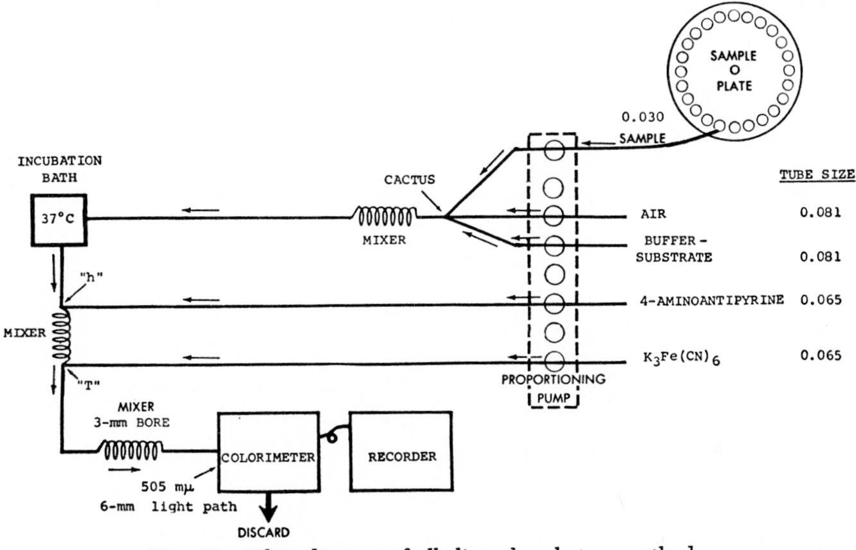

FIG. 22. Flow diagram of alkaline phosphatase method.

been adapted is a modification (K1, P1) of the King-Armstrong (K-A) procedure. Alkaline phosphatase activity is measured by determining the amount of free phenol produced by enzymatic hydrolysis of phenyl phosphate. This particular modification is simplified by virtue of the fact that color development occurs readily at room temperature in the

[34] Personal Communication: Dr. E. E. Baird, University of Texas Medical Branch, Galveston, Texas.

presence of protein. Free phenol condenses readily with 4-aminoantipyrine in the presence of the oxidant, ferricyanide, to form a red product. Each sample is run twice, once with substrate and once without. The difference in phenol between the two is related to alkaline phosphatase activity in conventional K-A units. It also seems feasible, in most cases, to deduct an average control value from all samples and to run the controls only in the cases which are on the borderline between normal and abnormal.

The schematic diagram shows that, after mixing of sample and buffered substrate, and segmenting with air, the mixture is passed into an incubating bath at 37°C. On leaving the bath the 4-aminoantipyrine reagent is added and mixed. The enzyme activity is stopped by the addition of potassium ferricyanide, which is also necessary for color development. The color is sufficiently developed after passage of the stream through one mixing coil to be read and recorded in the usual manner. When controls are run, the sample is mixed with buffer not containing phenyl phosphate. In this case, since incubation is unnecessary, the tubes may be hooked up to bypass the incubation bath. With any group of samples all assay (with substrate) determinations are made in sequence. The process is repeated for all controls (without substrate).

The incubation time has to be measured with a stopwatch each day of use. It should be measured when the system is full. This eliminates possible changes in pumping rate due to changes in back pressure in the system. Although any dye might be used, the yellow color of the ferricyanide reagent is sufficiently strong to use as a visual aid. This reagent is temporarily added through the buffered substrate line with the other reagent lines pumping as in the test. Timing is started when the yellow color intersects the sample line, and timing is stopped when the yellow column reaches the regular ferricyanide line. For most accurate work the buffered substrate should be prewarmed to 37°C, although use of buffered substrate at room temperature (ca. 22°C) does not cause much difference in the results.

The incubation time is generally about seven minutes, although longer and shorter times can be obtained by changing the air intake or buffered substrate volume or both. With these variables the usable scale can be conveniently adjusted to individual needs. After the exact incubation time has been determined there is a simple formula conversion to K-A units.

The recommended sampling rate is forty samples per hour, which requires about 0.3 ml of specimen for assay as well as control. At this rate there is some evidence of contamination between samples of high and low phenol content; however, it is reported that the results by the automatic device compare most favorably with the corresponding manual procedure.

2.5.3.7. *Carbon Dioxide.* A method has been adapted for the determination of carbon dioxide in blood. It can be seen from Fig. 23 that after

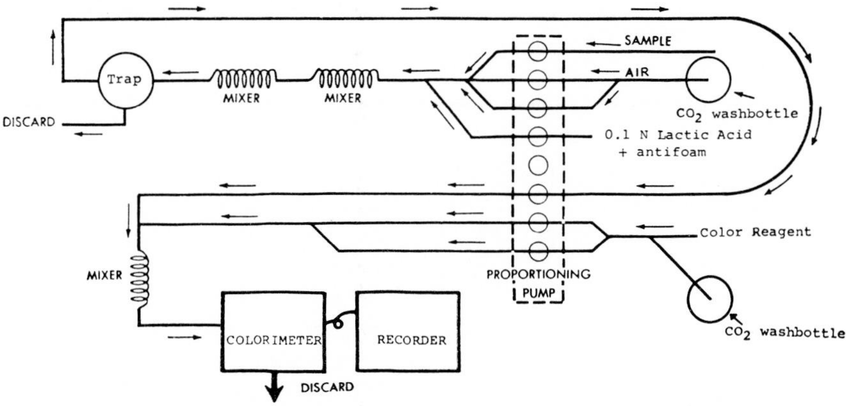

FIG. 23. Flow diagram of carbon dioxide method.

the sample has been segmented with CO_2-free air, lactic acid is added. Following the evolution of carbon dioxide, the gas phase composed of air plus carbon dioxide is separated from the liquid phase. The separation is accomplished in a small glass chamber where the liquid portions of the stream under the influence of gravity drop to a lower orifice and run to waste. A small portion of the gas phase also leaves with the liquid stream. The rest of the gas phase is drawn by the pump into a buffered aqueous stream containing the indicator, phenolphthalein. This indicator is sensitive to carbon dioxide and, after equilibration, the degree of color change, which is proportional to the carbon dioxide content of the initial sample, is read and recorded.

The method may be run with either plasma or serum. The recommended rate is forty determinations per hour, which utilizes about 0.3 ml of sample. It has been shown[35] that the method has given thoroughly adequate correlation with the results of the manometric Van Slyke technique.

The samples must be kept under oil to prevent losses of CO_2 from occurring during the time the specimens are on the sample plate. This appears to offer no major problem in the conduct of the analysis.

[35] Personal Communication: Dr. R. Gambino, St. Luke's Hospital, Milwaukee Wisconsin.

2.5.4. *Tests in the Final Stage of Adaptation*

Space considerations prevent the complete discussion of a number of tests which have already been proven adaptable to the instrument.[36]

An acid phosphatase method has been developed corresponding to that of the alkaline phosphatase already described.

A uric acid procedure has been devised which uses color development in the presence of cyanide and phosphotungstic acid.

A total protein method is available which employs an adaptation of the conventional biuret color development. The albumin/globulin ratio may also be determined by determination of albumin using the anionic dye, 2-(4′-hydroxybenzeneazo)benzoic acid.

A method for the determination of cholesterol has been adapted. The color developed in the presence of ferric chloride, sulfuric acid, and glacial acetic acid has been utilized for the analysis. The use of such strong reagents has required employment of tetrafluorethylene tubing. These acid reagents are not pumped directly from reagent bottles but are displaced from reservoirs by fluid pumped into the reservoir behind a flexible diaphragm.

Finally, the flame photometric methods for determination of sodium and potassium have been adapted to the device. In this case, after the addition of lithium as internal standard, the electrolytes are dialyzed into the recipient stream, which is then pumped into an atomizer-burner of more or less conventional design. A colorimeter has been designed to allow for simultaneous recording of sodium and potassium, utilizing a two-channel recorder as optional equipment. This arrangement would tend to conserve sample. Without the dual recorder, the recorder supplied as part of the basic instrument is used, and the specimens are sampled once for sodium and once for potassium.

2.5.5. *General Considerations and Certain Restrictions in Adapting Tests to the Autoanalyzer*

Should one attempt to adapt a test to the Autoanalyzer, certain characteristics of the instrument become apparent. Each one of the individual units—sample plate, pump and tubing, dialyzer, heating bath, and recording colorimeter—has certain aspects of optimal performance and others of minimal or restrictive performance. The relationships of these units to a flowing stream must be considered along with contamination effects.

2.5.5.1. *Sampling and Flow.* In the absence of a timing marker on the record, it is possible to provide a visual reference of the division between

[36] Personal Communication: Technicon Instruments Corp., Chauncey, New York.

samples by breaks in the recorded curve. These abrupt color changes are supplied by the admission of air through the sample tube before the uptake of each sample. Since air is not a liquid diluent but merely maintains flow rate during these periods, there is a reduction of volume and an increase in concentration of the reagent flowing through the upper level of the dialyzer. Although this undoubtedly affects both dialysis from the sample stream as well as back dialysis from the recipient stream, the changes of reagent color values as a result of this have generally been kept quite small. It must be considered, however, in the choice of reagents in both the sample stream as well as the acceptor stream.

The "in-and-out" time of the sample pick-up tube which in turn determines the length of time between samples is a compromise. At present, the "in" time is twice the "out." In a system which would show no contamination or lag, the time between samples could be considerably shortened. This is not the actual case. The sample must be picked up for a sufficient length of time and volume to overcome a lag of about thirty seconds in the dialyzer before dialysis (S5) proportional to concentration of test substance is achieved. On the other hand, the sample tube must remain in air for a sufficiently long time to allow adequate reagent to pass to remove any traces of residual contamination. These factors differ from test to test. In the specific tests described above, the average approximate volumes metered at forty determinations per hour are as follows: total volume of specimen, 0.3 ml; total diluted volume dialyzed, 2.7 ml; total wash-out volume between specimens, 1.2 ml. Although prediluted specimens, as in the micromodifications described, allow different sample and diluting volumes to be employed, there is a limited range allowed in adapting these variables.

2.5.5.2. *Volume and Flow; Pump and Tubing.* It is obvious that the plastic composition of the tubing under the pump rollers and throughout the system precludes the use of many solvents other than water. At the present time this prevents extractions with nonpolar solvents or use of mixtures containing these solvents. The tubes under the pump rollers are the critical ones since flexibility is an essential here. As previously indicated, it is possible in the rest of the system to use tubing which has a greater degree of chemical inertness. In order to pump liquids which would destroy the flexible tubing under the pump, a displacement principle must be employed. The pump may pump water into a system containing a diaphragm or piston. This in turn will displace an equal volume of a reagent which would normally destroy the utility of the pump tubing in a short time.

At present, the volumes which can be accurately metered with the pump have a lower limit beyond which tubing wear plus small changes in patency or inaccuracies in tubing internal diameter tolerances can make relatively large errors in volumes pumped. It is not known whether this places a greater restriction in approaching microscale work than does contamination or some of the other factors considered below.

In a flowing system of this design, it is not possible to make a reagent addition or to perform a dilution by increasing the volume of one of the reagents without also increasing the flow rate at all subsequent points. Since some of the subsequent steps are sensitive to flow rate (e.g., dialysis, color development) this volume and flow rate relationship must be taken into account when attempting to modify a test method for adaptation to the Autoanalyzer.

One method of controlling flow rate without altering the fluid volumes, is by altering the air volume. This technique has its limitations since there is an optimal size of the liquid segment between the two air compartments moving in the stream. Either too great or too small a subdivision of the liquid segment is usually accompanied by erratic or noisy records. The method of varying air volume to gain varying flow rates, however, can be utilized within these restrictions. With most of the methods, the instrument accomplishes a number of proportional processes, possibly no one of which goes to equilibrium or completion. In adapting any method, it may be advantageous, therefore, by increasing the time interval to allow the over-all rate-limiting step to proceed further toward completion. There are two ways of gaining time. With a constant volume being pumped, time can be gained by increasing the path length. This has conveniently been done by such devices as adding mixing coils or doubling the dialyzer path length. There is of course an upper practical limitation on total path length due to creation of back pressure (resistance to flow), which will affect pumping efficiency. As mentioned above, time can also be influenced by the volume introduced using either liquid or air volume, or both, for the purpose.

2.5.5.3. *Dialysis and Flow.* Controlling the relationship between dialysis rate and flow and the factors which affect dialysis is probably the most complex of all the unit processes carried out by the automatic device. Two alternate streams of air and liquid enter the dialyzer, one on each side of and in contact with the cellophane membrane. Due to the interspersing of air and liquid, the effective surface area of the membrane is less than the actual surface area. In some of the tests previously outlined, wetting agents have been added to the stream flowing through the dialyzer. These possibly increase the effective area through which dialysis may occur.

It was seen that the two readily diffusible substances, glucose and urea, did not dialyze at equivalent rates and that protein binding of calcium had to be overcome to obtain a dialysis rate proportional to its concentration. The instrument usually provides a dialysis rate (and therefore an amount dialyzed) which is proportional to the concentration of the substance being tested. However, this proportionality would need to be checked for any test adapted to automatic analysis. It was also seen that dialysis was influenced by the presence of protein, salts, membrane charges, etc., in that dialysis rate from standards was not always identical with that from plasma or serum. To date these differences have been corrected for by a constant factor.

As mentioned earlier, a simple dilution effect may not be achieved by increasing the solution volume. The increase in flow rate on one side of the dialyzing membrane may effect an increased efficiency of dialysis. In this case, part of the diluent effect would be negated by obtaining a larger proportion of the test material in the recipient stream. On the other hand, increased flow rate might decrease the dialysis efficiency and the result obtained would be out of proportion to the simple dilution to be expected.

At the present time, in those tests run at forty determinations per hour and also using dialysis for separation of the test material from protein, the average total volume dialyzed per sample has been 2.7 ml. The corresponding recipient volume has averaged about 2.8 ml. The flow rates of the two streams which are established by the combined air plus liquid volumes pumped per minute, have also been kept approximately equal. Although these values may be altered to some extent in adapting any test, it is considered that there are certain restrictions which must be placed on variation of these liquid volumes and differential flow rates. Countercurrent flow of the two fluid streams has not given good results for reasons which are obscure.

Although insufficient knowledge is currently available to predict dialysis efficiency, there must always be some dilution of the test material on passing into the recipient stream. This may ultimately limit the degree to which microanalysis or analysis of trace substances may be adapted to the current design of the instrument. However, two of the tests, glucose and urea, have been adapted to a microscale. The suggested differences in the microprocedures as compared to the macro- indicates what may be done in cases of analysis of trace substances. In the glucose method, the effective dialysis area was doubled by doubling the dialysis path. In this particular case, but not in the urea method, a significantly smaller proportion of the initial test material was transferred to the recipient stream.

In both micromethods, a sample pick-up tube of larger dimensions was employed. This provides a more even pumping rate and eliminates small variations which do not affect results in the macroprocedures, but become relatively important when analyzing trace substances. Finally, in both of the micromodifications, the analysis rate is decreased to twenty determinations per hour. This halving of the analysis rate is accomplished by doubling the time the sample is being metered into the system. This provides a longer time interval and larger volume to overcome the lag in the dialysis system. The time that the sample tube is metering air between specimens is also doubled. During this time relatively greater volumes of reagent are being pumped through the instrument to remove contaminants. The situation is roughly similar to that of the pipet-buret mentioned in an earlier section. The leading volume of the specimen as it migrates through the system is diluted by the residuum of preceding reagent. This volume in the pipet-buret is that which is rejected. In the Autoanalyzer, the color produced by this leading volume is absorbed in the leading shoulder of the chart tracing. When a definite volume of test solutions has passed, the shoulder levels off at a plateau of color which is representative of the optimal or proportional color for that sample. After this point is reached, further addition of sample is not needed to produce accuracy, and the machine commences to pump only reagents into the system. The residual color from the test material is then washed out by a definite volume of reagents and the process may then be repeated.

2.5.5.4. *Heating and Flow*. The purpose of heating is to facilitate the production of color. In a number of tests, the rate of color formation is sufficiently rapid at room temperature not to require additional heat. It is not necessary that color development proceed to completion. The conditions needed for accuracy in the instrument are met when the rate of color development is proportional to the concentration of test substance over a wide range of concentration. This condition frequently exists. In adapting a method to automated analysis, the amount of color produced may be varied by changing the path length between the final reagent added and the colorimeter. When the reaction occurs at room temperature, glass mixing coils may be added to give any desired path length. When the reaction requires heating to 95°C, the total path length in the bath is fixed, and length of time between the final reagent addition and colorimeter must generally be controlled by the total volume of air and reagents admitted to the system. In the tests which have been outlined, although the average liquid volume in the recipient stream going through the dialyzer is 2.8 ml per minute, the average final volume amounts to 5.5 ml. It is the final volume which will determine the flow rate and therefore the reaction time in the heating bath.

2.5.5.5. *Recording Colorimetry and Flow.* The principles of continuous flow colorimetric recording have been well worked out. In the present instance, the only point worthy of additional mention is the interrelation of flow rate to light-path length of the cell and wash-out volume. If it were possible to introduce a flow cell of the same volume as the ones currently employed but containing a light path longer than 10 mm, it might be expected that dilute solution or microchemical procedures could be more readily adapted to this instrument.

2.5.6. *Summary: Autoanalyzer*

The essential components of the Autoanalyzer have been covered in some detail. The clinical chemical methods which have successfully been or possibly will be adapted for use by this device have been described. Finally some of the problems encountered in adapting test methods for use with the analyzer and some current limitations were discussed. It is to be expected that the manufacturer if not the users will explore every means of increasing the versatility of the machine and that with time and development some of the current restrictions will be removed. It is certain that the instrument has the potentiality of performing a greater variety of tests than has currently been discussed. It has demonstrated its capability in a sufficient number of different tests to make one optimistic that, even though the present test methods may undergo modification, the instrument will ultimately (and does now) produce reliable chemical results in many laboratories.

3. Conclusions

The discussion has centered around the growth of the clinical chemistry laboratory as an incompletely defined entity but one faced with an ever-increasing number and complexity of test requests. The need for greater over-all accuracy and reproducibility of test results coming from different laboratories has been mentioned as well as the trend toward microchemical examination of blood as a conservation measure. Eight major activities needed for the performance of most of the clinical chemical tests have been presented with some description of devices which have reduced or eliminated manual techniques in the performance of these activities.

For the activities under discussion, the word "automation" was used in its broadest sense, and no clear distinction was attempted between those devices which were primarily developmental, were used for the application of a new technique, or merely changed one form of manual manipulation into another more convenient or more accurate one.

The eight activities chosen for presentation were: glassware washing; filtration and centrifugation; pipetting and transferring; titration and buretting; mixing and shaking; weighing; evaporation and distillation; and colorimetric reading and recording. Automation in certain other areas has not been covered because the activities are not widely performed in clinical laboratories. The development of automation in chromatography (both paper and gas phase), electrophoresis, ion exchange, and radioactivity has not been investigated.

The answer to the question of whether to automate or not depends on whether there is need of increasing the output in numbers and/or accuracy of tests and whether there is available an adequate number of trained personnel. Many institutions are faced with the prospect of increased demand for test results with a relative shortage of trained personnel or with a high turnover of trainable personnel. In these cases the answer to whether or not to automate might well be yes. The decision of what, and when, to automate is so much a problem of individual circumstance that only generalities can be attempted. There is no doubt that relatively small expenditures, particularly in the field of simple automatic pipets and pipet-buret combinations, can go a long way toward increasing the output and probably the accuracy of most laboratories. It has even been reported that savings in washing and breakage have more than paid for the initial outlay. Another advantage of these devices is that they never go out of style and can always be kept one step ahead of the more fully automatic equipment.

The use of motor-driven automatic pipets can certainly be justified in institutions requiring multiple pipettings. Such operations as adding anticoagulants to containers can be done rapidly and accurately without a trained operator. The pipets can also be used for accurately pipetting many of the conventional test reagents.

The use of automatic titrating devices is a subject of greater complexity. In the larger institutions, the acquisition of a single instrument (e.g. chloride titrator) limited to one test may very well be justified. The increased accuracy and saving of time alone could be ample reason for such a purchase. In the future, other tests may be adapted to electroanalysis. This could very well implement a decision to automate with any of the available automatic titrators.

Before purchase of a completely automated unit there must be a period of assessment of its virtues and faults. There may be one point of hesitation. After a chemical method has become automated, the chemical reagents and even the know-how for doing the determination manually will have vanished. What does one do when the machine breaks down? This is a prime consideration; however, there has been a precedent in most laboratories and it is not a new decision. Reference is made to the current wide use of the flame photometer. The passage of the chemical methods for sodium and potassium into the background may be mourned by some, yet the headaches of an additional instrument in the laboratory have become accepted. It may be thus with the fully automated device applicable to seven or more tests. There is yet another point to consider. Mention has already been made of the flame photometer, which is capable of rapid analysis of two and possibly three test materials. Earlier mention was made of a rapid and accurate chloride titrator. The question might be asked whether or not it would be feasible to wait long enough for the development of individual instruments for individual or small groups of tests. A corollary of this would be whether or not a fully automated central core of tests could be run on the multitest device, other tests being obtained with individual instruments. There are too many unknowns at the present time to do anything more than speculate. The trend of developments to date makes it a foregone conclusion that, as the clinicians request more chemical information about more patients, the clinical chemist will depend more heavily on accurate automated instrumental methods of analysis.

References

B1. Baginski, E. S., Williams, L. S., Jarkowski, T. S., and Zak, B., Polarographic and spectrophotometric micro serum determinations of chloride. *Am. J. Clin. Pathol.* **30**, 559-563 (1958).

C1. Cotlove, E., Trantham, H. V., and Bowman, R. L., An instrument and method for automatic, rapid, accurate and sensitive titration of chloride in biological samples. *J. Lab. Clin. Med.* **51**, 461-468 (1958).

C2. Cunningham, W. J., Automation. *Am. Scientist* **45**, 74-78 (1957).

D1. Dern, R. S., and Pullman, T. N., The accuracy, precision and utility of the syringe as a pipetting device. *J. Lab. Clin. Med.* **36**, 494-500 (1950).

F1. Fearon, W. R., The carbamido diacetyl reaction, a test for citrulline. *Biochem. J.* **33**, 902-907 (1939).

F2. Fiske, C. H., and Subbarow, Y., The colorimetric determination of phosphorus. *J. Biol. Chem.* **66**, 375-400 (1925).

H1. Hamilton, R. H., Graphic determination of prothrombin time (abstract). *Clin. Chem.* **4**, 549 (1958).

H2. Hoffman, W. S., A rapid photoelectric method for the determination of glucose in blood and urine. *J. Biol. Chem.* **120**, 51-55 (1937).

J1. Janney, N. W., The hospital chemical laboratory. *J. Lab. Clin. Med.* **1**, 747-752 (1916).

K1. Kind, P. R. N., and King, E. J., Estimation of plasma phosphatase by determination of hydrolyzed phenol with amino-antipyrine. *J. Clin. Pathol.* **7**, 322-325 (1954).

K2. Kirman, D., Morgenstern, S. W., and Feldman, M. A., A modified microtechnique for determining chloride. *Am. J. Clin. Pathol.* **30**, 564-566 (1958).

L1. Lingane, J. J., "Electroanalytical Chemistry." Interscience, New York, 1953.

M1. Marsh, W. H., Fingerhut, B., and Kirsch, E., Determination of urea nitrogen with the diacetyl method and an automatic dialyzing apparatus. *Am. J. Clin. Pathol.* **28**, 681-688 (1957).

M2. Marsh, W. H., Fingerhut, B., and Kirsch, E., Alkaline phosphatase-adaptation of a method for automatic colorimetric analysis. *Clin. Chem.* **5**, 119-126 (1959).

M3. McGuckin, W. F., and Power, M. H., Determination of blood sugar by an automatic method (abstract). *Clin. Chem.* **4**, 541 (1958).

M4. Meites, L., "Polarographic Techniques." Interscience, New York, 1955.

O1. Ormsby, A. A., Direct colorimetric method for determination of urea in blood and urine. *J. Biol. Chem.* **146**, 595-604 (1942).

P1. Powell, M. E. A., and Smith, M. J. H., The determination of serum acid and alkaline phosphatase activity with 4-aminoantipyrine (A.A.P.). *J. Clin. Pathol.* **71**, 245-248 (1954).

R1. Rowe, A. H., An automatic pipette. *J. Lab. Clin. Med.* **1**, 439-441 (1916).

S1. Saifer, A., Gerstenfeld, S., and Zymaris, M. C., Rapid system of microchemical analysis for the clinical laboratory. *Clin. Chem.* **4**, 127-141 (1958).

S2. Schales, O., and Schales, S. S., Simple and accurate method for determination of chloride in biological fluid. *J. Biol. Chem.* **140**, 879-884 (1941).

S3. Seligson, D., An automatic pipetting device and its application in the clinical laboratory. *Am. J. Clin. Pathol.* **28**, 200-207 (1957); see also Seligson, D., and Marino, J., Automatic pipetting attachment for the Van Slyke manometric apparatus for CO_2 determination. *Clin. Chem.* **4**, 120-126 (1958).

S4. Seligson, D., McCormick, G. D., and Sleeman, K., Electrometric method for determination of chloride in serum and other biologic fluids. *Clin. Chem.* **4**, 159-169 (1958).

S5. Skeggs, L. T., An automatic method for colorimetric analysis. *Am. J. Clin. Pathol.* **28**, 311-322 (1957).

S6. Stauffer, R. E., Dialysis and electrodialysis . *In* "Technique of Organic Chemistry" (A. Weissberger, ed.), Vol. 3, pp. 313-361. Interscience, New York, 1950.

W1. Williams, M. B., and Moser, J. H., Colorimetric determination of calcium with ammonium purpurate. *Anal. Chem.* **25**, 1414-1417 (1953).

W2. Wootton, I. D. P., Standardization in clinical chemistry. *Clin. Chem.* **3**, 401-405 (1957).

Z1. Zall, D. M., Fisher, D., and Garner, M. Q., Photometric determination of chlorides in water. *Anal. Chem.* **28**, 1665-1668 (1956).

AUTHOR INDEX

Numbers in parentheses are reference numbers and are included to assist in locating references in which authors' names are not mentioned in the text. Numbers in italics refer to pages on which the references are listed.

A

B

D

E

F

G

H

L

M

N

O

P

Q

R

S

T

Y

Z

SUBJECT INDEX